D1296343

GEOMETRY

Third Edition

Ron Tagliapietra

Kathy D. Pilger

BJU PRESS

Greenville, South Carolina

Note:

The fact that materials produced by other publishers may be referred to in this volume does not constitute an endorsement of the content or theological position of materials produced by such publishers. Any references and ancillary materials are listed as an aid to the student or the teacher and in an attempt to maintain the accepted academic standards of the publishing industry.

GEOMETRY
Third Edition

Ron Tagliapietra, EdD
Kathy D. Pilger, EdD
Larry Hall, MS
Larry Lemon, MS

Editors	**Composition**	**Project Manager**
Nathan Huffstutler	Anthology, Inc.	Kevin Neat
Rebecca Moore	Melba Clark	
Suzette Jordan		

Design	**Photo Acquisition**	**Cover Design**
Anthology, Inc.	Terry Latini	Anthology, Inc.
	Susan Perry	Elly Kalagayan
	Drew Fields	

Produced in cooperation with the Bob Jones University Division of Mathematical Sciences of the College of Arts and Science, the School of Education, and Bob Jones Academy.

Photograph credits appear on pages 660-61.

© 2006 BJU Press
Greenville, South Carolina 29614
First Edition © 1985 BJU Press
Second Edition © 1999 BJU Press

Printed in the United States of America
All rights reserved

ISBN 1-59166-347-4
ISBN 978-1-59166-347-8

15 14 13 12 11 10 9 8 7 6 5 4 3 2 1

Congratulations

Your search for the best educational materials available has been completely successful! You have a textbook that is the culmination of decades of research, experience, prayer, and creative energy.

The facts

Nothing overlooked. Revised and updated. Facts are used as a springboard to stimulate thoughtful questions and guide students to broader applications.

The foundation

Nothing to conflict with Truth and everything to support it. Truth is the pathway as well as the destination.

The fun

Nothing boring about this textbook! Student (and teacher) might even forget it's a textbook! Brimming with interesting extras and sparkling with color!

BJU PRESS
Learning for Life

1.800.845.5731   www.bjupress.com

Using This Book

Every chapter contains biblically based material, providing a scriptural foundation.

Questions about the Bible text help you grasp what is important to the chapter.

The Bible as truth makes what you learn personal and practical.

The Bible as the basis for geometry sheds light on the purpose of the subject.

$4 \cdot \left(\frac{8}{9}\right)^2$ or 3.16 and Babylon used $3\frac{1}{8}$ or 3.125. Because some civilizations had more accurate estimates, some modern critics have called this an error in the Bible.

God knows everything. He knows more than the Egyptians, the Babylonians, and the modern critics. Let's look more closely at what God said.

8. Reread the diameter measurement. It measures from where to where?
9. Reread the circumference measurement. What did it measure around?
10. Does the sea itself have the same diameter as the brim according to verse 26? What is the width of the brim?

11. Use the conversions from Chapter 3 (p. 113) to give these measurements in inches: AD, AB, AC, BC, and the sea's circumference.
12. Using problem 11, what do you get for the value of π? Is it better or worse than the ancient values listed above?

This value of π is the only irrational number approximated by Bible measurements. The theme verse lays the basis for the real number system that includes both rational and irrational numbers.

Line upon Line

AND HE MADE a molten sea, ten cubits from the one brim to the other: it was round all about, and his height was five cubits: and a line of thirty cubits did compass it round about.

1 KINGS 7:23

GEOMETRY AND SCRIPTURE **411**

The theme verse helps you to link the temporary with the eternal in your studies.

Different colors help you to distinguish **Postulates,** Definitions, **Theorems,** and Constructions.

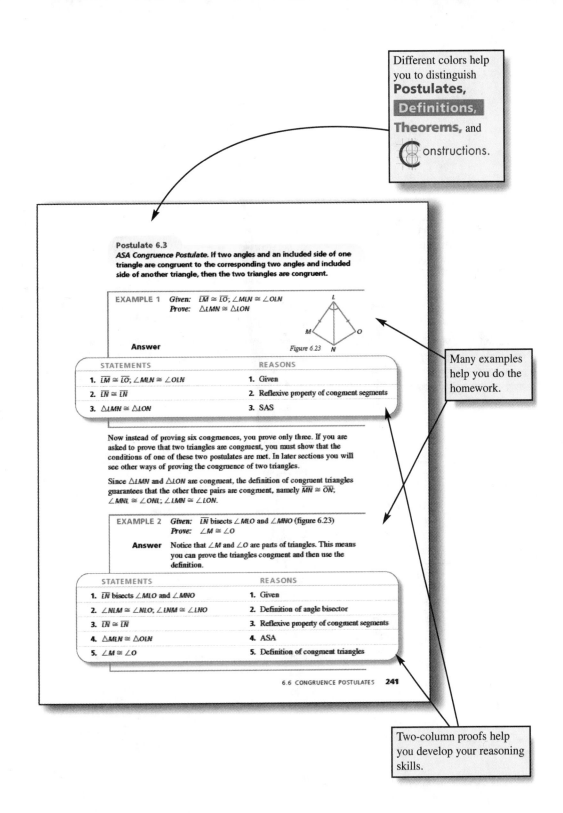

Postulate 6.3

ASA Congruence Postulate. If two angles and an included side of one triangle are congruent to the corresponding two angles and included side of another triangle, then the two triangles are congruent.

EXAMPLE 1   *Given:* $\overline{LM} \cong \overline{LO}$; $\angle MLN \cong \angle OLN$
　　　　　　　Prove: $\triangle LMN \cong \triangle LON$

Answer

Figure 6.23

STATEMENTS	REASONS
1. $\overline{LM} \cong \overline{LO}$; $\angle MLN \cong \angle OLN$	1. Given
2. $\overline{LN} \cong \overline{LN}$	2. Reflexive property of congruent segments
3. $\triangle LMN \cong \triangle LON$	3. SAS

Now instead of proving six congruences, you prove only three. If you are asked to prove that two triangles are congruent, you must show that the conditions of one of these two postulates are met. In later sections you will see other ways of proving the congruence of two triangles.

Since $\triangle LMN$ and $\triangle LON$ are congruent, the definition of congruent triangles guarantees that the other three pairs are congruent, namely $\overline{MN} \cong \overline{ON}$; $\angle MNL \cong \angle ONL$; $\angle LMN \cong \angle LON$.

EXAMPLE 2   *Given:* $\overline{LN}$ bisects $\angle MLO$ and $\angle MNO$ (figure 6.23)
　　　　　　　Prove: $\angle M \cong \angle O$

Answer   Notice that $\angle M$ and $\angle O$ are parts of triangles. This means you can prove the triangles congruent and then use the definition.

STATEMENTS	REASONS
1. $\overline{LN}$ bisects $\angle MLO$ and $\angle MNO$	1. Given
2. $\angle NLM \cong \angle NLO$; $\angle LNM \cong \angle LNO$	2. Definition of angle bisector
3. $\overline{LN} \cong \overline{LN}$	3. Reflexive property of congruent segments
4. $\triangle MLN \cong \triangle OLN$	4. ASA
5. $\angle M \cong \angle O$	5. Definition of congruent triangles

6.6 CONGRUENCE POSTULATES   **241**

Many examples help you do the homework.

Two-column proofs help you develop your reasoning skills.

Exercises with different levels of difficulty help guide your progress.

▶ **B. Exercises**

16. Explain why there are no other cases in exercise 6.
17. Explain why there is no third case in exercise 5.

Prove the following theorems.
18. Theorem 9.18
19. Theorem 9.17b
20. Theorem 9.17c

Level C exercises challenge your thinking skills.

▶ **C. Exercises**

21. Prove Theorem 9.19.

Dominion Thru Math helps you apply technology to problem solving.

▶ **Dominion Thru Math**

One room of a museum includes archeological artifacts in a circular cylinder case. Three hidden security cameras evenly placed around the room constantly monitor the case. The case has a radius of 9 feet, and each camera lens has a view angle of 40°.

22. How many degrees of the case can each camera view, and how many degrees of overlap are there? (*Hint*: Use exterior tangents.)

23. How many feet of the circumference of the case does each camera view that the other cameras cannot view?

■ **Cumulative Review**

Use the quadrilateral shown for the following questions.
24. Name the type of quadrilateral.
25. How do the diagonals relate?

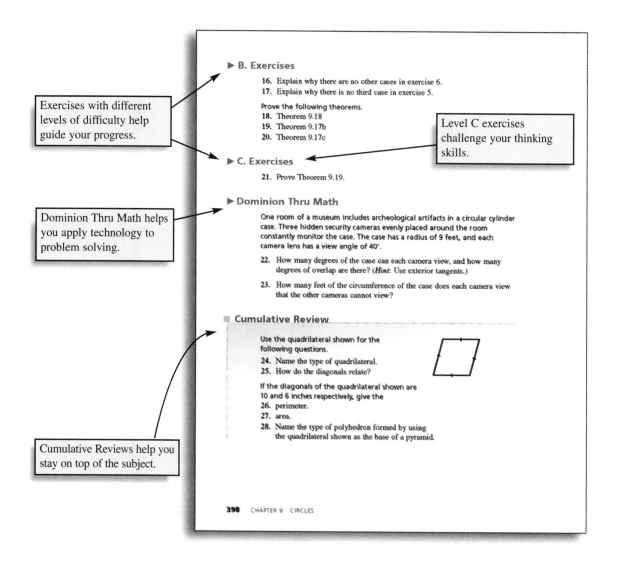

If the diagonals of the quadrilateral shown are 10 and 6 inches respectively, give the
26. perimeter.
27. area.
28. Name the type of polyhedron formed by using the quadrilateral shown as the base of a pyramid.

Cumulative Reviews help you stay on top of the subject.

roof. The foundation for a house usually consists of one or more rectangles that must be laid out perfectly. A carpenter cannot simply measure the lengths of the sides of the quadrilateral because that would guarantee only a parallelogram. Using theorems from geometry, he can either check that one of the angles is a right angle, or he can measure the lengths of the diagonals to see if they are the same.

A carpenter must also be able to read an architect's drawing. This is essential if he is going to build a house exactly the way the future owner wants it. If he makes

A carpenter nails boards for the roof with a pneumatic (air-driven) tool.

Geometry Around Us reveals some of geometry's secret hideouts.

9 Analytic Geometry

Graphing Circles

A circle is a conic section formed by the intersection of a right circular cone with a plane that is perpendicular to the axis. A circle is also a locus of points that are a given distance from a given point in a plane. From this latter definition we can develop an equation for a circle.

Analytic Geometry helps you make the algebra-geometry connection.

MIND OVER MATH

Find the perimeter and area of the shaded region in the marked figure.

The Mind over Math brain teasers exercise your gray matter.

Geometry Through History

Geometry Through History brings the subject to life.

GEORG FRIEDRICH BERNHARD RIEMANN

Georg Bernhard Riemann was born on September 17, 1826, in the village of Breselenz, in Hanover, Germany. His father was a Lutheran pastor. The Lutheran congregation of the small village could not sufficiently support the pas-

Chaos theory uses fractals to show the pattern in what otherwise appears random to man. Fractals have become popular as artwork, but chaos theory also describes populations in ecology, randomness in statistics, fluctuating market values in economics, earth's magnetic field in physical science, and other topics from chemistry to astronomy.

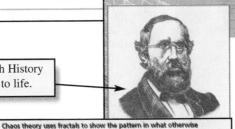

After this chapter you should be able to
1. classify and use real numbers.
2. apply properties of real numbers.
3. find distances using absolute values.
4. apply distances to betweenness, midpoints, and congruent segments.
5. find perimeters and circumferences.
6. distinguish inscribed and circumscribed figures.
7. justify the midpoint and circumference formulas.
8. construct congruent segments and bisectors.

Chapter openers are your ticket to new achievements.

Contents

Introduction

What Is Geometry?

All mathematics has its roots in man's efforts to understand and describe the world around him. Since Creation, man has measured objects, described shapes, and used reason as tools in exercising dominion over the earth. In Egypt, the subject was especially important for building pyramids and re-establishing property boundaries along the changing floodplain of the Nile River.

The Greeks organized the ideas of several centuries into principles and properties. In fact, our word *geometry* comes from two Greek words and means "earth measure." About 300 B.C. the Greek mathematician Euclid set forth the known principles in an orderly and systematic presentation. His system is called Euclidean geometry.

Columns at Temple of Hera, Olympia, Greece

Euclid's system is a model. Some of you have built model cars. Just as model cars represent key features of real cars, geometry represents key features of God's creation. Model cars represent the shape and scale of real cars even though they are too small to be used for transportation. Likewise geometry shows the relationships among measurements such as lengths and areas as well as the reasons for such relationships. Thus geometry is a mathematical model.

Geometry is also abstract. Something that is not a visible, concrete object is abstract. Numbers are abstractions describing "how much." For example, the idea of "fiveness" is exemplified by many things, like your fingers and toes, and is represented by many different symbols, such as V, 5, ⊞, and so on. "Fiveness," then, is an abstraction.

Basswood stem under a microscope (magnified 63X)

Euclidean geometry is an abstract model of our orderly physical universe. Geometry expresses properties that describe the structure of the physical universe. According to the Euclidean model, light from the stars travels to us in a straight line—without bending. While this is a useful model, does light really behave this way? Only God knows the full truth. The Christian can rejoice that God created both the visible universe and its invisible aspects such as mathematics.

> *For by him were all things created, that are in heaven, and that are in earth, visible and invisible, whether they be thrones, or dominions, or principalities, or powers: all things were created by him, and for him: and he is before all things, and by him all things consist. (Col. 1:16–17)*

Why Is Geometry Important?

Geometry is valuable because its applications are useful to people in many occupations: surveying, engineering, science, architecture, building, trades, designing, navigating, the military, and others. In fact, mathematical training, or the lack of it, has become a technological filter to eliminate those who do not have the skills to enter the various technical fields.

King Khalid International Airport, Riyadh, Saudi Arabia, is the biggest airport complex in the world.

Snowflake under a microscope

Geometry also plays a role in making things beautiful. From artists to architects and from designing clothing to making patterns for wallpaper, geometry is all around us. The French mathematician Blaise Pascal was so enamored with geometry in nature that he concluded, "God is the great Geometer." Maybe he had been studying the designs of seashells, pine cones, nuts, flowers, snow-flakes, frost, or honeycombs. Or he may have been looking into the heavens on a clear, dark night:

> *The heavens declare the glory of God; and the firmament sheweth his handywork. Day unto day uttereth speech, and night unto night sheweth knowledge. (Ps. 19:1–2)*

Besides these direct applications, geometry develops a comprehension of spatial relationships. Nearly every intelligence test has some questions on spatial relationships. A well-known industrial mathematician claims that people trying to solve problems in industry are greatly hindered by poor geometry backgrounds. If a picture is worth a thousand words, it should be obvious that geometrical representations will help a person solve problems.

Most importantly, geometry has trained young minds to think logically and clearly even in nonmathematical situations. As a young lawyer Abraham Lincoln worked diligently through Euclid's books *The Elements* to improve his mind for the practice of law. Geometry helped him learn the principles of deductive reasoning, which he in turn used to present forceful arguments in a court of law. All of us will reason through problems and issues throughout our lives, and the most "practical" courses are those that train us to come to the correct conclusions to complex or controversial problems.

In fact for centuries young men studying for the ministry had to study Latin and geometry. They probably wondered, "Why should a preacher study geometry?" The famous preacher C. H. Spurgeon enjoyed doing geometry for fun to keep his mind sharp. He found that the skills learned in geometry also helped him to reason through issues in theology and to organize his sermons.

Your geometry studies should help you to make better judgments about the statements that other people make and should help you to present your own ideas more skillfully. As Lincoln and Spurgeon found, geometry should improve your skills in drawing correct conclusions and organizing essays and speeches. Good reasoning and geometric concepts will also help you solve problems and understand concepts in other math and science classes and will prepare you for college as well. Most college entrance exams include a large amount of mathematics, including geometry and problems that involve reasoning.

But Is It Worth the Effort?

To answer this question, think about other skills you have learned already. Have you learned how to water ski, snow ski, ice skate, roller skate, swim, or drive a car or motorcycle? Have you played team sports such as volleyball, soccer, basketball, baseball, or lacrosse? What is true about your skills in each case? You may start out very poorly and even fail many times to perform the skills correctly. But if you keep trying, you finally succeed. Others may learn more quickly than you do, and many may be better than you are, but you can still do it. Remember, too, that a sports skill is not nearly as much fun while you are learning it as it is after you have mastered it. Like sports skills, thinking logically and doing proofs are skills that everyone can learn. Some will learn faster, and some will be better than others, but everyone can enjoy these skills at his own pace. An attitude of determination and diligence is important for every skill we learn in life.

How many sizes of spheres can you find in this sporting goods display?

Questions to Discuss

1. Name an abstract mathematical model that you have studied in previous math courses.
2. Make a list of geometric designs and/or applications that you see at home.
3. Explain how mathematical concepts have influenced the way people think about the world.

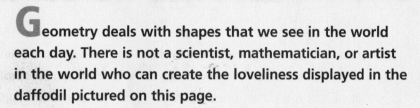

Geometry deals with shapes that we see in the world each day. There is not a scientist, mathematician, or artist in the world who can create the loveliness displayed in the daffodil pictured on this page.

The flower of the daffodil displays a hexagon pattern. For you see, the greatest master of geometry is our God.

Do you think you are sitting in geometry class today because of chance? You're not. Through the discipline of geometry, God has something to teach you about yourself and, more importantly, about Him. As you glimpse God's order through geometry, you should marvel at His wisdom. And you should apply yourself with all diligence to "do all to the glory of God," which includes geometry class—today.

After this chapter you should be able to

1. express sets, subsets, and elements of sets with proper symbols.

2. perform set operations: unions, intersections, complements.

3. define terms and list the three undefined terms.

4. discuss points, lines, and their relationships, using proper symbols.

5. distinguish between postulates and theorems.

6. explain the criteria for an ideal geometric system.

7. state incidence postulates.

8. prove incidence theorems.

1.1 Sets and Subsets

In mathematics, collections of objects are called *sets*. Set symbols enable us to write statements about sets faster just as number symbols enable us to write numerical statements quickly. A brief review of set symbols appears on the next few pages. Make sure that you understand the symbols and definitions.

Do the objects shown have anything in common? A set is denoted by braces, { }, and is named with a capital letter. You can use any capital letter to name this set of objects. Let's use C for cleanup.

The objects of the set are called *elements* or members of the set. There are two ways to describe elements: the list method or set-builder notation. In the list method the elements are listed inside set braces.

The basket represents the set. What seven elements does the set contain?

EXAMPLE 1   Use the list method to describe set C.

Answer   C = {comb, brush, toothbrush, paste, soap, razor, cloth}

This same set can be described using set-builder notation. The general form of set-builder notation is {x|x is . . . }, where x is an arbitrary element of the set, and the vertical line, |, indicates the words *such that*. This vertical line is followed by a description of a representative element of the set.

EXAMPLE 2   Use set-builder notation to describe set C.

Answer   C = {x|x is an object used in your morning cleanup}

This notation is read "The set C is the set of all elements x such that x is an object used in your morning cleanup." You can see that set-builder notation describes the set without listing the objects.

To symbolize the fact that the comb is an element of set C, you can use the symbol ∈. This symbol means "is an element of."

$$\text{comb} \in C$$

This notation says that the comb is an element of the set C.

Remember that the symbol for *equals* means "is the same as." This means that two sets are *equal* if they are exactly the same set.

EXAMPLE 3 Let $A = \{1, 3, 5, 7, 9\}$ and
$B = \{x \mid x \text{ is an odd number less than } 10\}$.
Symbolize a relation between A and B.

Answer $B = \{1, 3, 5, 7, 9\}$ 1. Convert set B into a list.
$A = B$ 2. Compare B to A and notice that
                   they are the same set.

If set A contains set B, then set B is a subset of set A and we write $B \subseteq A$. Each element of B must also be an element of set A. In particular, every set is a subset of itself: $A \subseteq A$.

The *empty set*, or *null set*, denoted by { } or ∅, is the set that has no elements. The empty set is considered to be a subset of every set. Therefore, $\emptyset \subseteq A$.

EXAMPLE 4   If $P = \{2, 7, 12, 17\}$ and $Q = \{2, 12\}$, name three subsets of P.

Answer $Q \subseteq P$, $P \subseteq P$, and $\emptyset \subseteq P$

A is a proper subset of B if it is a subset of B other than B itself. A proper subset is denoted by the symbol ⊂. In example 4, $Q \subset P$ and $\emptyset \subset P$.

Do you know what the slash means in these signs? In the same way, a slash can be used to add a "not" to any set relations.

$\notin$ not an element of

$\not\subseteq$ not a subset of

$\not\subset$ not a proper subset of

$\neq$ not equal to

Sets of the same size are called *equivalent sets*. The terms *equivalent sets* and *equal sets* do not mean the same thing. Equal sets are sets with the same elements, whereas equivalent sets are sets that are in *one-to-one correspondence*.

The idea of one-to-one correspondence of sets can be seen in this illustration. Consider the set of boys called B = {Jack, Fred, Ron} and the set of girls called G = {Rhonda, Heather, Joy}. In a relay race, each element of set B is paired with one and only one element of set G. Therefore, the two sets are considered to be equivalent sets. But notice that the sets are not equal sets.

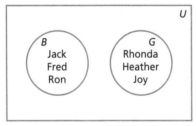

Figure 1.1

Another special set is the *universal set,* denoted by U. A universal set is a set that contains the elements in the context of a given problem. For instance, in the example above about sets B and G, the universal set might be all the students in the youth group. The universal set that you choose for a particular problem depends on the extent and type of problem.

A *Venn diagram* will help you illustrate the set relations. The above example can be illustrated as figure 1.1 shows.

In previous math classes, you studied the set of real numbers, $\mathbb{R}$, which includes both rational and irrational numbers. In this course you will be working with this set as well as sets of points in space.

▶ A. Exercises

Write the following sets with both listing statements and set-builder notation.

1. States that border your state
2. Students in your class
3. Your favorite foods
4. Letters in your last name
5. Classes you are taking

Tell whether the following pairs of sets are equivalent sets, equal sets, or neither. Give the most specific answer possible.

6. $A = \{a, b, c\}$ $B = \{1, 2, 3\}$
7. $K = \{5, 7, 8\}$ $M = \{1, 9, 2, 7\}$
8. $G = \{3, 1, 9\}$ $H = \{9, 3, 1\}$
9. $L = \{man, son, brother\}$ $N = \{woman, daughter, sister\}$
10. $Q = \{Santa Fe, Boise\}$ $R = \{New Mexico, Idaho, Virginia\}$

Use the proper notation to describe these statements.

11. Set A is a subset of set L.
12. b is an element of set K.
13. The empty set is a subset of the universal set.
14. c is not an element of set M.
15. Set K is equal to set F.
16. Set J is a proper subset of set M.
17. Set N is not a subset of L.
18. The set with elements a and f is a subset of set P.
19. The set with elements k, l, and m is not a subset of the set consisting of elements k, l, and n.
20. The null set is a subset of set B.

▶ B. Exercises

Draw Venn diagrams to illustrate the following sets. Give the universal set in each case.

21. $C = \{Jill, Joy, Gary\}$
    $D = \{Susan, Jean, Judy, Ann\}$
22. $L = \{Kentucky, Kansas\}$
    $M = \{Arkansas, Alabama, Arizona\}$
23. $N = \{1, 2, 3, 4, 5, 6\}$ $P = \{1, 3, 5\}$
24. $A = \{1, 3, 9\}$ $B = \{2, 4, 6\}$

▶ C. Exercises

25. Show how the set of even positive integers is equivalent to the set of all counting numbers.

1.2 Set Operations

You already know operations with real numbers, such as addition and multiplication. In this section you will learn three operations on sets.

American bison roam in sets called herds. This herd roams Custer State Park in the Black Hills of South Dakota.

Look at these sets. The universal set is the set of all geometric figures.

A = {triangle, square, rectangle, circle}
B = {rhombus, rectangle, triangle, parallelogram}

The *union* of sets A and B, written $A \cup B$, is the set combining all the elements of the given sets.

$A \cup B$ = {triangle, square, rectangle, circle, rhombus, parallelogram}

Venn diagrams can be used to show operations on sets. In the Venn diagram, geometric figures are substituted for the names of the figures. The shaded area represents $A \cup B$.

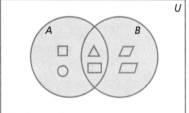

Figure 1.2

The second basic operation on sets is *intersection*. The intersection of sets A and B, denoted by $A \cap B$, is the set that contains the elements belonging to both A and B. What do sets A and B have in common?

$A \cap B$ = {triangle, rectangle}

A Venn diagram can also illustrate the intersection of sets. The shaded area in figure 1.3 represents $A \cap B$.

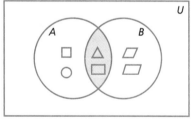

Figure 1.3

When two sets have nothing in common (their intersection is empty), the sets are called *disjoint sets*. In the sets

K = {algebra, history, music, chemistry}
L = {English, Spanish, German}
$K \cap L = \varnothing$

Therefore, *K* and *L* are disjoint sets. The Venn diagram on page 4 provides another example of disjoint sets.

Two sets are necessary for the operations of union and intersection, so they are called *binary operations*. An operation on a single set is called a *unary operation*.

The last operation on sets that you will see in this section is the *complement* of a set. This operation is different from the other two operations because it is a unary operation. The complement of a set *M*, denoted by *M'*, is the set of all elements in the universal set that are not in the set *M*. If the universal set is the set of whole numbers, {0, 1, 2, 3, 4, . . .}, and *M* = {1, 3, 5, 7, . . .}, then *M'* = {0, 2, 4, 6, . . .}.

EXAMPLE 1   Draw a diagram of *C'*.
C = {1, 4, 5, 7} and the
universal set is the digits
from one to seven.

Answer   *C'* is shaded in the
Venn diagram.
C' = {2, 3, 6}

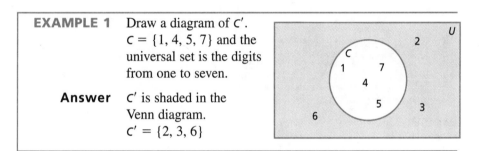

You can combine set operations too. Be careful to use the correct order of operations. Do operations in parentheses first, just as you learned to do with algebraic expressions. Then do complements, and finally do unions and intersections.

EXAMPLE 2   Find *A'* ∩ *B* and (*A* ∩ *B*)'. Let *A* = {1, 2, 3, 4} and
B = {1, 3, 5, 7}. Use the same universal set as in example 1.

Answer

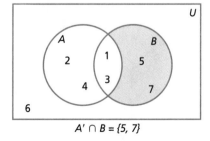

A' ∩ *B* = {5, 7}

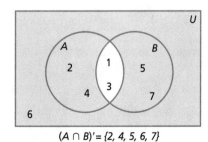

(*A* ∩ *B*)' = {2, 4, 5, 6, 7}

▶ A. Exercises

If A = {dog, bird, cat, snail, rabbit} and B = {snail, snake, bird}, find the following sets.

1. $A \cup B$ 2. $A \cap B$

Let the universal set, U, be the set of whole numbers, and
K = {1, 3, 6, 9, 12}
L = {2, 4, 6, 8, 10}
M = {1, 4, 8, 12}
N = {0, 5, 15}

Find the following sets and make a Venn diagram to illustrate each operation.

3. $K \cap M$ 9. $(K \cup L) \cap M$
4. $L \cup N$ 10. $(N \cap K) \cup L$
5. $K \cap L$ 11. K'
6. $M \cup N$ 12. $M \cap N$
7. $L \cap N$ 13. $L \cup (M \cup N)$
8. L'

▶ B. Exercises

14. $(L \cup M)'$ 17. $K' \cap M$
15. $(K \cap N)'$ 18. $(M \cap K') \cup (M' \cap K)$
16. $L \cap M'$ 19. $(K \cap L) \cup (K' \cap L')$

▶ C. Exercises

20. Use intersection, union, and complement to write a set statement that describes the shaded area.

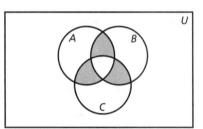

▪ Cumulative Review

A = {1, 3, 8} B = {1, 3, 9} C = {3, 8, 9} D = {3, 9}

True/False
21. $A \subseteq B$
22. $D \in B$
23. $D \subset C$
24. $9 \notin A \cap D$
25. $(A \cup D) \subset (B \cup C)$

1.3 Undefined Terms and Definitions

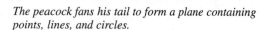

Where do you start a mathematical system such as geometry? To begin the study of geometry, you must understand what basic geometrical words mean. A good definition of some words is hard to find.

The peacock fans his tail to form a plane containing points, lines, and circles.

What is a good definition anyway? A definition is a statement of meaning. A good definition must have the following characteristics to successfully convey the meaning of the term.

1. **Clear**. The definition must communicate the point and state the term being defined. Avoid vague or ambiguous language.
2. **Useful**. The definition must use only words that have been previously defined or are commonly accepted as undefined.
3. **Precise**. The definition must be accurate and reversible. Identify the class to which the object belongs and its distinguishing characteristics.
4. **Concise**. The definition must be a good sentence and use good grammar. Stick to the point and avoid unnecessary words.
5. **Objective**. The definition must be neutral. Avoid emotional words, figures of speech, and limitations of time or place.

EXAMPLE Evaluate this definition. *Space* is the set of all points.

Answer To determine if this is a good definition of space, we need to analyze each of the five characteristics.

1. **Clear**. The object being defined is named in the definition. It is not ambiguous.

2. **Useful**. Look at all the words in the definition. Have they all been defined? At this point in our study, we must say no. The word *set* was discussed in the first two sections of the chapter, but has the word *point* been defined?

3. **Precise**. The definition describes space accurately. Space is classified as a set and is described as the particular set containing all points. The definition is also reversible. We could say, "The set of all points is space."

Continued ▶

---

4. Concise. This definition contains only necessary words and uses proper grammar.

5. Objective. This definition uses appropriate objective language.

We can conclude from this analysis that this definition of *space* would be a good one if the word *point* were defined.

Consider the word *point*. Can it be defined? Most of you certainly have some idea of what a *point* is. If we define *point*, some words in that definition would also need to be defined, and this process could continue forever. To avoid this needless waste of time and energy, mathematicians have agreed to accept some undefined terms as basic building blocks for a mathematical system. Of course, they desire that there be as few undefined terms as possible. Since the term *point* is one of these basic terms, the definition of space is a good definition.

In geometry there are three undefined terms that can be described but not defined: *point*, *line*, and *plane*.

Undefined term	Description	Illustration	Notation
Point	Spot; an object with no dimensions, length, width, or thickness; a location in space	pinpoint	denoted by a capital letter; location marked by a dot • P
Line	Straight; an object that extends infinitely in one dimension; has length but no width or thickness	laser beam	$\overleftrightarrow{CD}$; or a lowercase script letter
Plane	Flat; an object that extends infinitely in two dimensions; has length and width but no thickness	thin sheet of glass	plane *k*; denoted by a lowercase letter

These descriptions are not definitions. The descriptions should simply help you visualize concepts. In reality, points, lines, and planes cannot actually be seen.

The sheer cliff face, down which this man is rapelling, can be represented by a plane.

Likewise, there is a starting place for Christian doctrine. There are certain things that the believer must accept as truth although he may not fully understand them. For example, the believer must believe that the Bible is the verbally inspired, inerrant Word of God. He must believe in and accept the atoning death and shed blood of Jesus Christ for salvation. Although he may not thoroughly understand these things, his faith in these basic truths forms the foundation on which he can build a strong spiritual life.

▶ A. Exercises

Check all the characteristics of a good definition and determine which of the following definitions are good. If a definition is not good, explain why.

1. A frog is an amphibian.
2. A noun is a word used to name a person, place, thing, quality, or action.
3. A glove is a covering for the hand.
4. Plasma is the clear, yellowish liquid part of the blood.
5. Nine is the sum of six and three.
6. An empty set is any set.
7. An atheist is like when some crazy nut denies that God exists.
8. Love is unselfish concern for the best interests of another.
9. A retable is a reredos.
10. Braille is a system of writing and printing for blind people that consists of different patterns of raised dots that represent letters, words, numbers, and punctuation marks.

Illustrate each idea.
11. point K
12. plane p
13. line l
14. $\overleftrightarrow{AC}$

Write a sentence describing each illustration.

15.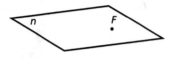

16.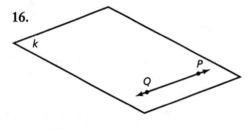

17.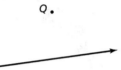

18.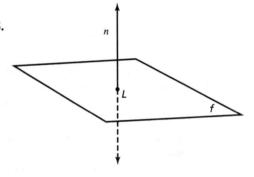

▶ B. Exercises

Define the following.
19. king
20. pen
21. mosaic
22. emu

▶ C. Exercises

23. Evaluate the definition of *element* of a set from page 2.
24. Correct any definitions in exercises 1-10 that were not good.

▶ Dominion Thru Math

Study the daffodils on the opening pages of this chapter and answer the questions.

25. How many petals are there, and what figure do their tips form?
26. What figure is formed by endpoints of every other petal?
27. What shape do you see near the center with the orange edges?
28. Even though the petals are formed by slightly curved pieces, what general shape are they?
29. What shapes are the little bundles of pollen pistils in the center?

Study the flowers and the butterfly on page 498 and answer the questions.

30. How many petals are there, and what figure do the tips form?
31. What is the basic shape of each petal?
32. What shape is in the white center of the flower?
33. What is different about these pollen pistils as compared to those of the daffodil?

■ Cumulative Review

State in words the most specific relationship that you can between each pair of sets. Express as many of the relations in symbols as you can. Consider the integers as the universal set.

34. $A = \{1, 3, 5\}$
    $B = \{x \mid x \text{ is an odd integer}\}$
35. $A = \{x \mid x \text{ is a prime number greater than 2}\}$
    $B = \{x \mid x \text{ is an even integer}\}$
36. $A = \{x \mid x \text{ is the square of an integer}\}$
    $B = \{0, 1, 4, 9, 16, 25, \ldots\}$
37. $A = \{1, 11, 21, 31\}$
    $B = \{-11, 1, 11, 121, 1331\}$
38. $A = \{x \mid x > 2\}$
    $B = \{x \mid x \leq 2\}$

Analytic Geometry

Graphing Points

You may be asking yourself, "How does any of this geometry relate to what I learned in algebra?" *Analytic geometry* makes a connection between the figures that you see in geometry and the equations that you saw in algebra. In each chapter of this book, you will find a feature on analytic geometry to help you see the connection.

Points are graphed on a *Cartesian plane,* named for René Descartes, who was the first mathematician to connect geometric figures and equations. Every point in a plane is identified by an *ordered pair,* such as (3, −2), which is measured from a reference point called the *origin*. The origin is the intersection of two perpendicular lines called the *axes*. The horizontal line is called the *x*-axis, and the vertical line is called the *y*-axis. The axes divide the Cartesian plane into four *quadrants*.

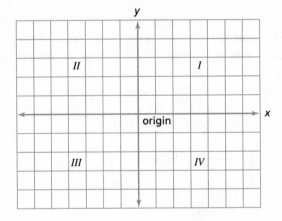

The origin is the ordered pair (0, 0). You can graph any other ordered pair by counting from the origin. The first number in an ordered pair, such as the 3 in (3, −2), represents the *x*-value, or distance of the particular point in the horizontal direction. The second number is the *y*-value, or vertical distance.

To locate the point (3, −2), start at the origin and move 3 units right and 2 units down. Be sure to label the point.

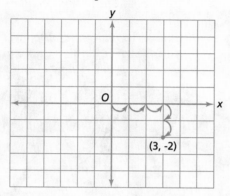

Each point of the plane has an *x*-value and a *y*-value. By relating geometric points to these algebraic variables, you can derive geometric truths. In this way, analytic geometry uses algebra to study geometry.

▶ Exercises

Give the coordinates of each point.

1. *A*
2. *B*
3. *C*
4. *D*
5. *E*

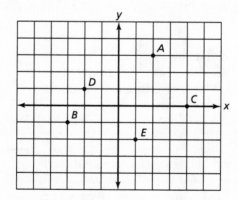

1.4 A Framework with Definitions

What is a skeleton? Can you give a good definition of *skeleton?*

A skeleton of Albertosaurus looms over its kill, Lambeosaurus, at the Field Museum of Natural History in Chicago, Illinois.

Most of you will think of a human skeleton. But a good definition of *skeleton* in general is "a supporting structure or framework." This describes not only the human skeleton but also the geometric skeleton. The geometric skeleton, or framework, is illustrated in the diagram to the right.

The system of geometry builds step by step up the ladder. There are three undefined terms, many definitions, two dozen postulates, and many theorems. Each upper level of the framework is built from one or more of the lower levels.

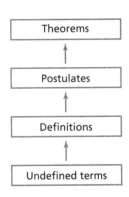

Before discussing postulates and theorems, we need to establish a few more definitions in order to have a wider base on which to build.

Collinear points are points that lie on the same line.

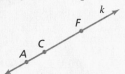

Figure 1.4

Points *A*, *C*, and *F* are collinear points because they lie on line *k*.

Noncollinear points are points that do not lie on the same line.

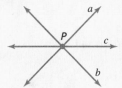

Figure 1.5

Points *X*, *Y*, and *W* are noncollinear points because no line could contain all of them.

Concurrent lines are lines that intersect at a single point.

Figure 1.6

Lines *a*, *b*, and *c* are concurrent lines because they intersect at point *P*.

Coplanar points are points that lie in the same plane.

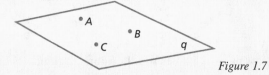

Figure 1.7

Points *A*, *B*, and *C* are coplanar since they all lie in plane *q*. Can you give the definition for *noncoplanar points?*

Coplanar lines are lines that lie in the same plane.

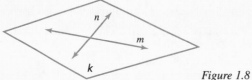

Figure 1.8

Lines *m* and *n* are coplanar lines.

Parallel lines are coplanar lines that do not intersect.

Can you illustrate parallel lines? If two lines *l* and *m* are parallel, we write *l* ∥ *m*. The symbol " ∥ " is read "is parallel to."

Skew lines are lines that are not coplanar.

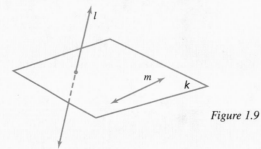

Figure 1.9

Lines *l* and *m* are skew lines. No plane could contain both lines.

Parallel planes are planes in space that do not intersect.

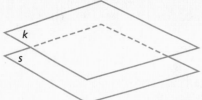

Figure 1.10

The symbol " ∥ " can be used for parallel planes too. In figure 1.10, planes *k* and *s* are parallel: *k* ∥ *s*.

The last two building blocks in the framework of geometry are postulates and theorems. Both postulates and theorems describe relationships among the terms (whether defined or undefined). In contrast to definitions, postulates and theorems are not always reversible.

A statement that can be shown to be true by a logical progression of previous terms and statements is a *theorem*. The process of justifying a theorem is called *proving* a theorem.

Just as undefined terms provide a starting point for defining other terms, postulates are the basic statements from which theorems are proved. Postulates, then, are assumed to be true without proof. Most postulates (sometimes called axioms) are obvious truths observed in God's created order.

In the next two sections, you will learn the most basic postulates and prove your first theorems. For now, just remember that postulates are assumed, but theorems must be proved from definitions, postulates, and previous theorems.

► A. Exercises

Use the diagram for exercises 1-5.
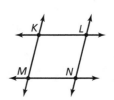
1. Name the lines that contain point *K*.
2. Name the lines that are concurrent at point *N*.
3. Name all lines shown.
4. Which lines appear to be parallel?
5. Name three sets of collinear points.

Use the diagram to the right for exercises 6-14. Assume that the edges of the boxlike figure are lines that continue infinitely.
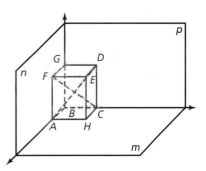
6. Name two planes.
7. Name three concurrent lines that intersect at point *H*.
8. Name two pairs of skew lines.
9. Name the intersection of $\overleftrightarrow{HC}$ and $\overleftrightarrow{CB}$.
10. Name two intersecting planes. What is their intersection?
11. Give four noncoplanar points.
12. Name two pairs of lines that appear to be parallel.
13. Name three coplanar lines.
14. What planes intersect in $\overleftrightarrow{BC}$?

► B. Exercises

15. Do you ever have to prove a postulate?
16. Do concurrent lines always intersect?
17. Do skew lines ever intersect?
18. Name three things you can use to prove a theorem.
19. What is logic?

20. Can you find three noncoplanar points?

Cumulative Review

21. Define *space*.
22. Define *subset* (assume *set* and *element* as undefined terms).
23. If $A \subseteq B$ and $B \subseteq A$, what can you conclude?
24. If $A \subseteq B'$, what can you conclude about sets A and B?
25. Draw a picture to illustrate $\overleftrightarrow{AB} \cap \overleftrightarrow{CD} = \{P\}$.

1.5 An Ideal Euclidean Model

A geometry in general is a system of definitions, postulates, and theorems that is built in a logical progression. Just what makes up a good geometry? The key is its set of postulates or basic assumptions. Mathematicians desire three qualities for a system of postulates. The system should be

1. consistent

2. independent

3. complete

Consistent postulate systems contain postulates that do not contradict one another. If two postulates of the system had opposite meanings, the system would be inconsistent and therefore defective. For example,

Postulate 1. If two lines intersect, then every other line in the plane intersects one of them.

Postulate 2. If two lines intersect, then there is exactly one line parallel to both.

According to Postulate 1, every line in the plane must intersect one of the two intersecting lines. However, Postulate 2 requires a line that is parallel to both and so does not intersect either. These postulates lead to contradictory conclusions. Inconsistent postulate systems must be exposed so that they do not lead us astray.

An *independent* postulate system is a system in which no postulate can be deduced or proved from the other postulates in the system. Therefore every postulate is independently necessary. An example of a system that is not independent is given here. Can you tell why it is not independent?

> *Postulate 1.* If the animal is a parakeet, then it is a bird.
> *Postulate 2.* If the animal is named Polly, then it is a parakeet.
> *Postulate 3.* If the animal is named Polly, then it is a bird.

The last qualification for an ideal postulate system is that it be *complete*. This means that every statement that can be expressed within the system can be proved or disproved from the postulates. This does not mean that the postulates must give every detail of the system but that they must form a sufficient base for a full description.

An ideal postulate system, then, is consistent, independent, and complete. But no ideal system can be constructed. Because man would tend to boast of a perfect system, God has limited man's logic and reason. In 1931, the modern mathematician Kurt Gödel proved mathematically that ideal postulate systems are impossible.

If we cannot have an ideal postulate system as a foundation for geometry, does geometry crumble? No. God is the foundation for geometry. God created geometry and He is consistent, independent, and complete.

How far can our understanding of God's geometry go? Our system must be consistent. God demands consistency: we are not to be hypocrites. The system can be independent—not independent of God, but of repetition. Since Jesus Himself repeated some things in His teaching, we can see that repetition has a place in teaching. However, no system is ever complete. Man cannot know everything about geometry (or any other subject). In summary, every useful postulate system in mathematics is consistent, some are independent, but none are complete.

The system presented in this book is consistent (so as not to lie or lead to contradictions), but it is not complete since no possible presentation of geometry could be, as Gödel proved. It is sufficiently complete, however, to help you understand God's creation, to develop your God-given thinking skills, and to prepare you for college. This book does not present an independent system. Some of the postulates overlap in order to make learning easier. A truly independent system for geometry is beyond the scope of this book, but you may study one in college.

The first five postulates in the system of geometry are called *Incidence Postulates*. Know them! The entire structure of geometry is built on these postulates (and twenty-three other postulates that will be introduced throughout this course). The word *incidence* comes from *in,* meaning "in" or "on," and *cadere*, meaning "to fall." Therefore, a point that is incident with a line falls on, or lies on, the line.

Incidence Postulates

Postulate 1.1

Expansion Postulate. A line contains at least two points. A plane contains at least three noncollinear points. Space contains at least four noncoplanar points.

Postulate 1.2

Line Postulate. Any two points in space lie in exactly one line.

Postulate 1.3

Plane Postulate. Three distinct noncollinear points lie in exactly one plane.

Postulate 1.4

Flat Plane Postulate. If two points lie in a plane, then the line containing these two points lies in the same plane.

Postulate 1.5

Plane Intersection Postulate. If two planes intersect, then their intersection is exactly one line.

You will need complete understanding of these five postulates to prove the incidence theorems in the next section.

Is the foundation of a structure important? It certainly is. If the foundation crumbles, then the whole structure falls. Our spiritual lives also need the strong foundation of the Word of God, the Bible, and faith in the shed blood of Jesus Christ. Is our spiritual foundation sure? (See Luke 6:47-49; Prov. 14:27; Ps. 71:3; I Sam. 2:2.)

New York City has 185 skyscrapers over 500 feet tall, more than any other city in the world. Notice the many points, lines, and planes in the skyline.

▶ A. Exercises

1. Name the three characteristics of an ideal system of postulates and give a brief explanation of each.

Observing the Line Postulate, draw as many lines as possible through the following points.

2.

4.

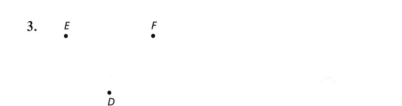

3.

Place the word *plane(s)*, *line(s)*, *point(s)*, or *space* in the blank to complete each sentence.

5. Three noncollinear _____ lie in a plane.
6. Two _____ either are parallel or intersect in a line.
7. An infinite number of lines can intersect in one _____.
8. Four noncoplanar points determine _____.
9. If a line lies in a certain plane, then there are at least two _____ that also lie in that plane.
10. Three distinct collinear points lie in a _____.

Draw a sketch for each figure described in exercises 11-15.
11. A line and a point that is not on the line
12. A plane that contains points *A*, *B*, and *C*
13. Three planes that intersect in one line
14. Two planes that do not intersect
15. A line and a plane that intersect in only one point

16. How many possible answers are there for the number of points of intersection between a line and a plane? Draw a line and a plane to illustrate each case.

Give the postulates (by name) that would verify the following statements.

17. Plane *k* contains points *A* and *C*; $\overset{\leftrightarrow}{AC}$ also lies in *k*.

18. If *m* is a plane, then *m* must contain three points *A*, *B*, and *C* that are noncollinear.

19. If point *K* lies on line *l*, then there exists another point that lies on *l*.

20. If plane *ABCD* intersects plane *ABEF*, then the planes intersect in $\overset{\leftrightarrow}{AB}$.

Identify each statement as true or false based on the five Incidence Postulates. If the answer is false, draw a diagram to illustrate.

21. Any three points lie in exactly one plane.

22. If three planes intersect, they intersect in exactly one line.

23. Given two points, there is exactly one line running through those two points.

24. A line determines exactly one plane.

▶ **C. Exercises**

25. Which postulate guarantees that a camera tripod set on a level floor will remain steady? Explain.

■ **Cumulative Review**

Identify each statement as true or false.

26. If $5 \in A$, and $A \subseteq B$, then $5 \in B$.

27. If $A \subseteq B$, and $B \subseteq C$, then $A \subseteq C$.

28. If $8 \in A \cap B$, then $8 \in A$ and $8 \in B$.

29. If $7 \in A \cup B$, then $7 \in A$ and $7 \in B$.

30. If $C \subseteq A \cup B$, then $C \subseteq A$ or $C \subseteq B$.

1.6 Incidence Theorems

Theorems are statements that can be proved by a logical progression of definitions, postulates, and previously proven theorems. The Incidence Theorems are theorems derived from the Incidence Postulates.

Sheepeater Cliff in Yellowstone National Park, Wyoming, displays the parallel lines of volcanic basalt.

Figure 1.11

These pictures suggest the first incidence theorem.

Theorem 1.1
If two distinct lines intersect, they intersect in one and only one point.

The pictures above are not sufficient proof of this theorem. You will not study formal proofs for a few chapters, but you can understand the reasoning behind the proof of this theorem.

Suppose two distinct lines did intersect in more than one point. If they intersect in at least two points, what do these two points determine? According to the Line Postulate, they determine exactly one line. How can we have two distinct lines and exactly one line at the same time? If both were true, the system would not be consistent. Something must be wrong. Since the only questionable statement supposes that two distinct lines intersect twice, the assumption must be false, and the lines must intersect in exactly one point.

The approach used in this proof is called "indirect proof." This approach will be discussed in Chapter 9 in more detail. The proofs of "only one" in the following theorems also require indirect proof.

Next consider the following diagram.

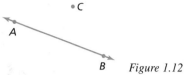

Figure 1.12

How many lines pass through both *A* and *B*?
How many planes can pass through $\overleftrightarrow{AB}$?
How many planes can pass through points *A*, *B*, and *C*?
How many planes pass through any three given noncollinear points?
How many planes pass through point *C* and $\overleftrightarrow{AB}$?

Your answers to these questions have helped you discover the following theorem.

Theorem 1.2

A line and a point not on that line are contained in one and only one plane.

The reasoning used in proving this theorem follows. Given a line and a point *P* not on the line, we know the line contains two points, since the Expansion Postulate states that every line contains at least two points. Call the two points that are on the line *A* and *B*. Because *P* does not lie on the line, points *A*, *B*, and *P* are noncollinear. Since we have three noncollinear points, we know that exactly one plane passes through these points (Plane Postulate). Since *A* and *B* are both in this plane, the Flat Plane Postulate guarantees that $\overleftrightarrow{AB}$ also lies in this plane. Thus there is exactly one plane that passes through a line and a point not on that line.

Theorem 1.3

Two intersecting lines are contained in one and only one plane.

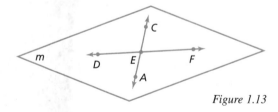

Figure 1.13

See whether you can supply the reasoning for the proof of Theorem 1.3. (See exercise 20.)

Theorem 1.4
Two parallel lines are contained in one and only one plane.

To give reasons for this theorem, we must first draw a picture to see whether we recognize that the theorem is true.

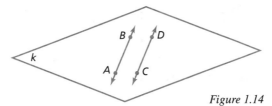

Figure 1.14

If $\overleftrightarrow{AB}$ is parallel to $\overleftrightarrow{CD}$, denoted by $\overleftrightarrow{AB} \parallel \overleftrightarrow{CD}$, then we know by the definition of parallel lines that they are coplanar and do not intersect. Since the lines are coplanar, some plane k contains them. Since the lines do not intersect, C is not on $\overleftrightarrow{AB}$. Since the three points A, B, and C are not collinear and plane k contains them, there can be no other plane containing them (Plane Postulate).

Be sure you understand the phrase "one and only one." This phrase means the same as the phrase "exactly one," which you saw in the postulates. "Exactly" means no more and no less.

Each domino has two numbers on it. In a double-six set of dominoes, the numbers range from blank (zero) to six. No two dominoes are alike.
1. How many dominoes are in a double-six set?
2. a double-nine set?
3. a double-twelve set?

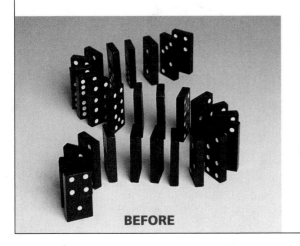

BEFORE

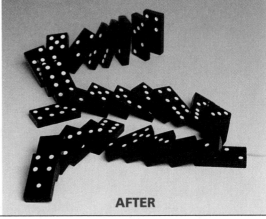

AFTER

For exercises 1-10 refer to the figure below.

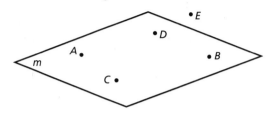

Answer each question, explain your answer, and state a definition, postulate, or theorem that supports your answer.

1. Does $\overleftrightarrow{AB}$ lie in plane *m*?
2. Does $\overleftrightarrow{EC}$ lie in plane *m*?
3. How many planes pass through points *A*, *E*, and *D*?
4. How many planes pass through $\overleftrightarrow{EA}$ and $\overleftrightarrow{AB}$?
5. How many planes pass through $\overleftrightarrow{CD}$ and point *E*?
6. If $\overleftrightarrow{AD}$ and $\overleftrightarrow{CB}$ are parallel lines, what guarantees that only one plane contains these lines?
7. What is the intersection of $\overleftrightarrow{ED}$ and $\overleftrightarrow{DB}$? Is there more than one point of intersection?
8. Can you draw any skew lines using the points in this diagram?
9. If $\overleftrightarrow{AD}$ is parallel to $\overleftrightarrow{CB}$, will these two lines ever intersect?
10. How many planes pass through $\overleftrightarrow{AB}$?

► **B. Exercises**

Give examples of the postulates and theorems stated in exercises 11-19, based on the diagram. All six faces of the solid are rectangles.

11. A line and a point not on that line are contained in one and only one plane.
12. Three noncollinear points are contained in one and only one plane.
13. Two parallel lines are contained in one and only one plane.
14. If two lines intersect, then they intersect in one and only one point.

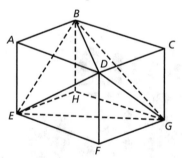

15. If two lines intersect, then the lines are contained in one and only one plane.
16. If two planes intersect, then they intersect in a line.
17. Are $\overleftrightarrow{BC}$ and $\overleftrightarrow{EF}$ skew lines?
18. What plane intersects all the other planes shown?
19. Give the reasoning behind this statement: "There are at least three lines in any plane."

▶ C. Exercises

20. Prove Theorem 1.3. Use Figure 1.13 (shown).

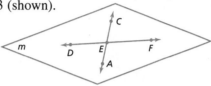

▶ Dominion Thru Math

Look at the world around you, both man-made and created. Notice the geometric shapes and concepts everywhere.
21. Give examples of lines that do not look parallel but are.
22. What causes parallel lines not to look parallel?
23. Give examples of lines that look parallel but are not.
24. What makes them (answers to exercise 23) intersect?
25. Give examples of lines that look straight but are curved.
26. What makes them (answers to exercise 25) curved?

■ Cumulative Review

27. Name three undefined terms.
28. Name three defined terms.
29. Name a postulate.
30. State a theorem.
31. According to the postulates and theorems so far, must there be an infinite number of points in space? Explain.

EUCLID

Euclid (ca. 330-275 B.C.) became a teacher of geometry at the great ancient school in Alexandria, Egypt. He encountered two attitudes among his students that are still common today. Some wished to learn geometry overnight, without having to study and discipline themselves. Tradition says his response to this attitude was, "There is no royal road to geometry." A Christian teacher would add that God promises to bless diligence (Gen. 3:19). Another student complained to Euclid: "What do I gain by learning all this stuff?" Euclid replied that knowledge is valuable in itself but gave the student some coins so that he would "profit" from his learning. Euclid understood that there is more to life than making a living. Christian teachers would qualify his remark by explaining why knowledge is valuable—since the source of knowledge is God, knowledge helps us understand the Creator and His creation.

Euclid disciplined his students' minds with a thirteen-volume work on geometry called *The Elements*, which is the earliest existing systematic treatment of geometry. Although others had written on geometry, Euclid can be called the Father of Geometry because he presented geometry systematically with proofs of theorems based on postulates and definitions.

Here is a list of the contents of his thirteen volumes:

Vol. I—Plane geometry including the Pythagorean theorem

Vol. II—Geometry for solving algebra problems, including the quadratic formula

Vols. III, IV—Theorems on circles

Vol. V—Discussion of proportions

Vol. VI—Proportions for studying similar figures

Vols. VII, VIII—Properties of numbers

Vol. IX—Prime and perfect numbers, including that there are infinitely many primes

Vol. X—Geometry of irrational distances expressed as square roots

Vols. XI, XII, XIII—Three-dimensional (solid) geometry

Euclid saw the importance of careful reasoning in the development of geometry. He had five basic postulates, although only one of them matched a postulate in this book: Two points determine a unique line. He considered some of the postulates to be obvious common notions. He called these special postulates *axioms*, but today the terms *postulate* and *axiom* are synonymous. Euclid did not understand the need for undefined terms. He uselessly defined *point* as "that which has no part." While the system has been refined since Euclid's time, the system established is still rightfully named in his honor. Euclidean geometry is the main subject of this whole book!

Euclid's works, like the Scriptures, would not exist today if they had not been hand copied many times. The oldest known copy of *The Elements* is dated A.D. 380, at least 650 years after it was written. In contrast, the oldest copies of the New Testament are from the second century A.D., less than 100 years after its writing. No scholars dispute the reality of Euclid, his existence 300 years before Christ, or his teachings. Liberals, however, attack the apostles on these grounds, even though there are many times more ancient manuscripts of the New Testament and archaeological finds to confirm Bible history.

Euclid saw the importance of careful reasoning in the development of geometry.

1.7 Sketches and Constructions

Diagrams are a very important part of geometry. In this section you will learn three different methods of making diagrams: sketching, drawing, and constructing geometric figures.

A *sketch* is a careful freehand picture made only with a writing instrument such as a pencil. If you sketch a rectangle, it may look like this:

Figure 1.15

Notice that the figure is not accurate in angle measure or side measure, but it roughly represents a rectangle. Such a sketch can be a valuable aid when you are working geometric word problems or proving theorems.

A *drawing* of a geometric figure is made with the aid of tools. These tools may include a ruler, a protractor, a compass, a parallel bar, a T-square, drafting triangles, or French curves. With rulers and protractors you can measure lengths and angles. By correctly using these instruments, you can make drawings that are quite accurate.

Since the development of computer-assisted drawing (CAD), few professional draftsmen still use hand-held instruments like plastic triangles and French curves. The computer actually draws by a coordinate plane system that allows points to be connected by the computer either as straight lines or various curves. CAD drawings are not any faster to produce, but they are very accurate and neat and changes can be made easily. They also make it possible to include intricate details that would require too much expensive labor using hand-held instruments.

To ancient Greek mathematicians, only special kinds of drawings were acceptable. They would not accept drawings made with measuring devices. Yet as far back as the time of Euclid, Greek mathematicians were making accurate constructions and learning much geometry from them.

The Parthenon on the Acropolis in Athens is the most famous example of architecture from ancient Greece.

A **construction** is a drawing made with the aid of only two instruments: an unmarked straightedge and a compass.

The straightedge is used only for making straight lines through two points. The straightedge cannot be used in any way for measuring. The compass is used to make circles and arcs and to mark off segments of a certain length.

Constructions are important in the study of geometry because they help you see clear physical relationships that are models of mathematical theory. If you understand these physical relationships, you will better understand related mathematical concepts. Constructions must always be accurate and neat.

Remember the differences between these three instructions.

Sketch

permits you to make a neat freehand picture.

Draw

allows you to use any tools you desire, including rulers and protractors, but freehand is forbidden.

Construct

requires a drawing using only a straightedge and a compass.

onstruction 1

Circle

> *Given:* Two points *A* and *B*
> *Construct:* A circle with a radius having a length
> equal to the distance between *A* and *B*

1. Place the point of the compass on point *A*.
2. Adjust the pencil to rest on point *B*.
3. Without changing the compass width, rotate the compass completely around point *A*.

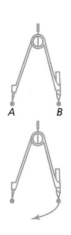

Unless otherwise specified, every diagram in this book is a drawing but not a scale drawing. View these drawings as you would view your own sketches. They are convenient visual aids, but they are not intended to convey accurate lengths or angle measures.

▶ A. Exercises

Do all work carefully and neatly.
1. Sketch $\overleftrightarrow{AB}$.
2. Draw $\overleftrightarrow{AB}$.
3. Construct $\overleftrightarrow{AB}$.
4. Which two of the above amounted to the same thing?
5. Sketch a circle.
6. Draw a circle using a soup can, protractor, or template.
7. Construct a circle.

Identify which diagram methods could use each tool.
8. Ruler
9. Pencil
10. Compass
11. Protractor
12. Template
13. Colored pens
14. Straightedge

▶ B. Exercises

15. Sketch a two-inch segment.
16. Draw a two-inch segment.
17. Can you construct a two-inch segment?

Label two points X and Y on a piece of paper. Measure the distance between them with your compass. Use that distance as a radius and construct a circle with its
18. center at X.
19. center at Y.

▶ C. Exercises

20. How many lines are determined by n coplanar points, no three of which are collinear? Explain your answer.

► Dominion Thru Math

Construct a daffodil. You must construct only arcs of circles in each step.

21. On a clean piece of paper make a circle with a 6 cm radius. Aligning your straightedge with the center, mark the ends of a diameter.

22. From each end of the diameter, mark two more points on the circle by making an arc with the radius of the circle. The six points form a hexagon at the tips of the petals. Connect every other point with a light line to see two overlapping equilateral triangles forming a six-pointed star. (Note the smaller hexagon in the middle.)

23. Connect each vertex of the inner hexagon to the center of the circle with a light line. Notice each petal forming as a diamond-shaped rhombus. Construct the center circle (end of the cone) with a 1.8 cm radius.

24. Construct every other petal (the other three petals are behind these). Construct arcs of circles with the centers as the middle two vertices of the rhombuses and the radii as the width of the rhombus. The arc starts at the center circle (edge of the flower's conical center) and goes to the end of the petal on the original circle. Now draw the last three petals. Be careful to stop at the edge of each of the first three petals, since the last three are behind them. Add color.

■ Cumulative Review

25. Which term describes skew lines: intersecting, complementary, disjoint, equal, empty?

26. Sketch the set statement $A \cap B \cap C$ using a Venn diagram.

27. Sketch a pair of parallel planes and a pair of intersecting planes.

28. How many planes are determined by four noncoplanar points where no three are collinear?

29. If $A = \{2, 4, 6\}$ and $B = \{2, 4, 8\}$, describe the relationship between sets.

Geometry and Scripture

God's Consistency

Do you remember the three historic goals for an axiomatic system (a system based on postulates)? All three are qualities of God.

Give the letter of the characteristic that each verse illustrates.

1. Isaiah 40:13-14

2. Romans 9:18

3. Ephesians 4:6

4. II Timothy 2:13

A. Consistent—no contradictions in God

B. Independent—God's knowledge, power, and will do not depend on anything or anyone else

C. Complete—nothing missing from God

We have seen that man can attain only some of these goals. One attainable goal is consistency. In fact, we are commanded to be consistent and not to contradict ourselves. Identify the letter of the command that best describes each verse. Use each letter once.

5. Exodus 20:16

6. Matthew 6:5

7. Ephesians 4:25

8. I John 2:21

9. James 3:10-11

A. Speak the truth only.

B. Pure sources produce only purity.

C. Don't intermix false and true testimony.

D. Falsehood is not part of the truth.

E. Don't be inconsistent hypocrites.

Independence is attainable but not always desirable. Repetition is valuable in teaching even though it may be less efficient. An independent postulate system is beyond the scope of this book.

10. Turn to Luke 6. In which two verses does Christ repeat the command to love our enemies?

> **HIGHER PLANE:** Use a concordance to find this command in another book of the Bible.

Completeness in a postulate system is impossible for men according to Gödel's theorem. In spite of man's limitation, God does want us to study and reason and thus develop our skills. For each verse, identify which side of this balance is emphasized: limitation (L) or responsibility (R).

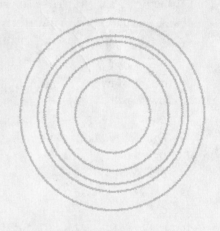

11. II Timothy 2:15

12. Job 39:1-2, 26; 40:1-2

13. Proverbs 23:23

14. Matthew 6:27

15. I Corinthians 1:20

16. II Corinthians 10:5

According to the principles above, God is consistent, independent, and complete. Geometry exalts God for all three qualities but reflects consistency most clearly. God's consistency is stressed in the theme verse.

> ### Line upon Line
>
> **I**F WE BELIEVE not, yet he abideth faithful: he cannot deny himself. 🐚
>
> **II TIMOTHY 2:13**

Chapter 1 Review

Write the following sets, both by listing and using set-builder notation.

1. The states of the United States whose names start with a C
2. The students in the class who have at least one brother

Write a phrase that describes each mathematical phrase.

3. $K \subset L$
4. $\varnothing \subset M$
5. $x \in B$
6. $\{l, m\} \subseteq P$
7. $k \notin M$
8. Give an example of two sets that are equal.
9. Give an example of two sets that are equivalent but not equal.
10. Draw a Venn diagram to illustrate these sets.

   $A = \{blue, black, brown\}$
   $B = \{orange, black, red, purple\}$
   $C = \{brown, gray, black\}$

Let the set of even natural numbers be the universal set. Use the sets below to answer exercises 11-19.

   $D = \{2, 6, 8\}$ $E = \{2, 8, 10, 12\}$ $F = \{4, 14\}$

11. Draw a Venn diagram.
12. $D \cap E$
13. $E \cup F$
14. $\varnothing \cap D$
15. $D \cap F$
16. $(E \cup F) \cap D$
17. D'
18. F'
19. $(E \cap F)'$

20. Is every set equal to itself? Is every set equivalent to itself?
21. If set A is equivalent to set B, and set B is equivalent to set C, what can you conclude? What if $A = B$ and $B = C$?
22. If set A is equivalent to set B, is set B equivalent to set A? What if $A = B$?

Use the lines and points shown for exercises 23-28. Use three or four points to name a plane. All six faces are rectangles.

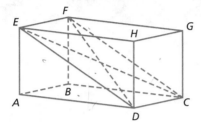

23. Name the planes.
24. Name six pairs of parallel lines.
25. Name three pairs of parallel planes.
26. Give four noncoplanar points.
27. Name two pairs of intersecting lines.
28. What planes intersect in $\overleftrightarrow{BD}$?
29. Define *construction.*
30. Explain the mathematical significance of II Timothy 2:13.

One of the great architectural feats of history was the construction of the Pyramid of Cheops, the Great Pyramid.

In order to build that edifice—it took 100,000 men thirty years to complete—the Egyptians had to be experts in the field of geometry.

How would you like the job of building the Great Pyramid? No problem, right? After all, you need only to build a square pyramid out of 2,300,000 separate blocks averaging $2\frac{1}{2}$ tons each. How do you suppose the pharaoh, who claimed to be a god, would react if you made an error in the calculations for his sacred burial place?

The head engineer did very well for Cheops. On a base of 13.1 acres with almost perfect right angles, he built the pyramid 481.4 feet high. The difference between the longest and the shortest side is only 7.9 inches. Each side is aligned with the true compass points—north, south, east, west. The inside corridors have almost exactly the same gradients, and the exterior inclines are each 51°52′.

There is one other thing to consider. How would you like to trust your eternal life to a god who hired you to build his tomb? Isn't it reassuring to have placed your confidence in the God who left behind an empty tomb? Christ said, "Because I live, ye shall live also." "And I give unto them eternal life; and they shall never perish, neither shall any man pluck them out of my hand" (John 10:28).

After this chapter you should be able to

1. define various subsets of lines, planes, and space.

2. use proper notation for rays, half-lines, segments, angles, circles, and betweenness.

3. state the separation postulates and the Jordan Curve theorem.

4. classify curves and surfaces.

5. classify polygons by number of sides and convexity.

6. classify polyhedra based on the number of faces.

7. distinguish among closed regions, closed surfaces, and solids.

2.1 Subsets of Lines

What is a subset of a line? What are lines made from? A line is made up of points. If this is true, then a subset of a line is a set of points. When you think of a line and a point on that line, what relationships exist between that point and the line? Look at the following representation of a line with a particular point marked.

This roller coaster in Georgia is supported by many segments.

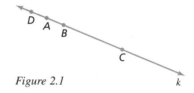

Figure 2.1

Points *B* and *C* are on the *same side* of point *A* on the line.

Definition

A **half-line** is the set of all points on a line on a given side of a given point of the line.

The notation for half-line *AB* is $\overset{\circ}{AB}$. Notice the open circle at the left end of the arrow. The open circle means that point *A* is not included in the half-line.

Keeping this definition in mind, you should see that point *A* divides line *k* into three sets.

Postulate 2.1

Line Separation Postulate. **Every point divides any line through that point into three disjoint sets: the point and two half-lines.**

According to the Line Separation Postulate, point *A* in figure 2.1 separates line *k* into three sets. The three subsets of line *k* are the sets containing point *A*, $\overset{\circ}{AB}$, and $\overset{\circ}{AD}$. The point *A* that divides the line into half-lines is called the *origin* of each half-line. Notice that the origin is not part of the half-line.

A **ray** is the union of a half-line and its origin. It extends infinitely in one direction from a point.

The rays form sunbeams, illuminating the Grand Canyon in Arizona.

Figure 2.2

Figure 2.2 represents ray *CD*. The notation for ray *CD* is $\overrightarrow{CD}$. Notice that the origin, or *endpoint,* of the ray is always named first in the notation and that the arrow always points to the right. What is the difference in the notation for a half-line and a ray?

▶ A. Exercises

Give the proper notation for the following figures.

1.   X   Y

2. C   D

3. S   T

Use the figure below for exercises 4-10.

4. Name three points that are on $\overrightarrow{AB}$.
5. How many disjoint sets are formed on $\overleftrightarrow{AD}$ by the point C? Name them.
6. What is the endpoint of $\overrightarrow{BC}$?
7. Explain why $\overrightarrow{BC}$ and $\overrightarrow{BD}$ are the same ray.

8. Are $\overrightarrow{AC}$ and $\overrightarrow{CA}$ the same ray? Explain.
9. Name $\overrightarrow{AC}$ in two other ways.
10. Name three half-lines.

▶ B. Exercises

Refer to the figure above exercise 4.
11. Find $\overleftrightarrow{AB} \cup \{A\}$
12. $\overrightarrow{BC} \cap \overrightarrow{BA}$
13. $\overleftrightarrow{CD} \cap \overleftrightarrow{CA}$
14. $\overrightarrow{BC} \cup \overrightarrow{BA}$
15. $\overrightarrow{AC} \cap \overleftrightarrow{BC}$
16. $\overrightarrow{DC} \cup \overrightarrow{DA}$
17. $\overrightarrow{BD} \cup \overrightarrow{CA}$
18. List all the rays.
19. Name two physical phenomena that are rays.

▶ C. Exercises

20. Name as many rays as possible in the figure.

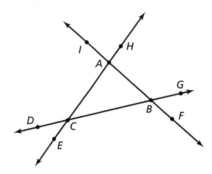

■ Cumulative Review

Identify the undefined term best illustrated in each example.
21. a piece of paper
22. an electron
23. a crease in a garment
24. a ramp for loading a truck
25. a pixel (smallest dot on a computer screen)

2.2 Betweenness

What does the word *between* mean? *Between* is a preposition that means "in an intervening position, separating." The Christian must be sure that he lets nothing separate him from fellowship with God. What is the one thing that can separate a Christian and God? *Sin*. By looking at Adam, we can see the broken fellowship that occurs between man and God because of sin (Gen. 3:23-24). God created Adam to have fellowship with Him, but sin came between Adam and God and broke the original fellowship between man and God. Because of Adam's sin, we all are sinners and all have broken fellowship with God (Rom. 5:12). How gracious God is to redeem man and restore fellowship by the precious blood of His only begotten Son, Jesus Christ (Rom. 5:9; I Pet. 1:18-19).

What does *between* mean in geometry?

Triangular trusses support this bridge over the Columbia River at Astoria, Oregon. At 1232 feet long, it is the longest truss bridge in North America.

Definition

B is **between** A and C if $\overrightarrow{BC} \cap \overrightarrow{BA} = \{B\}$ when A, B, and C are collinear. In symbols, you can write A-B-C.

$$\overset{\longleftrightarrow}{\underset{\textstyle A \quad B \qquad\qquad C}{\bullet\qquad\bullet\qquad\qquad\bullet}}$$

Figure 2.3

Show that C is not between A and B.

Answer If C is between A and B, then by the definition of *betweenness,* $\overrightarrow{CA} \cap \overrightarrow{CB} = \{C\}$, which is false. Notice that $\overrightarrow{CA} \cap \overrightarrow{CB} = \overrightarrow{CA}$.

The definition of betweenness is important in defining other geometric figures, such as *opposite rays* and *line segments.*

Definitions

$\overrightarrow{BA}$ and $\overrightarrow{BC}$ are **opposite rays** if and only if B is between A and C.

A **segment** is the set consisting of two points A and B and all the points in between. The symbol for segment AB is $\overline{AB}$. $\overline{AB} = \{A, B\} \cup \{X | A\text{-}X\text{-}B\}$.

A   B   *Figure 2.4*

A geometrical concept that is often confused with a ray because of the similarities in notation is a *vector.* Vectors will not be studied in this course, but you should be aware of the notation and the general meaning of a vector. A vector is a directed line segment. Vector AB, denoted by $\overrightarrow{AB}$, goes from A to B in the direction of the ray, but it has only the finite length of the segment. Do not confuse rays and vectors; they are different.

It is important for you to use precise and accurate notation in geometry. The following table summarizes the correct notation that you have learned so far.

Symbol	Meaning
A	point A
$\overleftrightarrow{AB}$	line AB
$\overset{\circ\!\!\rightarrow}{XY}$	half-line XY
$\overrightarrow{CD}$	ray CD
$\overline{AC}$	segment AC
$\overrightarrow{LM}$	vector LM

▶ A. Exercises

Use the correct notation for the following phrases.

1. line KM
2. ray OP
3. point X
4. half-line FA
5. vector LO

Use the figure for exercises 6-12.

6. Is *F* between *C* and *K*? Use the definition to explain.
7. What other points is *F* between?
8. Is *B* between *L* and *M*? Explain.
9. Name two pairs of opposite rays.
10. Are $\overrightarrow{FB}$ and $\overrightarrow{FM}$ opposite rays? Why?
11. Name four segments.
12. Name two pairs of rays that are not opposite.

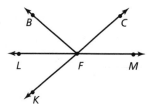

▶ B. Exercises

Use the figure below and correct notation to describe each set of points in exercises 13-21.

$$\overset{\leftrightarrow}{\underset{\overset{\bullet}{A}\ \ \overset{\bullet}{B}\ \ \overset{\bullet}{C}\ \ \overset{\bullet}{D}}{}}$$

13. $\overrightarrow{AB} \cup \overrightarrow{BC}$
14. $\overrightarrow{AB} \cup \overrightarrow{DC}$
15. $\overset{\circ\circ}{BC} \cap \overrightarrow{AC}$

16. $\overset{\circ\circ}{CB} \cap \overrightarrow{AB}$
17. $\{A\} \cap \overset{\circ}{\overrightarrow{AB}}$
18. $\{A\} \cup \overset{\circ}{\overrightarrow{AB}}$

19. $\overrightarrow{DC} \cap \overleftrightarrow{AC}$
20. $\overline{BC} \cup \overrightarrow{CD}$

21. Name two rays that have *B* as their endpoint. What is the special name that these two rays have?
22. Draw two rays that are not opposite but have the same endpoint.
23. What is the intersection of $\overrightarrow{AX}$ and $\overrightarrow{XA}$?

▶ C. Exercises

24. A segment without its endpoints is an *open interval*. What symbol should be used for an open interval? Why?
25. What is the intersection of $\overset{\circ}{\overrightarrow{AB}}$ and $\overrightarrow{BA}$?

■ Cumulative Review

Name the postulate that justifies each statement.
26. *B* is between *A* and *C*; therefore $\overset{\circ}{\overrightarrow{BA}} \cup \{B\} \cup \overset{\circ}{\overrightarrow{BC}} = \overleftrightarrow{AC}$.
27. *A*, *B*, and *C* are collinear. *B*, *C*, and *D* are also collinear. Therefore *A*, *C*, and *D* are collinear, and $\overleftrightarrow{AB} = \overleftrightarrow{BC} = \overleftrightarrow{CD}$.

Assume that *A-B-C* and *X-B-Y*. What kind of lines are $\overleftrightarrow{AC}$ and $\overleftrightarrow{XY}$, if
28. *A*, *X*, and *B* are not collinear.
29. *A*, *X*, and *B* are collinear.
30. Suppose *B* is between *A* and *C* and *C* is between *B* and *D*. Draw two conclusions (include a sketch).

2.3 Subsets of Planes

What is a plane? Recall that a plane in geometry can be described as a set of points that extends infinitely and has both length and width but no thickness. How could you describe a subset of a plane?

Cable-stayed bridges, such as the Tenpozan Watasi Bridge at Osaka, Japan, join the roadway to the towers with cables to form triangles.

$\overleftrightarrow{AB}$ divides plane *k* into three sets. This fact should sound familiar. Remember that a point divides a line into three sets: the point and two half-lines. Similarly, a line divides a plane into three sets: the line and two half-planes. In plane *k*, $\overleftrightarrow{AB}$ divides the plane into half-planes, p_1 and p_2. This discussion can be summarized by the Plane Separation Postulate.

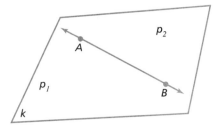

Figure 2.5

Postulate 2.2
Plane Separation Postulate. **Every line divides any plane containing the line into three disjoint sets: the line and two half-planes.**

From the Plane Separation Postulate, we can define several terms.

Definitions

A **half-plane** is a subset of a plane consisting of all points on a given side of a line in the plane. If points *P* and *Q* are in the same half-plane, then so is the segment joining them.

An **edge of a half-plane** is the line that separates the plane into two half-planes. The line is not part of either half-plane.

Opposite half-planes are the two half-planes that are separated by a particular line of the plane. If points *P* and *R* are in opposite half-planes, the segment joining them must intersect the edge.

Figure 2.6 illustrates these terms.

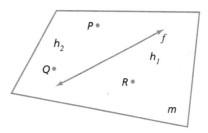

Notice that plane *m* is separated into three subsets: line *f* and half-planes h_1 and h_2. The edge of half-plane h_2 is *f*. What is the edge of half-plane h_1? Notice that h_1 and h_2 are opposite half-planes.

Since $\overline{PR}$ intersects *f* at one point, *P* and *R* are in opposite half-planes. Since *P* and *Q* lie in h_2, so does $\overline{PQ}$.

Figure 2.6

Another subset of a plane is an angle.

Definitions

An **angle** is the union of two distinct rays with a common endpoint.

The **sides of an angle** are the two rays that form the angle.

The **vertex of an angle** is the common endpoint (origin) of the two rays.

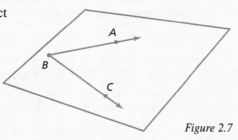

Figure 2.7

Figure 2.7 illustrates angle *ABC*, denoted by $\angle ABC$. Notice that $\overrightarrow{BC}$ and $\overrightarrow{BA}$ are two distinct rays, that $\overrightarrow{BC} \cap \overrightarrow{BA} = \{B\}$ and that $\overrightarrow{BC} \cup \overrightarrow{BA} = \angle ABC$. The vertex of $\angle ABC$ is *B*. Notice that the vertex is the middle point in the notation. $\overrightarrow{BC}$ and $\overrightarrow{BA}$ form the sides of the angle.

Now consider the lines containing the sides of ∠CDE. In figure 2.8 $\overleftrightarrow{CD}$ separates the plane into two half-planes, one of which contains point E. Likewise, $\overleftrightarrow{DE}$ separates the plane into two half-planes, one of which contains point C. The intersection of these two sets forms the interior of the angle.

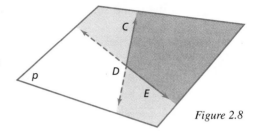

Figure 2.8

Definitions

The **interior of an angle** is the intersection of the two half-planes each determined by a side of the angle and containing the other side (except for the vertex).

The **exterior of an angle** is the complement of the union of the angle and its interior.

The interior of ∠CDE is shaded in figure 2.9.

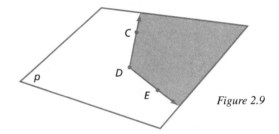

Figure 2.9

▶ A. Exercises

Use the figure shown for exercises 1-2.
1. Name two sets of collinear points.
2. Describe two pairs of half-planes and give the lines that determine them.

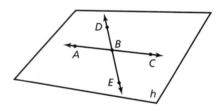

Use the angle in the figure for exercises 3-8.
3. Name the angle.
4. Give three other names for the angle.
5. Name three points in the exterior of the angle.
6. Name three points in the interior of the angle.
7. Name the sides of the angle.
8. Name the vertex of the angle.

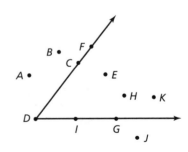

Use the next figure for exercises 9-15.

9. Name the two half-planes.
10. Name the edge of these two half-planes.
11. Find $s_1 \cup s_2 \cup k$.
12. What postulate justifies your answer to exercise 11?

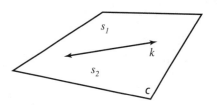

▶ B. Exercises

13. Find $s_1 \cap k$.
14. Find $(s_1 \cup s_2)'$.
15. Find $(s_2 \cup k)'$.

Use the figure for exercises 16-23.

16. Name ∠1 in another way.
17. Name a point in the interior of ∠CFE.
18. Name a point in the exterior of ∠AFC.
19. Name three points in the interior of ∠AFE.
20. Name all the angles shown, using letters.
21. Find ∠1 ∩ ∠3.
22. Find ∠1 ∪ ∠2.
23. Name three angles that have $\overrightarrow{FA}$ as a side.

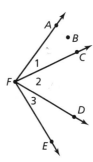

▶ C. Exercises

24. Draw two angles that have no interior region in common and whose intersection is $\overrightarrow{AB}$.
25. List all the angles determined by the figure. You should be able to list eighteen.

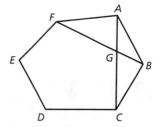

■ Cumulative Review

Use the diagram to decide whether each statement is true or false.

26. $\overline{AB} \in \overleftrightarrow{AB}$
27. $B \in \overrightarrow{AB}$
28. $A \subseteq l$
29. $\overleftrightarrow{CQ} \subseteq r$
30. $l \in r$

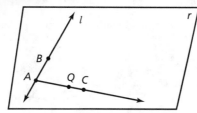

DESIGNERS AND THEIR PATTERNS

The oldest and best examples of designs come from the arts and crafts of Indians. Indian artists cleverly developed designs by rotating or reflecting some basic shape to cover a surface. In geometry we call such patterns *tessellations*.

Today clothing designers still use patterns in dresses, suits, and ties. Designers use geometric ideas to create the multitude of geometric designs used in these patterns. And such designs are not limited to clothing. The production of many commercial products like wallpaper, textiles, flooring, furniture, and decorator items requires the creative use of geometry. Interior design magazines show beautiful pictures of rooms decorated with patterns on expensive wallpaper or plush furniture. Stop by a store that sells carpet and vinyl flooring and look over their sample books.

Artists who design the layout work for advertising and public relations brochures use techniques of perspective and pattern to capture attention. Many eye-catching company logos are just ordinary geometric shapes. Those in graphic arts who prepare printed material must be aware of the geometry of moiré patterns. These are undesirable background patterns that form when multicolor halftone screens are not rotated correctly with respect to the other colors. An error as

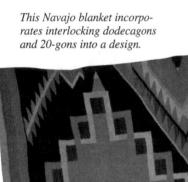

This Navajo blanket incorporates interlocking dodecagons and 20-gons into a design.

small as 0.1 degree between screen angles can cause undesirable moiré in areas where three or four colors are printed.

Designers use geometric ideas to create the multitude of geometric designs used in these patterns.

You may wonder where the Indians got their geometric ideas and why so many designs have a geometric look. The answer can be found by examining God's creation. God designed snowflakes in the shape of a hexagon with intricate symmetry and beauty. Birds and fish have an aerodynamic structure that God knew would help them soar through the air or glide smoothly through the water. Only in recent years have men begun to design planes and cars to provide the same low drag factors that birds and fish have enjoyed all along.

An entrance incorporates patterns in tiling and wallpaper.

Interior decorators use patterns in rugs, upholstery, curtains, and wallpaper.

2.4 Curves and Circles

The spokes of the mountain bike form radii for the circular wheels.

You will need to remember the definition of triangle before considering the main concepts of this section.

Figure 2.10

Definition

A **triangle** is the union of segments that connect three noncollinear points. A triangle is designated by the symbol △ followed by the three noncollinear points.

Each segment is a *side* of the triangle, and each of the three noncollinear points is a *vertex*. An *angle* of a triangle is one of three angles having its vertex at a vertex of the triangle and having rays containing sides of the triangle.

When there is no ambiguity, the angles can also be named by the vertex: $\angle A$, $\angle B$, and $\angle C$ respectively. Furthermore, each vertex (and therefore each angle) has a corresponding opposite side. The side opposite vertex A is the side that does not contain point A, namely $\overline{BC}$. What side is opposite $\angle ACB$?

The word *triangle* means "three-angled" from the Latin words *tri* (three) and *angulus* (angle). Similarly the word *trinity* comes from the Latin word *trinus*, which means "triune" or "threefold." The trinity describes our threefold Godhead: the Father, the Son, and the Holy Spirit. Matthew 28:19 lists all three as equals: "Go . . . and teach . . . baptizing them in the name of the Father, and of the Son, and of the Holy Ghost."

Place your pencil on a sheet of paper and, without lifting your pencil, move it so that you make a mark across the paper. The mark that you just made is called a *curve*. A curve is a continuous set of points.

(a)   (b)   (c)   (d)

Figure 2.11

Trace each curve shown above. Your pencil marked the same point twice on some of them. When you trace (*c*) and (*d*), you end on the same point where you began. These are *closed curves*. When you trace (*a*) and (*c*), you never cross your path. These are *simple curves*. Which curve could be called a simple closed curve?

Definitions

A **closed curve** is a curve that begins and ends at the same point.

A **simple curve** is a curve that does not intersect itself (unless the starting and ending points coincide).

A **simple closed curve** is a simple curve that is also a closed curve.

Curves need not lie in a plane. A cat can show you a very complex space curve all over your bedroom if you give it a ball of yarn. However, curves in this book will be *planar curves*—that is, curves lying in a plane.

The bridge over the Arkansas River at the Royal Gorge in Colorado is 1053 feet above the water. The cables supporting a suspension bridge form a curve that is not an arc.

EXAMPLE Classify a triangle as a curve.

Answer First verify that it is a curve. Since you can make a triangle without lifting your pencil, a triangle is a continuous set of points (or a curve).

Since you ended where you began, it is a closed curve. Since it intersects itself only at the end, it is a simple curve.

A triangle is a simple closed curve.

Choose a point in a plane and draw a simple closed curve around this point so that every point of the curve is the same distance from this point. The resulting curve is a circle, the most familiar simple closed curve.

Definitions

A **circle** is the set of all points that are a given distance from a given point in a given plane.

The **center** of the circle is the given point in the plane.

A **radius of a circle** is a segment that connects a point on the circle with the center. (The plural of *radius* is *radii*.)

A **chord of a circle** is a segment having both endpoints on the circle.

A **diameter** is a chord that passes through the center of the circle.

An **arc** is a curve that is a subset of a circle.

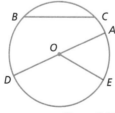

Semicircular arcs have long been used to support bridges. The bridge over New River Gorge in West Virginia has a main span 1700 feet long, making it the longest steel arch bridge in the world.

Circles are named by the center point, and the symbol ⊙ means "circle." The circle in figure 2.12 is ⊙O.

In ⊙O, $\overline{OA}$ is a radius; $\overline{OD}$ and $\overline{OE}$ are also radii. $\overline{DA}$ is a diameter of ⊙O, and $\overline{BC}$ is a chord of ⊙O. The red portion of ⊙O from *A* to *E* is arc *AE*, denoted by $\overset{\frown}{AE}$. The interior of the circle is shaded below.

Figure 2.12

Figure 2.13

The **interior of a circle** is the set of all planar points whose distance from the center of the circle is less than the length of the radius (*r*).

The **exterior of a circle** is the set of all planar points whose distance from the center is greater than the length of the radius.

Like the circle, the simple closed curve shown in figure 2.14 divides the plane into three sets of points: the curve itself, the interior of the curve, and the exterior of the curve. Which set is shaded?

The next theorem makes a generalization from these curves. Try to generalize on your own before you read it.

Figure 2.14

Theorem 2.1

Jordan Curve Theorem. **Any simple closed curve divides a plane into three disjoint sets: the curve itself, its interior, and its exterior.**

The general proof of this theorem is extremely difficult but here is the proof for circles.

Given any circle, let *C* be the center and the radius *r*. For any point *P* in the plane, its distance from the center is a number *d*. This number must be less than *r*, equal to *r*, or more than *r*. Therefore, according to the definitions, the point must be in the interior of the circle if the distance is less than *r*, on the circle if the distance equals *r*, or in the exterior of the circle if the distance is more than *r*. Thus any point in the plane must be in one of these three disjoint sets.

Because this theorem applies only to simple closed curves, it will help to have a name for a figure such as figure 2.14.

Definition

A **region** is the union of a simple closed curve and its interior. The curve is the *boundary* of the region.

▶ A. Exercises

Refer to this diagram for exercises 1-5. Give the name for each figure.

1. $\overline{XD}$
2. $\overline{AB}$
3. $\odot X$
4. $\overarc{CA}$
5. What is the intersection of the interior and exterior of the circle?

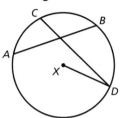

Use the figure for exercises 6-10.

6. Name the circle in this diagram.
7. Name three chords.
8. Name five radii. What do you notice about their lengths?
9. Name two arcs.
10. If the length of a radius is *x*, what is the length of a diameter? How many diameters of a circle are there?

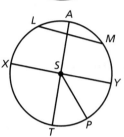

Classify each figure as a curve, a closed curve, a simple curve, a simple closed curve, or not a curve. Use the most specific term possible.

11.

12.

13.

14.

15.

16.

17.

▶ B. Exercises

Use the figure for exercises 18-22.

18. Name six triangles.
19. Name all the angles.
20. $\triangle ABC \cap \triangle ADE$
21. $\triangle ABD \cap \triangle ADE$
22. What side is opposite $\angle B$ in the large triangle?
23. If X, Y, and Z are noncollinear, find $\overline{XY} \cup \overline{YZ} \cup \overline{XZ}$.
24. Explain why the definition of *triangle* states that the three points are noncollinear.

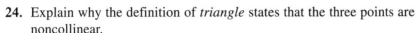

Use the figure shown for exercises 25-29. S is the region bounded by rectangle *AHFC*. Tell whether the statements are true or false.

25. $\triangle BCI \cap S = \triangle BCF$
26. $CDEF \cap S = \overline{CF}$
27. $S \cap BGED = BGFC$
28. $\triangle BGF \cap \triangle BCF = \{I\}$
29. $ABGH \cup BGFC = ACFH \cup \overline{BG}$

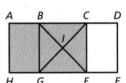

▶ C. Exercises

30. Explain logically why the following statement is true: Each triangle is contained in one and only one plane.

▶ Dominion Thru Math

31. Study the blanket and rug patterns on pages 52-53 and then make your own design for a rug, blanket, or wallpaper. Try to incorporate triangles, polygons, and circles or arcs of circles.

■ Cumulative Review

True/False

32. The intersection of two planes can be a single point.
33. The intersection of two opposite half-planes is their common edge.
34. A segment is a curve.
35. The Line Separation Postulate asserts that a line separates a plane into three disjoint sets.
36. If planes s and t are parallel, then every line in plane s is parallel to every line in plane t.

2 Analytic Geometry

Graphing Lines and Curves

You can use your analytic geometry skills to graph the subsets of the plane that you have studied. Make a table of x and y values from the equation. Plot pairs until you see the pattern and then connect the points from left to right. Remember that first degree equations are lines; other equations are curves.

EXAMPLE 1 $y = -2x + 1$

Answer

x	y
0	1
1	-1
-1	3
2	-3
-2	5
3	-5

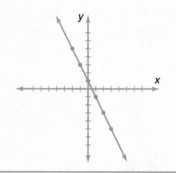

EXAMPLE 2 $y = \frac{1}{2}x^2$

Answer

x	y
0	0
1	$\frac{1}{2}$
-1	$\frac{1}{2}$
2	2
-2	2
3	$\frac{9}{2}$
-3	$\frac{9}{2}$

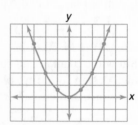

▶ Exercises

Graph.

1. $y = x - 5$
2. $y = 3x$
3. $y = x^2 + 1$
4. $y = 2x + 3$

2.5 Polygons and Convexity

All sets of points can be classified as *convex* or *concave*. It is easy to show that a set is concave. Concave sets always have some points that cannot be connected by segments without leaving the set. The region below is concave because the segment connecting A to B goes outside the region. $\overline{AB}$ is not completely contained in the region, so $\overline{AB}$ is not a subset of the region.

Figure 2.15

EXAMPLE 1   Show that an angle is a concave set.

Answer   Label two points on the angle so that the connecting segment is not contained in the angle. Notice that most of $\overline{AB}$ is in the interior of the angle instead of being part of the angle itself.

If no segment exists to show that the set is concave, then it is a *convex* set. The sets shown are examples of convex sets. Any segments joining points in these sets are always completely contained in the set.

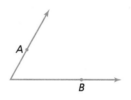

Figure 2.16

Definitions

A **convex** set has the property that any two of its points determine a segment contained in the set.

A **concave** set is a set that is not convex.

This bridge over the St. Lawrence River at Quebec City, Canada, is the longest cantilever bridge in the world. Note the two supporting structures, called cantilevers, and the connecting piece joining them.

Definitions

A **polygon** is a simple closed curve that consists only of segments.

A **side of a polygon** is one of the segments that define the polygon.

A **vertex of a polygon** is an endpoint of a side of the polygon.

An **angle of a polygon** is an angle with two properties: its vertex is a vertex of the polygon and each side of the angle contains a side of the polygon.

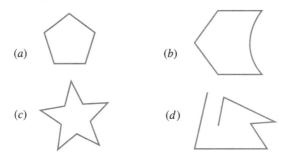

Figure 2.17

Figure (*d*) is a *polygonal curve* since it consists of segments, but it is not a polygon because it is not closed. Figure (*b*) is a simple closed curve, but it is not a polygon because not all the sides are segments. Both figures (*a*) and (*c*) are polygons. A polygon together with its interior forms a *polygonal region*.

EXAMPLE 2   Classify these polygonal regions as convex or concave.

Answer   Polygonal regions (*b*) and (*c*) are convex, while (*a*) and (*d*) are concave. Can you show segments to prove that (*a*) and (*d*) are concave?

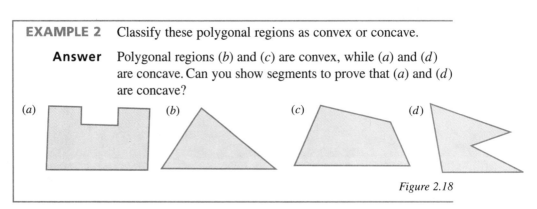

Figure 2.18

Some polygons have special names, like the triangle. Polygons are classified according to the number of sides they have.

Polygon Classification		
Number of sides	Name of polygon	Example of polygon
3	triangle	
4	quadrilateral	
5	pentagon	
6	hexagon	
7	heptagon	
8	octagon	
9	nonagon	
10	decagon	
11	hendecagon	
12	dodecagon	
n	n-gon	

Polygons are curves; the interior is not part of the polygon. This means that technically polygons are always concave (and have no area). However, normal usage is not confusing. When you are asked whether a polygon is convex (or to find the area of a polygon), you will know that you are to consider the polygonal region. Here are a few more terms to help you describe polygons.

Definitions

An **equilateral polygon** is a polygon in which all sides have the same length.

An **equiangular polygon** is a convex polygon in which all angles have the same degree measure.

A **regular polygon** is a polygon that is both equilateral and equiangular.

A **diagonal** of a polygon is any segment that connects two vertices but is not a side of the polygon.

You should now be able to classify polygons by the number of sides, as convex or concave, and as regular or not regular. The notation for a polygon is a list of the vertices in consecutive order (clockwise or counterclockwise). Polygon *ABCDE* is a regular pentagon. $\overline{BD}$ and $\overline{BE}$ are diagonals of the pentagon. Can you name three more diagonals?

Figure 2.19

Definitions

The **interior** of a convex polygon is the intersection of the interiors of its angles.

The **exterior** of a convex polygon is the union of the exteriors of its angles.

For instance, the interior of △*ABC* is shaded in the figure.

The exterior of a triangle consists of points that are neither in the interior nor on the triangle itself. (In other words, the exterior can be described as the complement of the union of the triangle and its interior.)

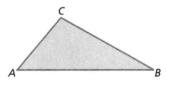

Figure 2.20

▶ A. Exercises

Assume that each set below includes its interior if it is bounded by a closed curve. 1) Classify each set as convex or concave; 2) classify polygons according to the number of sides; 3) if it is not a polygon, explain why.

1.

5.

2.

6.

3.

7.

4.

8.

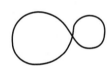

Refer to the diagram for exercises 9-10.

9. Name the sides of this polygon. What is its special name?

10. Using vertex C as one endpoint, name five diagonals of this polygon.

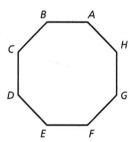

▶ B. Exercises

Draw the following.

11. A regular quadrilateral
12. An equilateral pentagon that is not convex
13. An equiangular hexagon that is not regular
14. A convex heptagon that is neither equilateral nor equiangular

Exercises 15-18 refer to the figure at right.

15. Name two points in the exterior of △EFH.
16. Name four points on △EFJ.
17. Name two points in the interior of ∠EFG.
18. Name a point in the interior of △EFJ.
19. Is the interior of an angle a convex set? the exterior?

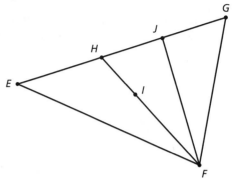

For each polygon in the table below, give

20. the name of each polygon based on the number of sides.
21. the number of vertices.
22. the numbers of diagonals from each vertex.
23. the total number of diagonals.

Number of sides *n*	Name of polygon	Number of vertices *V*	Number of diagonals from each vertex *P*	Total number of diagonals *D*
3				
4				
5				
6				
7				
8				
9				
10				

▶ C. Exercises

24. Look at the pattern in the previous table. Make a generalization about each column. Let *n* represent the number of sides for an *n*-gon.
    a. How many vertices does an *n*-gon have?
    b. How does the number of diagonals from each vertex compare to the number of sides?
    c. How many diagonals from each vertex does an *n*-gon have?
    d. Find the total number of diagonals.

25. Given △LMN and point *F*, which is not on △LMN but is in the same half-plane as point *N* and in the same half-plane as point *L*, answer the following:
    a. Is *F* in the interior of ∠LMN?
    b. Is *F* in the exterior of △LMN?

▶ Dominion Thru Math

The Circular Academy lays out buildings equally spaced around a circle with a grassy area in the center. To minimize steps, the faculty wants to connect each building to every other building by sidewalks. For a cost analysis, the planner needs to know how many sidewalks must be constructed for different numbers of buildings.

26. How many sidewalks are needed for two to eight buildings?

27. Frustrated by the indecision on the number of buildings, the planner asks you for the number of sidewalks needed for *n* buildings. What formula do you give him?

■ Cumulative Review

True/False

28. If two lines do not intersect, then they are parallel.

29. Skew lines may be coplanar.

30. A diameter is a chord of a circle.

31. If *l* and *m* are skew lines and *m* and *n* are skew lines, then *l* and *n* are skew lines.

32. Is the union of an angle and its interior a region? Explain.

A mathematical mosaic is an arrangement of regular polygons. These polygons are arranged so that each vertex has the same number of polygons of the same type around it. Three mathematical mosaics are shown below. The third example has two types of vertices. Draw three more.

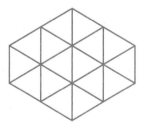 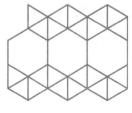

2.6 Subsets of Space

You already know that a plane is a subset of space. A piece of paper can represent a plane, but a plane has no thickness. If you bend the paper, it is no longer a plane, but it still represents a geometric surface.

The Leaning Tower at Pisa, Italy, incorporates circles and cylinders.

Definition

A **surface** is a connected set of points in space having only the thickness of a point.

A sphere is an important surface in space.

Definitions

A **sphere** is a surface in space consisting of the set of all points at a given distance from a given point.

The **center** of a sphere is the given point.

A **radius** of a sphere is a segment that connects a point of the sphere with the center.

A circle provides a simple example of a simple closed curve bounding a region. Similarly, a sphere is a basic example of a closed surface bounding a solid.

A **closed surface** is a surface with a finite size that divides other points in space into an interior and an exterior.

A **solid** is the union of a closed surface and its interior.

Cones and cylinders are two other important examples of closed surfaces.

Picture a region in a plane. The region is bounded by a simple closed curve. Consider a point not in the plane. If line segments connect the point to each point on the curve, the result is called a cone. The noncoplanar point is the *vertex*.

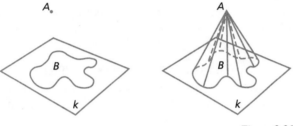

Figure 2.21

Definition

A **cone** is the union of a region and all segments that connect the boundary of the region with a specific noncoplanar point.

Take two regions of the same size and shape and place them in parallel planes. Connect corresponding points of the simple closed curves with segments. This result is a *cylinder.* Two cylinders are shown below.

Definition

A **cylinder** is the union of two regions of the same size and shape in different parallel planes, and the set of all segments that join corresponding points on the boundaries of the regions.

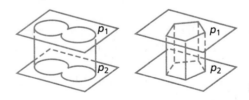

Figure 2.22

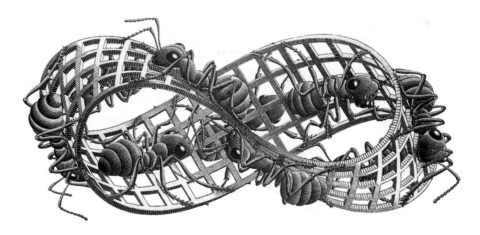

Follow the ants on this woodcut by M. C. Escher to see that the Moebius band is a surface with only one side.

The parts of cones and cylinders also have names. Each region in the definition is called a *base*. Base applies to these regions even when the cone or cylinder is sideways. The connecting line segments together form the *lateral surface*.

There are several types of cylinders and cones.

Definitions

A cylinder or a cone is **circular** if each base is a circle.

A **prism** is a cylinder with polygonal regions as bases.

A **pyramid** is a cone with a polygonal region as its base.

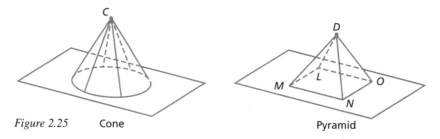

Figure 2.25   Cone   Pyramid

The vertex of the pyramid in the figure is *D*, and the base of the pyramid is the quadrilateral *LMNO*. The lateral surface of the pyramid consists of the four triangular regions. Each triangle is a *lateral face* of the pyramid. The intersection of two lateral faces is called a *lateral edge*. Like pyramids, prisms also have *lateral edges* and *lateral faces*.

Prisms and pyramids are classified by their bases. A square prism is a prism with a square for its base. A hexagonal pyramid is a pyramid with a hexagon for its base.

The term *right* refers to right angles (or perpendicular lines). A cylinder or a prism is *right* if the segments forming the lateral surface stand at right angles to the base. A cone or pyramid is *right* if the vertex is centered above the base. Cones and cylinders that are not right are *oblique*. Right cylinders, right cones, right pyramids, and right prisms will be important when you study surface area. The term *regular* is used of solids as well as of polygons. A pyramid or a prism is *regular* if it is right and its base is a regular polygon.

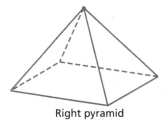

Right pyramid   Oblique pyramid

Figure 2.24

▶ A. Exercises

Identify each geometric figure with the most specific designation possible.

1.

2.

3.

4.

5.

6.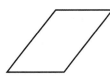

7.

8.

Use the diagram for exercises 9-11.

9. Name the figure. Is it a surface?
10. Name the vertex and the base.
11. Name the lateral faces. What kind of polygon bounds each lateral face?

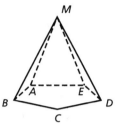

Use the next diagram for exercises 12-14.

12. Name the figure. Is it a closed surface?
13. Name the bases. What kind of region are they?
14. Name the lateral faces. What kind of region are the lateral faces?

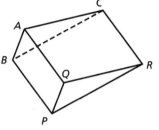

Name physical objects that could be represented by each figure.

15. cylinder
16. sphere
17. right circular cone

▶ B. Exercises

Sketch each surface in exercises 18-20 without showing the planes containing the bases.

18. A prism with a square region as base
19. An oblique circular cylinder
20. A cone with a nonconvex polygonal region as base

True/False. If the statement is false, explain why.

21. Every cone has a circular region for its base.
22. All prisms are cylinders.
23. Every pyramid has at least three lateral faces.
24. All spheres are solids.
25. All right pyramids are regular.

26. Reread the Line Separation Postulate and the Plane Separation Postulate. To obtain a similar separation postulate for space, complete the sentence below.
    Space Separation Postulate: Every plane in space separates space into . . .

27. Use the definition of closed surfaces to write a statement similar to the Jordan Curve theorem for three-dimensional figures.

28. What is the intersection of a right circular cone and a plane parallel to the base of the cone?
29. If S is a sphere, C is the center of S, and P is a point on S, then is $\overline{CP} \subseteq S$? How many points are in $\overline{CP} \cap S$?
30. Do the radii of a sphere have equal lengths? Explain.

Cumulative Review

True/False
31. A circle is a polygon.
32. A hexagon is a region.
33. A regular polygon is convex.
34. An angle can consist of two opposite rays.
35. The radii of a circle must have equal lengths.

2.7 Polyhedra

Prisms and pyramids differ from spheres because they have flat faces. Such closed surfaces are called polyhedra.

What polygon appears in this aerial view of Fort Jefferson in Dry Tortugas National Park west of Key West, Florida?

Definitions

A **polyhedron** is a closed surface made up of polygonal regions.

A **face of a polyhedron** is one of the polygonal regions that form the surface of the polyhedron.

Like other closed surfaces, a polyhedron separates space into three disjoint sets. These sets are the polyhedron itself and the interior and exterior of the polyhedron. Further, you will recall that a solid is the union of a closed surface and its interior. A polyhedral solid is bounded by a polyhedron.

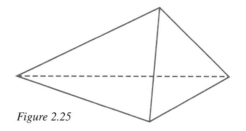

Figure 2.25

Polyhedra are named according to the number of faces. A four-sided polyhedron, such as the one shown, is called a *tetrahedron*. The table lists the special names given to particular polyhedra.

Number of Faces	Names
4	tetrahedron
5	pentahedron
6	hexahedron
7	heptahedron
8	octahedron
10	decahedron
12	dodecahedron
20	icosahedron

The following diagrams are illustrations of polyhedra.

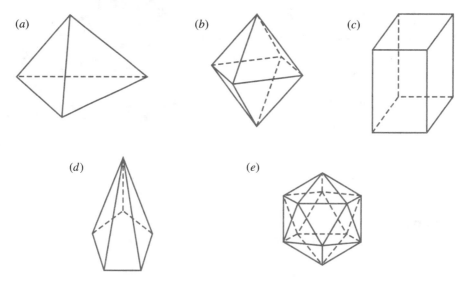

Figure 2.26

> **EXAMPLE** Classify the five polyhedra in figure 2.26.
>
> **Answer** Count the faces of each and use the correct label from the table. (*a*) tetrahedron (*b*) octahedron (*c*) hexahedron (*d*) hexahedron (*e*) icosahedron.

The intersection of adjacent faces of a polyhedron is called an *edge* of the polyhedron. The endpoints of the edges are called the *vertices*. The polyhedra that you will study in this course are *simple polyhedra*. A polyhedron that is not simple is one that has a hole in it, such as the dodecahedron below.

Polyhedron classification is similar to the classification of polygons in a plane. As a polygon is classified by the number of line segments that form it, so a polyhedron is classified by the number of faces that form it. Also, the sides of a polygon bound a region, while the faces of a polyhedron bound a solid. The least number of faces that a polyhedron can have is four.

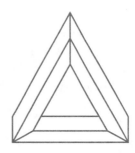

Figure 2.27

Have you observed how the concepts in this chapter have built upon one another? The concept of betweenness provided a foundation for segments, rays, and half-lines. From segments you learned definitions of angles, triangles, polygons, and circles. Polygons, in turn, enabled you to discuss special surfaces including bases for prisms and pyramids. These special polyhedra can be convex or concave, right or oblique. You will continue to build on these ideas throughout the book as you study surface area, volume, and various spatial relationships.

In similar manner, you as a Christian should progress from a general knowledge of Christ in salvation to a more specific knowledge of Him. When a person is first saved, he dwells on the milk of the Word and gains a broad overview of the Christian's life (I Pet. 2:2). But the Christian must strive to mature in the Word and not be as the people of Corinth who were not able to understand the details of Scripture (I Cor. 3:1-3). Make sure that you build carefully on your foundational knowledge of the Lord Jesus. There is no other sure foundation (I Cor. 3:9-11).

Now build another step on your knowledge of polyhedra.

A **regular polyhedron** is a convex polyhedron having two properties.

1) All faces are identical (same size and shape), and
2) The same number of edges meet at each vertex.

A cube is an example of a regular polyhedron. Its identical faces are squares. How many edges meet at each vertex?

▶ A. Exercises

Tell whether the statements are true or false.
1. Every polyhedron is a cone.
2. Every prism is a polyhedron.
3. Some cones are polyhedra.
4. A pyramid with a quadrilateral base has five faces.
5. A prism has only one base.
6. A circular cone is a polyhedron.
7. The smallest number of vertices of a polyhedron is four.
8. A cone with a heptagonal base is an octahedron.
9. A prism has the same number of faces as vertices.
10. A polyhedron can have no more than twelve faces.

Classify each polyhedron according to the number of faces.

11.

14.

16.

12.

15.

17.

13.

Use the figure for exercises 18-20.

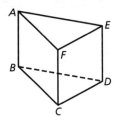

18. Name the faces of the polyhedron.
19. Name the vertices of the polyhedron.
20. Name the edges of the polyhedron.

▶ B. Exercises

Give another name for each figure.
21. A prism with a decagonal base region
22. A pyramid with a quadrilateral base region
23. The base of a solid is a regular n-gon. How many faces does the solid have if it is a pyramid? a prism?

A diagonal of a prism joins two vertices that do not lie in the same face. (This also means that a diagonal must intersect the interior of the prism.)

24. Complete the table.

Bases of Prism	Diagonals per Vertex	Total Diagonals
quadrilaterals		
pentagons		
hexagons		
heptagons		
octagons		

▶ C. Exercises

25. Generalize based upon your answers to exercise 24. How many diagonals does a prism have if its bases are n-gons?

■ Cumulative Review

For solids bounded by the given surfaces, decide whether each is convex or concave.
26. sphere
27. regular polyhedron
28. torus (doughnut-shape)
29. oblique circular cone
30. What geometric figure represents the core of a roll of paper towels? What shape results if you flatten the roll?

Geometry and Scripture

Variables

Many of the figures you have studied in this chapter play a role in the Bible. Let's find a few. For each passage below identify the figure.

1. Exodus 39:9
2. Joshua 18:11-20
3. Jeremiah 18:3

Notice that the territory of Benjamin forms a region. All 26 cities lie in the interior of the boundary curve (Josh. 18:21-28).

Scripture also refers to surfaces in space, such as cylinders. Most cylinders in the Bible are pillars.

4. Find a reference to pillars in Jeremiah. Since these pillars are hollow, they are indeed cylinders.

5. Find the reference to the pillars that Samson pulled down.

HIGHER PLANE: Some cylinders are so important that they had names. What were the names of the pillars in the porch of the temple that Solomon built?

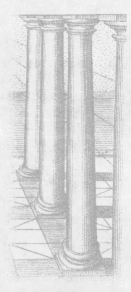

Now review the Line Separation Postulate.

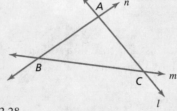

Figure 2.28

You know that *A* divides line *l* into the point and two half-lines. Can we also say that *B* divides *m* into the point and two half-lines? Both statements are true because the Line Separation Postulate uses the words *every* and *any*. These words act as variables or placeholders. We can replace "Every point" with the name of a point and "any line" with the name of a specific line as long as it meets the condition "through that point." In this sense the phrases "every point" and "line through that point" are variables.

6. Using figure 2.28, write as many applications of the Line Separation Postulate as you can. How many are there?

Just as algebra has variables having sets of numbers for domains, geometry has phrase variables having sets of points or figures for domains. The Bible also uses phrase variables having sets of people, creatures, or things for domains. You have noticed this before using John 3:16. Let's look further at this principle.

7. Read John 1:3*a*. What word is a placeholder that you could replace with things in its domain? Apply it to three things in its domain.

8. Name two things that are not in the domain of the variable in John 1:3*a*.

9. Read Romans 3:23 and Isaiah 53:6. What word is a "variable" in these verses? Put your name in each. What is the domain?

10. Read Romans 10:13 and find the variable. Apply it to yourself.

Line upon Line

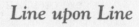

FOR WHOSOEVER shall call upon the name of the Lord shall be saved. ❧

ROMANS 10:13

Chapter 2 Review

Answer exercises 1-10 based on the diagram shown.

1. Name six rays.
2. Name three half-lines.
3. Name four angles.
4. Find $\overleftrightarrow{CE} \cap \overrightarrow{AD}$.
5. Find $\overrightarrow{DB} \cap \overrightarrow{AB}$.
6. Find $\overleftrightarrow{BD} \cap \overset{\circ}{BA}$.
7. Find $\overrightarrow{CB} \cup \overrightarrow{EB}$.
8. Is C between A and D?
9. Is B between C and E?
10. Name two pairs of opposite rays.

Sketch each figure listed below.

11. octagon
12. pentagon
13. triangle
14. hexagon
15. simple closed curve
16. sphere
17. cylinder with a hexagonal base region
18. cone with a pentagonal base region

Answer exercises 19-21 according to the figure shown.

19. Name all triangles.
20. Name the sides of $\triangle LMP$.
21. Name the vertices of $\triangle MPN$.

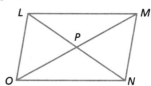

Answer exercises 22-26 based on the following diagram.

22. Name the circle.
23. Name three radii.
24. Name two diameters.
25. Name a triangle.
26. Name three chords.

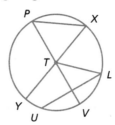

Classify the following polyhedra.

27.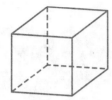

28.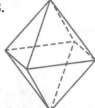

29.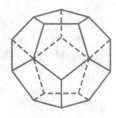

30. Explain the mathematical significance of Romans 10:13.

▶ Dominion Thru Math

The Circular Academy decides to build twelve circular buildings, each having its center on the large circle, which has a radius of 600 feet. The buildings lie at the vertices of a regular dodecagon, and each building has a 100-foot diameter.

31. If the face of each building is set back 50 feet from the edge of the property, how many acres are needed?

32. How many sidewalks are needed for the twelve-building campus? Use the formula from page 67.

33. If each proposed building is three stories, how many square feet will the combined buildings contain?

34. To park cars, the planner adds an 80-foot-wide band around the edge of the acreage (exercise 31). How many acres does this add to the land requirements? How many square yards of asphalt are needed if only the outside 5 feet of the parking perimeter is grass instead of pavement?

3 Segments and Measurement

How long is the coast of Great Britain?

Benoit Mandelbrot addressed this question and discovered *fractals*. If you sail around Britain without entering sounds and bays, you will travel about 2,520 miles of coastline. However, by hugging the shore of every bay, one may well travel 10,000 miles. Of course, a canoe can hug the shore closer than a tanker and would travel even farther. A person on foot would follow many coastal indentations within the bays that a canoe would bypass, while an ant would have to go around even smaller "bays" that a person would not notice. This process continues (an amoeba would find microscopic bays ignored by the ant) with bays inside bays until the shoreline grows astronomically large.

Such patterns occur in the ragged edge of a maple leaf or fern frond, mountain ranges, weather fronts, tree branches, water currents, and patterns of blood vessels. As Mandelbrot learned to describe such natural phenomena mathematically, he found that his formulas could generate fractals on a computer. Like the coast, each part of the picture can be viewed in greater and greater detail. The Mandelbrot set, shown here, is his most famous fractal.

Chaos theory uses fractals to show the pattern in what otherwise appears random to man. Fractals have become popular as artwork, but chaos theory also describes populations in ecology, randomness in statistics, fluctuating market values in economics, earth's magnetic field in physical science, and other topics from chemistry to astronomy.

After this chapter you should be able to

1. classify and use real numbers.

2. apply properties of real numbers.

3. find distances using absolute values.

4. apply distances to betweenness, midpoints, and congruent segments.

5. find perimeters and circumferences.

6. distinguish inscribed and circumscribed figures.

7. justify the midpoint and circumference formulas.

8. construct congruent segments and bisectors.

3.1 Real Numbers

The best way to describe the real number system is to build from the foundation of the natural numbers.

The *natural numbers* are the counting numbers that you learned as a small child. This set of numbers is symbolized by $\mathbb{N}$ for natural.

$\mathbb{N} = \{1, 2, 3, \ldots\}$

The concept of "none" is not represented by a natural number. For this reason, you soon added the smaller number zero to your counting skills. The new set formed from the natural numbers by including zero is called the *whole numbers* and is symbolized by $\mathbb{W}$.

$\mathbb{W} = \{0, 1, 2, 3, \ldots\}$

As you learned to measure temperature or keep checkbook balances, you learned the value of indicating numbers less than zero. These were called negative numbers. The natural numbers together with their negatives and zero form the set of *integers,* which is symbolized by $\mathbb{Z}$.

$\mathbb{Z} = \{\ldots -3, -2, -1, 0, 1, 2, 3, \ldots\}$

The integers are divided into the positive integers, zero, and the negative integers. You can see that $\mathbb{Z}$ can also be viewed as the whole numbers together with the negative integers.

The Fahrenheit and centigrade temperature number lines match at 40 degrees below zero.

Buying and cooking showed you the need for another kind of number. Fractions name amounts between integers to express portions of a dollar or parts of a measuring cup. Fractions are ratios of integers and should always be reduced to lowest terms such as $\frac{1}{8}$, $-\frac{2}{3}$, $\frac{13}{5}$, and $-\frac{7}{4}$. The set of all (meaningful) integer ratios is the set of *rational numbers.* The set of rational numbers is symbolized by $\mathbb{Q}$ (for quotients) and includes all the integers as well as other fractions.

$\mathbb{Q} = \{\frac{p}{q} \mid p \text{ and } q \text{ are integers and } q \neq 0\}$

Mushers experience such cold temperatures in polar regions.

Finally, you realized that some portions of things cannot be represented as ratios. Decimals enable you to express any portion. The set of all possible decimal numbers is called the *real numbers,* $\mathbb{R}$. Decimals that cannot be written as fractions are called *irrational numbers.* Irrational numbers in decimal form neither terminate nor repeat. Since they are so hard to write as decimals, they often have special names such as $\sqrt{3}$ or π. A decimal example of an irrational number is $2.7272272227\ldots$

The real number system obeys certain properties or laws. These properties are summarized in the following chart and should be familiar to you. In each of these properties, a, b, and c are real numbers.

Real Number Properties		
Property	Addition	Multiplication
Commutative	$a + b = b + a$	$a \cdot b = b \cdot a$
Associative	$(a + b) + c = a + (b + c)$	$(a \cdot b) \cdot c = a \cdot (b \cdot c)$
Distributive		$a(b + c) = ab + ac$
Identity	$a + 0 = a$ and $0 + a = a$	$a \cdot 1 = a$ and $1 \cdot a = a$
Inverse	$a + (-a) = 0$	$a \cdot (1/a) = 1$

You are also familiar with certain properties of equality.

Equality Properties	
Property	Meaning
Addition	If $a = b$, then $a + c = b + c$
Multiplication	If $a = b$, then $ac = bc$

Do you see why the subtraction and division properties are not necessary? Subtracting is accomplished by adding the additive inverse, so the addition property of equality can apply whenever subtraction is necessary. What property covers necessary divisions? Can you explain why?

More Equality Properties	
Reflexive	$a = a$
Symmetric	If $a = b$, then $b = a$
Transitive	If $a = b$ and $b = c$, then $a = c$

In algebra you learned that a relation is a set of ordered pairs. Equality ($=$) is a very important example of a relation. When a relation has the three properties listed on page 85, it has a special name.

Definition

An **equivalence relation** is a relation that is reflexive, symmetric, and transitive.

Equality ($=$) is an important equivalence relation. You will study other examples of equivalence relations as you progress in your study of geometry.

The Order of Operations tells you which arithmetic operations get first priority.

Order of Operations. Perform inside parentheses first, multiplications next, and additions last.

This is the easiest way to remember the order of operations, but it requires that you properly classify other operations. Addition includes subtraction. Multiplication includes division and exponentiation. Parentheses include absolute values. Parentheses can also be implied by fraction bars, exponents, or radicals. Within each level, perform the operations left to right.

Two other properties that you have used, although you may not know their names, are the substitution property and the trichotomy property.

Substitution property. If $a = b$, then a can replace b in any mathematical statement.

Trichotomy property. For any two real numbers a and b, exactly one of the following is true: $a = b$, $a > b$, or $a < b$.

We have reviewed many properties in this section. These are the rules for working with real numbers, and we must abide by these rules. Rules are given for our benefit so that our answers will be correct and our understanding will be enhanced. The Bible gives us rules to live by. These rules help us to live properly and to increase our spiritual understanding. The law of the Bible is our spiritual schoolmaster (Gal. 3:24), which shows us our need of salvation and of proper Christian living (Rom. 8:3-4).

Finally, you should review absolute value. The absolute value of a is denoted by $|a|$. Absolute value is a unary operation on real numbers. The absolute value is always positive. The formal definition of *absolute value* follows.

$$|a| = \begin{cases} a \text{ if } a \geq 0 \\ -a \text{ if } a < 0 \end{cases}$$

> **EXAMPLE** Find $|3|$ and $|-3|$.
>
> **Answer** $|3| = 3$ **1.** Since $3 \geq 0$, use the first rule: $|a| = a$.
>
> $|-3| = -(-3) = 3$ **2.** Since $-3 < 0$, use the second rule: $|a| = -a$ (read $-a$ as opposite of a).

Remember that absolute value bars imply grouping. As with parentheses, operations inside absolute values must be done first.

▶ A. Exercises

1. Draw a number line and graph the set of whole numbers less than 5.
2. Draw a number line and graph the set of real numbers between 3 and 6.
3. Classify the following real numbers as natural, whole, integer, rational, or irrational. Be as specific as possible.

$$5, -3, 0, -6, \tfrac{1}{2}, \sqrt{3}, 8, -5, \tfrac{2}{5}, \tfrac{1}{3}, \tfrac{7}{1}, \sqrt{7}, \pi$$

State the property described by each example.

4. $3(2 + 7) = 3 \cdot 2 + 3 \cdot 7$
5. $5 \cdot 8 = 8 \cdot 5$
6. $x = 6$ and $6 = y$, so $x = y$
7. $11 = 11$
8. $3 + (9 + 6) = (3 + 9) + 6$
9. $a > 5$; therefore $a \neq 5$
10. $3a = 9$; therefore $\tfrac{1}{3} \cdot 3a = \tfrac{1}{3} \cdot 9$

Compute the following.

11. $|-8|$
12. $|12|$
13. $|7| + |10|$
14. $|12| - |27|$
15. $|189 - 207|$
16. $|62 + 9|$

▶ B. Exercises

Tell whether the following statements are true or false.

17. Each natural number is a whole number.
18. Some irrational numbers are integers.
19. The addition property of equality is used to solve an equation such as $a - 5 = 10$.
20. If a is a negative number, then $-a$ is a negative number.
21. If b is a positive number, then $-b$ is a negative number.
22. If $a < 0$, then $|a| = -a$.

23. Explain why $|x| = -x$ when $x < 0$.
24. Give the reason (appropriate property) for each step in the solution of
$$5x + 7 = y$$
$$y = 9(x - 5).$$

a.	$5x + 7 = 9(x - 5)$	
b.	$5x + 7 = 9x - 45$	
c.	$5x + 7 - 5x + 45 = 9x - 45 - 5x + 45$	
d.	$5x - 5x + 7 + 45 = 9x - 5x - 45 + 45$	
e.	$0 + 52 = 4x + 0$	
f.	$52 = 4x$	
g.	$\frac{1}{4}(4 \cdot 13) = \frac{1}{4}(4x)$	
h.	$\left(\frac{1}{4} \cdot 4\right)13 = \left(\frac{1}{4} \cdot 4\right)x$	
i.	$1 \cdot 13 = 1 \cdot x$	
j.	$13 = x$	
k.	$x = 13$	
l.	$5(13) + 7 = y$	
m.	$72 = y$	
n.	$y = 72$	

25. Solve for x: $|x| = 4$

■ Cumulative Review

List the possible solutions for the intersection of
26. a line and a plane.
27. two planes.
28. a line and a convex polygon.
29. a line and a circular region.
30. Discuss possible relations among four points.

3.2 Segment Measure

Much of geometry deals with measurement. *Measure* refers to a basis of comparison. Christians should measure their spiritual lives by God's standards found in the Bible (II Cor. 10:12-18). Similarly, physical objects are measured by physical standards using an appropriate unit. Weight can be measured by an electronic scale or a two-arm balance in pounds or grams. Length can be measured by a ruler, a yardstick, a meter stick, or a pace. Volume can be measured by a quart, a pint, a liter, or a gallon.

Carpenters constantly apply measurement skills.

Linear measurement involves finding the length of a segment, or the distance between two points. You will study other types of measurement soon, but all measurement begins with the definitions and postulates for linear measure. The first postulate in our study of linear measurement is the Ruler Postulate.

Postulate 3.1
Ruler Postulate. **Every point of a line can be placed in correspondence with a real number.**

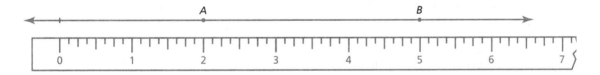

Definition

The **coordinate** of a point on a line is the number that corresponds to the point.

Lowercase letters are used for coordinates. The coordinate of point *A* is the number *a*. So in figure 3.1, *a* = 2. The point on a number line that corresponds to a number is called the *graph* of the number. In figure 3.1, point *A* is the graph of the number 2.

Number lines are rulers in the form of a coordinate system. If *A* and *B* are any two points of a line, coordinates can be positioned such that *A* corresponds to 0 and *B* corresponds to a positive real number. Figure 3.2 shows the coordinate system with 0 at *A* and *B* as a positive real number (namely 3). These coordinates make it easy to see that *A* and *B* are three units apart. However, you could also have subtracted to find this from figure 3.1, even though neither point was on zero.

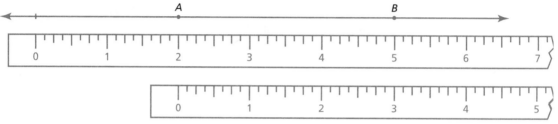

Figure 3.2

Definition

The **distance** between two points *A* and *B* is the absolute value of the difference of their coordinates. Distance between points *A* and *B* is denoted by *AB*, given by $AB = |a - b|$.

According to the definition of *distance,* we should be able to find the distance by subtracting the coordinates. But we can subtract in two possible orders.

$$5 - 2 = 3 \text{ or } 2 - 5 = -3$$

Distance is considered a positive number. By finding the absolute value, you will get a positive answer regardless of the order in which you subtract. Since the definition includes the absolute value symbol, you can subtract in either order.

EXAMPLES Find the distances.

      1. *BD*
      2. *AE*
      3. *CE*

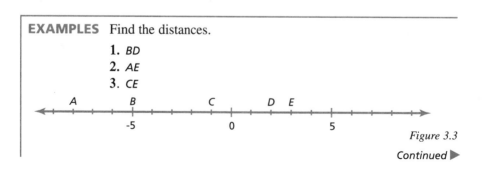

Figure 3.3

Continued ▶

Answers

1. $BD = |-5 - 2| = |-7| = 7$
2. $AE = |-8 - 3| = |-11| = 11$
3. $CE = |-1 - 3| = |-4| = 4$

Consider the number line in the example. According to the definition of *betweenness* given in Chapter 2, we can say that D is between B and E. A more refined definition of *betweenness* is given now.

Definition

A point M is **between** A and B if $AM + MB = AB$. The correct notation is A-M-B.

Remember that A-M-B and B-M-A mean the same thing. Also, notice that if $AM + MB = AB$, then the points are collinear.

We know that we can measure any segment on $\overrightarrow{AB}$ by the Ruler Postulate. However, how do we know that for every distance there will be a corresponding point? The Completeness Postulate tells us this; it is just the reverse of the Ruler Postulate.

Postulate 3.2

Completeness Postulate. Given a ray, $\overrightarrow{AB}$, and any positive real number r, there is exactly one point C on the ray so that $AC = r$.

This postulate guarantees a point at every distance. The line has no holes.

Paul ran the 100-yard dash in 9 seconds. Jim ran the 100-meter dash in 10 seconds. Who ran faster?

▶ A. Exercises

Draw a number line and graph each number.

1. $A = 3$
2. $B = -4$
3. $C = 0$
4. $D = -\sqrt{7}$
5. $E = \frac{1}{2}$
6. $F = \frac{-3}{2}$

Use the number line below for exercises 7-19.

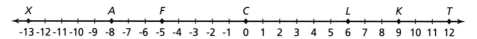

State the coordinate of each point.

7. *L*   9. *F*
8. *X*   10. *C*

Find each distance.

11. *CK*   13. *TX*   15. *AK*
12. *AF*   14. *LF*   16. *FC*

▶ B. Exercises

17. Show that *C* is between *A* and *T*.
18. Show that *F* is not between *L* and *K*.
19. Show that *A* is between *X* and *F*.

Find the distance between two points with the given coordinates.

20. 8 and 39   23. -1275 and 1384

21. 47 and -12 24. 5762 and 389

22. -18 and -197 25. $136\frac{3}{5}$ and $\frac{248}{3}$

$\frac{18}{215}$

▶ C. Exercises

Disprove each statement below by providing a counterexample (example with numbers to show that the statement is false).

26. $|a + b| = |a| + |b|$ 27. $|a - b| \le |a| - |b|$

▪ Cumulative Review

28. If *B* is between *A* and *C*, which is longer: *AB* or *AC*?
29. Can a half-line contain a ray?
30. For sets *A* and *B*, which statements are always true?
    a. $A \subseteq A$
    b. If $A \subseteq B$, then $B \subseteq A$
    c. If $A \subseteq B$ and $B \subseteq C$, then $A \subseteq C$
31. Is the subset relation reflexive? transitive?
32. Is the subset relation an equivalence relation?

The Distance Formula

Every point has coordinates in analytic geometry. Using the Pythagorean theorem, you can find the distance between the points to derive the Distance Formula.

Distance Formula.
The distance, d, between two points A (x_1, y_1) and B (x_2, y_2) is
$$d = \sqrt{(x_1 - x_2)^2 + (y_1 - y_2)^2}.$$

Make a right triangle by showing the point C (x_1, y_2).

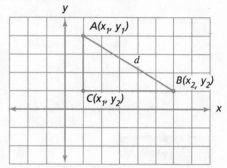

STATEMENTS	REASONS
1. $BC^2 + AC^2 = AB^2$	**1.** Pythagorean theorem
2. $\left(\lvert x_1 - x_2 \rvert\right)^2 + \left(\lvert y_1 - y_2 \rvert\right)^2 = d^2$	**2.** Substitute
3. $(x_1 - x_2)^2 + (y_1 - y_2)^2 = d^2$	**3.** Simplify (squares are positive)
4. $\pm\sqrt{(x_1 - x_2)^2 + (y_1 - y_2)^2} = d$	**4.** Square root both sides
5. $\sqrt{(x_1 - x_2)^2 + (y_1 - y_2)^2} = d$	**5.** Distance must be positive

EXAMPLE Find the distance between $(1, 4)$ and $(-2, 3)$.

Answer $d = \sqrt{(1 - [-2])^2 + (4 - 3)^2} = \sqrt{3^2 + 1^2} = \sqrt{10}$

▶ Exercises

Find the distance between each pair of points.
1. $(2, 5)$ and $(1, -2)$
2. $(0, 4)$ and $(3, 0)$
3. $(3, 3)$ and $(-1, 2)$
4. $(0, 0)$ and $(-4, -3)$
5. $(6, 5)$ and $(2, 7)$

3.3 Segment Bisectors

The distance between the two endpoints of a segment is called the *length* or the *measure* of the segment. You found some segment measures in the last section. When you look at a segment such as the one in figure 3.4, you can see that *M* is between *K* and *L* because they are collinear and $KM + ML = KL$.

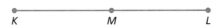

K M L

Figure 3.4

If $KM = ML$, then *M* is called the midpoint of *KL*.

Definition

The **midpoint** of $\overline{AB}$ is *M* if *A-M-B* and $AM = MB$.

The following theorem about midpoints can be proved.

Theorem 3.1
***Midpoint Theorem.* If *M* is the midpoint of $\overline{AB}$, then $AM = \frac{1}{2} AB$.**

Since *M* is the midpoint of $\overline{AB}$, *M* must be between *A* and *B* and also $AM = MB$. Betweenness guarantees that $AM + MB = AB$, and by substitution $AM + AM = AB$. Solving for AM, $2AM = AB$ and $AM = \frac{1}{2} AB$.

You can see that there is logical reasoning behind this theorem; every step has a reason. This reasoning process can be carefully organized in columns.

STATEMENTS	REASONS
1. *M* is the midpoint of $\overline{AB}$	1. Given
2. $AM = MB$ and *A-M-B*	2. Definition of midpoint
3. $AM + MB = AB$	3. Definition of betweenness
4. $AM + AM = AB$	4. Substitution (step 2 into 3)
5. $2AM = AB$	5. Distributive property
6. $AM = \frac{1}{2} AB$	6. Multiplication property of equality

From this theorem, you can prove the formula for the coordinate of the midpoint, $m = \frac{a+b}{2}$. You should learn the terms *bisector* and *congruent*.

Definition

A **bisector** of a segment is a curve that intersects the segment only at the midpoint.

A bisected apple

The bisector of a segment can be a line, a ray, another segment, or some other curve. The key is that it intersects the midpoint as shown below.

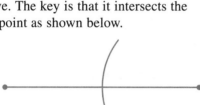

Definition

Congruent segments are segments that have the same length. The symbol ≅ is used for congruent segments.

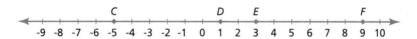

Figure 3.5

EXAMPLE Is $\overline{CD} \cong \overline{EF}$?

Answer $CD = \left|1 - (-5)\right| = 6$

$EF = \left|9 - 3\right| = 6$

Since $CD = EF$, $\overline{CD} \cong \overline{EF}$

(read "segment CD is congruent to segment EF").

$\overline{AB} = \overline{PQ}$ describes equal segments, meaning that the segments are the same set of points; but $\overline{AB} \cong \overline{PQ}$ describes congruent segments, meaning that the segments have equal lengths.

Correct notation is very important in geometry. Learn to use correct notation.

▶ A. Exercises

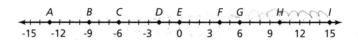

Use the number line for exercises 1-10. Find the indicated lengths.
1. *FC*   4. *CI*
2. *HG*   5. *EH*
3. *BD*   6. *AH*

7. Find the coordinate of the midpoint of $\overline{FH}$.
8. Find the coordinate of the midpoint of $\overline{CE}$.
9. Which point is the midpoint of $\overline{CG}$?
10. Find a segment congruent to $\overline{BE}$.

Tell whether the following statements are true or false.
11. If $\overline{AB} \cong \overline{CD}$, then $AB = CD$.
12. If $AX = XC$, then *X* is the midpoint of $\overline{AC}$.
13. If *T* is the midpoint of $\overline{SR}$, then *S*, *T*, and *R* are collinear points.
14. If *B* is the midpoint of $\overline{PQ}$ and $PQ = 24$, then $PB = 12$.
15. If $AX + XR = AR$, then *X* is the midpoint of $\overline{AR}$.

▶ B. Exercises

Given a segment, find the coordinates of the two endpoints *A* and *B*. Assume $\overline{XB}$ has the given length and a segment bisector passes through the point *X*, which has the given coordinate. It may help you to draw a picture.
16. $7, XB = 9$ 17. $-6, XB = 3$

Find *AB* if *X* is the midpoint of $\overline{AB}$ and $\overline{AX}$ has the given length.
18. 4   19. 157

Use the proper notation to say the following:
20. The distance between points *A* and *B* is equal to the distance between points *M* and *N*.
21. Segment *XD* is congruent to segment *FB*.
22. Point *P* lies on segment *KL*.

▶ C. Exercises

Give logical reasoning that supports each of the following statements.
23. A segment has one and only one midpoint.
24. If *L* is the midpoint of $\overline{MN}$, then $\overline{ML} \cong \overline{LN}$.
25. Let *M* be the midpoint of $\overline{AB}$; then $m = \frac{a+b}{2}$.

26. State the Ruler Postulate.
27. Where is the most convenient placement of a ruler to measure $\overline{AB}$?

28. Draw a hexagon.
29. Draw a hexahedron.
30. Draw a simple curve that bisects a segment.

3.4 Perimeter and Circumference

You have heard of the words *perimeter* and *circumference,* but what do they mean? These two words have similar meanings. First we will discuss the term *perimeter.* The word *perimeter* comes from the Greek words *peri* and *metron. Peri* means "around" and *metron* means "measure." So *perimeter* means "around measure."

Definition

Perimeter is the distance around a closed curve.

For example, the sum of the lengths of the sides of a polygon is its perimeter.

$\triangle ABC$ has sides whose measures are given in figure 3.6. Can you give the perimeter of $\triangle ABC$?

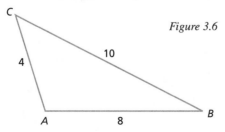

Figure 3.6

As runners race around the track, they travel the length of the sides plus the circumference of the two semicircles at the ends.

To find the *perimeter,* simply add the lengths of the sides of the triangle. So the perimeter of △*ABC* is 22.

If the lengths of the sides of any polygon are given, one can easily find the perimeter of a polygon. If the polygon is a regular polygon, what would be the easiest way to find the perimeter?

According to the definition of *perimeter,* the perimeter of regular pentagon *JKLMN* is *JK* + *KL* + *LM* + *MN* + *NJ* = 9 + 9 + 9 + 9 + 9 = 45. Isn't there an easier way to add five nines?

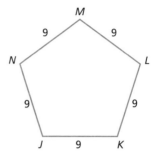

Figure 3.7

Theorem 3.2
The perimeter of a regular *n*-gon with sides of length *s* is *n* · *s*.

Length and distance refer to segments. The definition of perimeter extends the concept of distance to a polygon by adding the lengths of the segments that form its sides. Since circles do not have any sides, how can we find the distance around a circle?

By removing the label from the soup can and spreading it flat, we can find the length of the paper that represents the distance around the circular base of the can. If there is no label, the same idea works with a string. Wrap the string around the can, mark the length, pull it straight, and measure it with a ruler. You can see that this is a special type of perimeter. The name *circumference* will remind you that you cannot just add the lengths of the sides. First Kings 7:23 reports the circumference of a circle.

Definition

Circumference is the distance around a circle.

The diameter of a circle is a segment that connects two points of the circle through the center of the circle. A unique relationship exists in the comparison of the circumference of a circle to the diameter of the circle. The ratio that results is constant, regardless of the size of the circle. This ratio has been known for a long time.

The ratio of circumference to diameter has a special name, *pi,* represented by the Greek letter π, which is an irrational number and cannot be expressed as a repeating decimal. Approximately, π is 3.14, but 3.14159265358793 (to fourteen decimal places) is more accurate. Since we cannot write the entire decimal value, π is the quickest way to refer to the number. Plot π on a number line.

If we use c for the circumference of a circle and d for the diameter of the circle, the following equation is true.

$$\frac{c}{d} = \pi$$
Solving for c, $c = \pi\, d$.

Given the diameter of any circle, you can find the circle's circumference by using this equation. Could you find the circumference of a circle if you were given the radius of the circle? Since the diameter is twice the radius, another formula for circumference is

$$c = 2\pi r.$$

The skyline of Seattle, Washington, includes rectangular prisms in skyscrapers and the circular disk at the top of the Space Needle.

► A. Exercises

Find the perimeter of each polygon.

1.

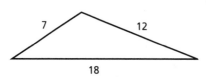

4.

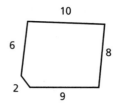

2.

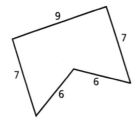

5.

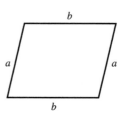

3.

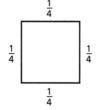

6.

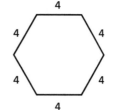

7.

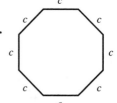

Find the perimeter of each polygon.

8. An equilateral triangle with sides 15 inches long
9. A quadrilateral with sides 3, 5, 15, and 8 centimeters long
10. A regular hexagon with sides 28 millimeters long

► B. Exercises

Write an equation to describe each situation and solve.

11. A regular quadrilateral *ABCD* has a perimeter of 72 meters. What is the length of each side?
12. The perimeter of an equilateral triangle is 78 centimeters. What is the measure of each side?
13. The perimeter of a triangle is 22 inches. Side *a* is twice as long as side *b*, and side *c* is 2 inches longer than side *a*. Find the length of each side.
14. The perimeter of a quadrilateral is 36 inches. Opposite pairs of sides have the same lengths, and one side is twice as long as its consecutive sides. What is the length of each side?

Complete the following table of circle measurements. The letter r represents radius, c represents circumference, and d represents diameter. Use 3.14 as an approximation for π.

	d	r	c in terms of π	c in decimal form
15.		6		
16.	13			
17.				28.26
18.			28π	
19.		3		
20.			15π	

▶ C. Exercises

Find answers to the nearest tenth.

21. How far does a bicycle travel in one revolution of its 26-inch tire?
22. Find the circumference of a roll of tape whose diameter is 5 centimeters.
23. The hands of a clock vary in length. If the hour hand is 4 inches long, the minute hand 6 inches long, and the second hand 7 inches long, how far does the point of each hand travel in one day?
24. How many feet does the second hand in exercise 23 travel in 24 hours? How many miles?
25. A circular track has a radius of 200 feet. What is the circumference of the track in yards?

■ Cumulative Review

26. What are the three criteria for an ideal postulate system?
27. What did Kurt Gödel prove about these three criteria in 1936?
28. What criterion is the most important?
29. Find the coordinate of N if $LM = 5$, $MN = 3$, M is between L and N, and the coordinate of L is -7.
30. If $x \neq 7$, then what can you conclude about x? Explain.

ARCHIMIDES

Archimedes was an ancient Greek mathematician. He set much of the mathematical foundation needed to build our present mathematical system. Archimedes was born about 287 B.C. in Syracuse on the island of Sicily. Little is known about his childhood. His father was an astronomer, and Archimedes undoubtedly gained his interest in mathematics from his father's study of the stars and planets and their orbits.

Archimedes spent some time in Egypt and probably attended the University of Alexandria. He was known for his mechanical war inventions, and he supposedly burned enemy ships by focusing the sun's rays through large reflecting glasses mounted on top of Alexandria's lighthouse, the Pharos.

Archimedes had a great ability to concentrate on the mathematical problem at hand. War could be raging around him, but he would calmly continue solving a problem. He often used sand or ashes as his scratch paper, drawing figures and writing equations with a stick. Sometimes his level of concentration resulted in quite embarrassing situations. While in the bathtub one day, he discovered the law of hydrostatics. He became so excited about his discovery that he jumped from his tub and ran through the streets of the city yelling, "Eureka!" which means "I found it!"

Archimedes also discovered many mathematical methods that earned him a reputation as the greatest mathematician of the ancient world. He invented a method for approximating the value of a square root and first stated the laws of exponents. He approximated π (in his work *On the Measurement of a Circle*) as being between $3\frac{10}{71}$ and $3\frac{10}{70}$. During the Middle Ages scholars considered *On Spirals* to be his greatest work because it showed how to trisect an angle using a spiral.

Today, we consider Archimides' study of area and volume to be his greatest achievement. His work *On the Sphere and the Cylinder* was one of several in which he derived volume formulas for solids with curved surfaces. His methods foreshadowed the development of methods now used in integral calculus to quickly and easily find the area of plane regions.

In 212 B.C. Rome conquered Syracuse, and at the culmination of the conquest Archimedes was killed. Tradition says that Archimedes was working on a problem when a Roman soldier came up. Archimedes told the soldier not to bother his drawing in the sand. The soldier thought Archimedes to be so presumptuous and insulting that he killed him on the spot where he was working his problem.

You can see that the many practical inventions made by Archimedes came from his years of fruitful theoretical research. His works *On the Equilibrium of Planes*, which first stated the law of levers, *Physics* (in eight volumes), and *On Floating Bodies*, which included his law of hydrostatics, earned him the title Father of Mathematical Physics.

Today, we consider Archimides' study of area and volume to be his greatest achievement.

3.5 Polygons: Inscribed and Circumscribed

On your paper make a sketch of the following figure.

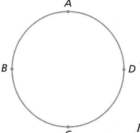

Figure 3.8

Connect the points in consecutive order with line segments. What kind of figure did you draw inside the circle?

Definition

An **inscribed polygon** is a polygon whose vertices are points of a circle.

Figure 3.9 shows an example of an inscribed pentagon.

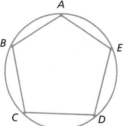

Figure 3.9

Notice that the sides of the inscribed polygon are chords of the circle. $\angle AED$ is called an inscribed angle; $\angle DCB$ is also an inscribed angle.

Definition

An **inscribed angle** is an angle whose vertex is on a circle and whose sides each contain another point of the circle.

Can you name three other inscribed angles in figure 3.9?

When a polygon is inscribed in a circle, the circle is *circumscribed about the polygon.* Do not be confused by the two terms *inscribed* and *circumscribed.* *Inscribed* means "written in," whereas *circumscribed* means "written around." So in figure 3.10 regular hexagon *LMNOPQ* is inscribed in ⊙*T*, while ⊙*T* is circumscribed about hexagon *LMNOPQ.*

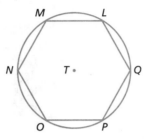

Figure 3.10

We can describe circumference in terms of the perimeter of inscribed polygons. Consider a set of regular polygons in which the number of sides steadily increases. The series of inscribed polygons shown in figure 3.11 will help you visualize the idea.

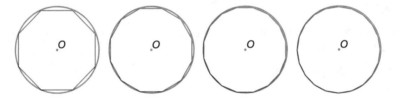

Figure 3.11

As the number of sides of the inscribed regular polygon increases, the perimeter of the polygon gets closer to the circumference of the circle. In other words, the perimeter of a regular *n*-gon approaches the circumference of its circumscribed circle as *n* increases.

Polygons can be circumscribed about a circle as well as inscribed in a circle.

Definition

A **polygon circumscribed about a circle** is a polygon whose sides each intersect the circle in exactly one point.

Figure 3.12 shows an example of a triangle that is circumscribed about a circle.

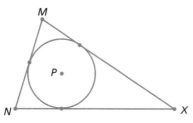

Figure 3.12

The hexagonal nut circumscribes the circular screw.

$\triangle MNX$ is circumscribed about $\odot P$. You can also say that $\odot P$ is inscribed in $\triangle MNX$. Notice that $\overline{MX}$ touches $\odot P$ in exactly one point.

Definitions

A **tangent line** (or *tangent*) is a line in the plane of a circle that intersects the circle in exactly one point.

The **point of tangency** is the point at which a tangent line and a circle intersect.

A **tangent segment** is a segment of a tangent line that contains the point of tangency.

Quadrilateral *ABCD* is circumscribed about $\odot O$. $\overline{AD}$ is a segment tangent to $\odot O$. The point of tangency is *Y*. Name three other tangent segments and give their points of tangency.

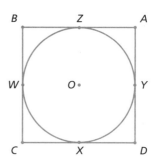

Figure 3.13

► A. Exercises

Look at the diagrams in exercises 1-5 and state whether the polygons are circumscribed about a circle, inscribed in a circle, or neither.

1.

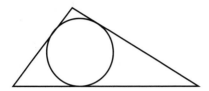

2.

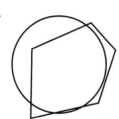

4.

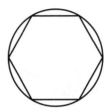

3.

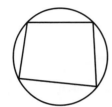

5.

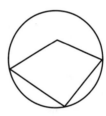

Use the figure for exercises 6-10.

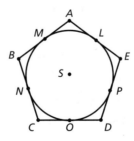

6. Is ⊙S inscribed in polygon *ABCDE*?
7. Is polygon *ABCDE* inscribed in ⊙S?
8. Name five tangent segments.
9. Name the point of tangency for each tangent segment.
10. Does *SP* = *SL*? Why?

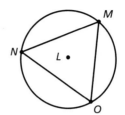

Use the figure for exercises 11-13.
11. Name three points on ⊙L.
12. Name three inscribed angles.
13. Is $\overline{MO}$ tangent to ⊙L? Why?
14. Draw an inscribed triangle.
15. Draw a hexagon inscribed in a circle.
16. Draw a circle circumscribed about a quadrilateral.

▶ B. Exercises

17. Draw a regular pentagon inscribed in a circle.
18. Draw another regular pentagon circumscribed about a circle.
19. Draw a regular triangle circumscribed about a circle.

▶ C. Exercises

20. Explain the steps you used to draw your answer to exercise 17.

▶ Dominion Thru Math

21. Olympic tracks consist of concentric ovals 1 meter wide. The inside lane has two 100-meter straightaways and two 100-meter semicircles. In the 400-meter dash, runners in successive lanes from the center start ahead of runners in previous lanes. (See photo on page 97.) Find the stagger needed for all runners to run the same distance.

▪ Cumulative Review

True/False. Correct any incorrect statements and justify correct statements with a definition, theorem, or postulate.
22. If $\overline{AB} \cong \overline{BC}$, then $AB = BC$.
23. If $\overline{AB} \cong \overline{BC}$, then $\overline{AB} = \overline{BC}$.
24. If $AB = BC$, then B is the midpoint of $\overline{AC}$.
25. If $AB = BC$, then $A = C$.
26. If B is the midpoint of $\overline{AC}$, then $\overline{AB} \cong \overline{BC}$.

3.6 Segment Constructions

Drawings are more accurate than sketches. Recall that a construction is a special drawing that permits use of only two tools. You will be amazed at the many things that you can construct using only a straightedge and a compass.

Before you begin making actual constructions, there are a few guidelines that you need to know and follow.

1. Make sure that your compass pencil and your free pencil are very sharp.
2. Lines and compass marks should look like eyelashes on the paper. These marks represent lines that have no width, so make the representations believable. Make light marks that can be erased if necessary.
3. Be neat. Carefully align your arcs and lines to pass through the correct points. Also, do not use dots for points. The marks of the compass where arcs cross are adequate and neater.

Raphael (1483–1520), The Marriage of the Virgin

Segments both visible and invisible are important to painters such as Raphael. The building in this painting contains many visible segments, while the segments undrawn between the bride, the groom, and the ring are important to the focus and balance of the piece.

To see the difference between a neat, precise construction and a messy construction, look at the two constructions below. Each shows the bisection of a given angle. Do you see how the bad construction disregards the guidelines?

For each construction the information that you are given will be listed under *Given,* and the figure that is to be constructed will be described under *Construct.*

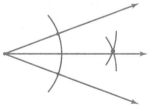
Bad construction

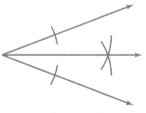
Good construction

Figure 3.14

onstruction 2

Copy a segment.

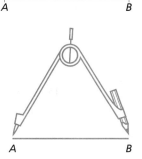

Figure 3.15

> **Given:** $\overline{AB}$
>
> **Construct:** A line segment congruent to $\overline{AB}$

1. Draw a line, using the straightedge. Mark any point on this line and label it A'.
2. Open your compass to measure the same length as the given segment, $\overline{AB}$.

3. Without changing the compass, place the point at A' and mark an arc on the line. Label the point of intersection B'.

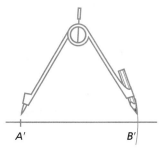

By repeating this basic construction, you can construct segments whose lengths are multiples of the given segment. The reason that construction 2 works is that the length you measure from the given segment corresponds to the radius of a certain circle. When you construct the segment, you are marking off a radius of equal measure; therefore, the segments are congruent.

onstruction 3

Bisect a segment.

> **Given:** $\overline{AB}$
>
> **Construct:** The bisector of $\overline{AB}$

1. Place the point of the compass at each endpoint, making intersecting arcs above and below the line segment.
2. Connect the two intersecting points to form the bisector of $\overline{AB}$. M is the midpoint.

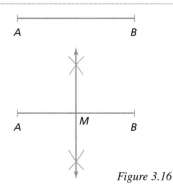

Figure 3.16

You will prove in Chapter 6 that this construction works. For now, verify that M is the midpoint using your compass to compare AM and MB.

▶ A. Exercises

Make the following constructions.

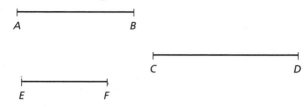

1. A segment congruent to $\overline{CD}$
2. A segment congruent to $\overline{EF}$
3. A segment congruent to $\overline{AB}$
4. A segment with length $AB + EF$
5. A segment with length $CD + EF$
6. A segment with length $AB + CD + EF$
7. A segment with length $CD - EF$
8. The bisector of $\overline{CD}$
9. The bisector of $\overline{AB}$
10. The bisector of $\overline{EF}$

▶ B. Exercises

Construct.
11. A circle with diameter $\overline{CD}$
12. A segment with length $3AB$
13. A segment with length $2(AB + EF)$
14. The bisector of a segment with length $AB + CD$

▶ C. Exercises

15. Construct a square inscribed in a circle.

■ Cumulative Review

16. Sketch a concave nonagon.
17. Sketch a circle. Show a chord, a diameter, and a tangent that are all parallel. Label each.
18. How many diagonals does a regular octagon have?
19. How many edges does a tetrahedron have?
20. Sketch a quadrilateral that has a diagonal that intersects the exterior of the quadrilateral.

Geometry *and* Scripture

Distance

Measurement is important in many areas of life. It is especially important for our understanding of the Bible. Since God commanded that men measure some things and some of those measurements are recorded in Scripture, it is important for us to understand them. What did God command men to measure in each verse?

1. Numbers 35:5   **2.** Deuteronomy 21:2   **3.** Revelation 11:1

HIGHER PLANE: Ezekiel 45:3

God is also concerned that our measurements be done fairly and accurately.

4. What does God command in Deuteronomy 25:15?

5. What does God prohibit (Prov. 20:10)?

So far, you have studied linear measures primarily. However, familiar units such as inches, feet, and meters are not used in the Bible. Furthermore, even the familiar mile and furlong refer to shorter distances in the Bible than they do today.

Complete the following table. Obtain missing conversions by using a ruler to measure the unit. (Your teacher will help you compare your answers against the accepted lengths of ancient units.) Also, find a Bible reference as an example for each unit.

	Unit	Measure	Conversion	Reference
6.	Finger	Width of your finger		
7.	Handbreadth	Width of your 4 fingers held together		
8.	Span	Thumb tip to pinky tip with your fingers spread		
9.	Cubit	Elbow to tip of your middle finger		Zech. 5:2
10.	Pace	Length between your left and right heels' footprints when walking		
11.	Reed	A stalk of sweet cane (calamus)	almost 11 ft.	
12.	Furlong	Length of race at Olympia	606.75 ft.	
13.	Mile	8 furlongs	4,854 ft.	

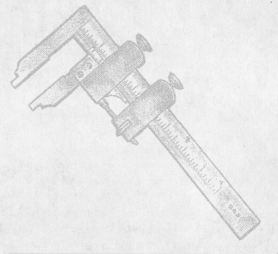

Convert the measurement in each reference below to modern units. Do not confuse ancient units with modern units. Express the size of the subject in a complete sentence.

14. I Samuel 17:4

15. Deuteronomy 3:11

16. Luke 24:13

17. Ezekiel 45:1

The Lord also used distance to teach us to go beyond the call of duty.

Line upon Line

AND WHOSOEVER shall compel thee to go a mile, go with him twain. ❧

MATTHEW 5:41

Chapter Review

State the property that is described by each equation. All letters represent real numbers.

1. $a + b = b + a$
2. $a(b + c) = ab + ac$
3. $a(bc) = (ab)c$
4. If $a = t$ and $a + b = c$, then $t + b = c$
5. For all a and b, $a > b$ or $a < b$ or $a = b$

Compute.

6. $|6| + |-27|$
7. $|12 - 19|$
8. $|-74| - |-70|$
9. $|18 - 12|$
10. $|136| \cdot |-9|$

Find the coordinate of each point on the figure shown.

```
          X          Z     Y         M
  -25 -20 -15 -10  -5  0   5  10  15  20  25
```

11. X
12. M
13. Z
14. Y

15. If the coordinate of A is 19 and the coordinate of B is -12, find AB.
16. If the coordinate of X is x and the coordinate of Y is y, find XY.
17. Find the perimeter of an equilateral triangle whose side measures 5 units.
18. Find the circumference of a circle whose diameter is 14 units.
19. Find the circumference of a circle whose radius is 17 units.
20. Draw a pentagon inscribed in a circle.
21. Draw a triangle circumscribed about a circle. Label the triangle $\triangle XYZ$.
22. In $\triangle XYZ$ label the points of tangency A, B, and C.

23. Find the perimeter of the polygon shown.

(Figure: a polygon with sides labeled 19, 10, 8, and 27)

Use $\overline{AB}$ for exercises 24-29.

$$\overset{\bullet}{\underset{A}{}}\text{———————}\overset{\bullet}{\underset{B}{}}$$

24. Sketch a segment congruent to $\overline{AB}$.

25. Draw a segment congruent to $\overline{AB}$.

26. Construct a segment congruent to $\overline{AB}$.

27. Sketch a circle with diameter $\overline{AB}$.

28. Draw a circle with diameter $\overline{AB}$.

29. Construct a circle with diameter $\overline{AB}$.

30. Explain the mathematical significance of Matthew 5:41.

▶ Dominion Thru Math

When plumbers install pipes from an architect's drawings, they calculate the needed lengths. The drawings give the distances between centerlines of pipes or fittings such as elbows and tees. The straight pieces must be cut shorter because the fittings are part of the length. The size and type of pipe determine how much to allow for the fittings.

31. Joe installs a 1-inch plastic water line branch with a tee, followed by $138\frac{3}{4}$ inches of pipe to a 45° elbow, then by $98\frac{1}{2}$ inches of pipe to a 90° elbow, and then by 42 inches of pipe to a valve. The pipe comes in 10-foot lengths. Couplings to join two straight lengths equal $\frac{1}{8}$ inches of pipe. Tees, 90° elbows, and valves have two sockets equal to $\frac{11}{16}$ inches of pipe each; 45° bends have two sockets equal to $\frac{3}{8}$ inches of pipe each. Find the lengths of pieces of pipe needed. *Hint:* Make a sketch.

32. In a 6-inch metal piping system, the discharge pipe from a pump goes 16 inches to a check valve, 12 feet to a 45° elbow, and then 8 feet to a tee; each side of the tee goes 5 feet to a 90° elbow at a tank. The pipe comes in 20-foot lengths. The equivalent lengths of pipe for these fittings are as follows: check valves have two ends at 4 inches each; 45° elbows have two ends at 4 inches each; 90° elbows have two ends at 6 inches each; and tees have three ends at 6 inches each. Find the lengths of pieces of pipe needed.

4 Angles and Measurement

The Dome of the Rock in Jerusalem, Israel, is the "Jewel of Islam" because it is the most famous and beautiful building in Islamic architecture. Completed in AD 691 with a lead dome, it was changed to the current gold color in 1965. The builders had to measure angles carefully in order to make the regular polygon and the 110-foot high wooden dome. Mosaics adorn the blue tiles and marble of the 36-foot base of the dome.

Muslims built the shrine to commemorate the place where Mohammed supposedly ascended to heaven. Yet even the Muslims recognize it as the place where Abraham offered his son on the altar (although they wrongly believe he offered Ishmael rather than Isaac). The shrine occupies the site of the temple of Israel.

After this chapter you should be able to

1. apply properties of inequality.
2. find angle measures with a protractor.
3. classify angles by measure.
4. identify postulates on angle measure.
5. define angle relations—congruence, bisection, and special pairs of angles.
6. classify triangles and quadrilaterals.
7. prove some theorems about angles.
8. construct congruent angles and angle bisectors.

117

4.1 Inequalities

In Chapter 3 you reviewed the properties of equality. The properties of inequality are similar, but remember to use caution when multiplying by a negative.

European castles, such as this one at Heidelberg, Germany, often incorporate inequalities by using towers of varying heights.

Inequality Properties	
Property	**Meaning**
Addition	If $a > b$, then $a + c > b + c$.
Multiplication	If $a > b$ and $c > 0$, then $ac > bc$. If $a > b$ and $c < 0$, then $ac < bc$.
Transitive	If $a > b$ and $b > c$, then $a > c$.

You should also recall that $a \geq b$ means $a > b$ *or* $a = b$. Do you remember the difference between "and" and "or"?

EXAMPLE 1   Graph $x \geq 3$ or $x < 2$.

Answer   1. Put a dot on 3 and shade to the right.

2. Put a circle over 2 to show that 2 is not included and shade left.

3. The word "or" requires the union of both parts above.

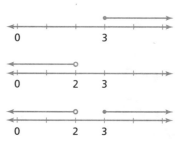

Figure 4.1

EXAMPLE 2   Graph $x \geq 3$ and $x < 2$.

Answer   You graphed both parts in example 1. "And" means intersection, so the answer should be where they overlap. Since there is no overlap, the solution is $\varnothing$.

EXAMPLE 3   Graph $|x| < 5$.

Answer   You know that $|x|$ represents a distance. The inequality says that the distance from zero must be less than five units. So $x < 5$ and $x > -5$. Since there is overlap, the solution can be summarized $-5 < x < 5$.

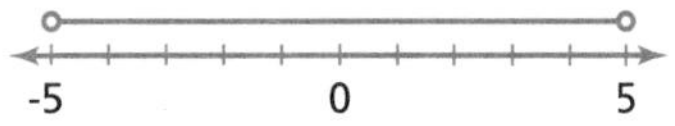

Figure 4.2

One other property of inequalities is especially important.

Definition

A real number a is **greater than** a real number b $(a > b)$ if there is a positive real number c so that $a = b + c$.

This definition shows that $5 > 3$ because $5 = 3 + 2$ and $2 > 0$.

▶ A. Exercises

Identify each property.
1. $x < 2$ and $2 < 5$, therefore $x < 5$.
2. Since $x \leq 3$, $-4x \geq -12$.
3. Since $x > 5$, $x = 5 + c$ for some positive constant c.
4. Since $x > 2$, $x - 3 > -1$.

Graph.
5. $x \geq 2$ and $x < 1$
6. $x \geq 2$ or $x < 1$
7. $x \leq 2$ and $x < 1$
8. $x \leq 2$ or $x < 1$
9. $x \leq 2$ and $x > 1$
10. $x \leq 2$ or $x > 1$
11. $x \geq 2$ and $x > 1$
12. $x \geq 2$ or $x > 1$

▶ B. Exercises

Solve each inequality.
13. $x + 6 > 5$ and $3x + 4 < 7$
14. $-6x \leq 12$ or $x + 8 < 5$
15. $|x| = 3$
16. $|x| < 3$
17. $|x| > 3$
18. $|x| = 0$
19. $|x| < -3$
20. $|x| > -3$

Consider the five inequalities: $<, >, \leq, \geq, \neq$.
Which inequalities have each property? Give examples to show which do not.
21. reflexive
22. symmetric
23. transitive
24. Are any of the inequalities equivalence relations?

▶ C. Exercises

25. Explain why $5 > 2$.
26. Explain why $-7 > -10$.
27. Show that $5 < 16$ and then conclude that $\sqrt{5} < 4$.
28. Show that $\sqrt{3} < \sqrt{11}$.

▶ Dominion Thru Math

Use the Dome of the Rock shown in the opening photo of the chapter to answer the questions.
29. Draw a footprint for the Dome of the Rock. (*Hint*: Compare front views to footprints for the four buildings shown in the chapter. Research on the Internet.)
30. Identify geometric shapes in the front view and the footprint.

■ Cumulative Review

State each postulate or theorem.
31. Line Separation Postulate
32. Theorem on perimeter of a regular *n*-gon
33. Ruler Postulate
34. Midpoint Theorem
35. Jordan Curve Theorem

4.2 Angle Measure

The Biltmore mansion in Asheville, North Carolina, includes angles of varying sizes in the roof.

According to the Ruler Postulate, each point on a line corresponds to a real number. Similarly, each angle corresponds to a real number.

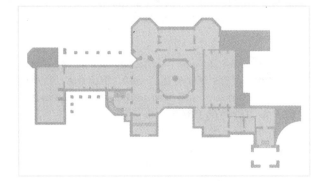

Postulate 4.1
Protractor Postulate. **For every angle *A* there corresponds a positive real number less than or equal to 180. This is symbolized $0 < m\angle A \leq 180$.**

Notice that the *m* in the symbolization indicates a real number, whereas the symbolization $\angle A$ without the *m* indicates the angle.

Definition

The **measure of an angle** is the real number that corresponds to a particular angle.

The *degree measure* is the real number that represents the measure of an angle. Angles are measured by an instrument called a *protractor.* A protractor is a semicircular device that is marked off in 180 equal divisions called *degrees.* The symbol for degrees is °, but the degree symbol will sometimes be omitted in this book.

You can measure an angle such as ∠BAC with a protractor by placing the hole or vertical slash mark of the protractor at the vertex of the angle and aligning one ray with the zero mark of the protractor. See figure 4.3. The place where the other ray intersects the protractor indicates the measure of the angle.

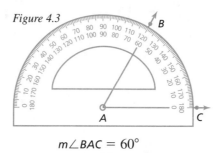

$$m\angle BAC = 60°$$

Do you see how you could draw a semicircle (half of a circle) with your protractor? Could you draw a circle? Notice that the degree markings go all the way around the semicircle. For every degree measure, you can make a corresponding angle. Otherwise an angle would be missing on the protractor and would leave a gap or a hole in our circle.

Postulate 4.2
Continuity Postulate. If *k* is a half-plane determined by $\overleftrightarrow{AC}$, then for every real number, $0 < x \le 180$, there is exactly one ray, $\overrightarrow{AB}$, that lies in *k* such that $m\angle BAC = x$.

This postulate is similar to the Protractor Postulate, but notice that the ideas appear in reverse order. When this happens, we call the statements *converses* of each other. The Continuity Postulate enables you to draw an angle of any measure by using a protractor.

Did you notice that these two postulates are similar to the Ruler Postulate and Completeness Postulate? The Ruler and Protractor Postulates tell us what can be measured: segments with a ruler and angles with a protractor. The Completeness Postulate tells us that rulers are number lines and have no missing numbers. Lines have no holes; they are complete. Similarly, the Continuity Postulate tells us that protractors have no missing numbers. Circles have no gaps; they are continuous. These two postulates enable us to draw segments and angles of any size.

EXAMPLE Draw a 42° angle.

Answer 1. Draw a ray, *DE*, to use as a
side of the angle.
2. Use a protractor to find
the other side of the angle
by placing the hole or vertical slash mark at *D* and
the zero mark along *DE*.
3. Draw *DF* to correspond
to 42 degrees.

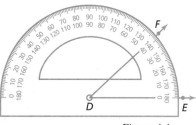

Figure 4.4

So $m\angle FDE = 42°$.

Angles are classified according to their angle measures. The following table defines special types of angles.

Angle name	Angle measure
Acute angle	$0° < x < 90°$
Right angle	$x = 90°$
Obtuse angle	$90° < x < 180°$
Straight angle	$x = 180°$

Definition

Congruent angles are angles that have the same measure.

The symbol for congruence is the same for angles as it is for segments. In figure 4.5 $m\angle ABC = 50°$ and $m\angle XYZ = 50°$, so $\angle ABC \cong \angle XYZ$.

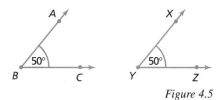

Figure 4.5

The arc in the interior of the angle indicates that the angles have the same measure. Notice that the equal sign is used for measure of angles. This is because $m\angle$ indicates a real number. The $\cong$ sign is used for the angles themselves but not for their measures.

▶ A. Exercises

Using a protractor, find the measure of each angle in the diagram.

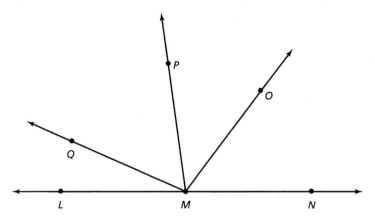

1. ∠LMQ
2. ∠OMN
3. ∠PMO
4. ∠QMN
5. ∠LMN

6. ∠PMN
7. ∠LMO
8. ∠QMP
9. ∠LMP

Draw angles having the given measures and label them correctly.

10. m∠TXY = 55°
11. m∠LMO = 117°
12. m∠XYZ = 150°
13. m∠ABC = 25°
14. m∠FYT = 10°
15. m∠LPQ = 60°
16. m∠KDP = 180°
17. m∠RSP = 120°

▶ B. Exercises

Draw an angle that is congruent to each given angle. Label each angle straight, right, obtuse, or acute.

18.

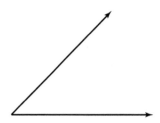

19.

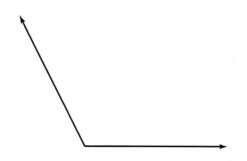

20.

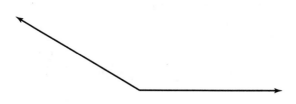

21.

22.

23.

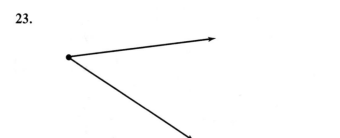

24. Use a protractor to determine which numbered angles below are congruent.

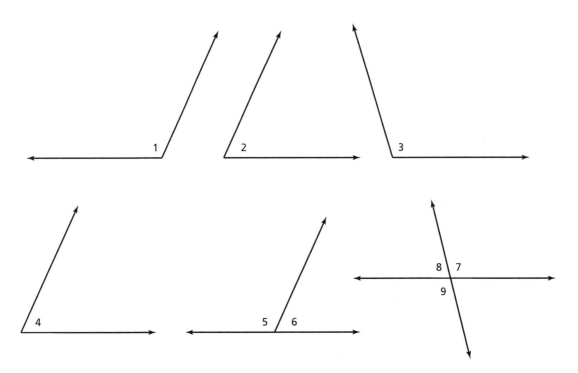

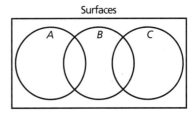

▶ **C. Exercises**

25. Is every angle congruent to itself?

■ **Cumulative Review**

Answer the questions about the Venn diagram below.

A = The set of cylinders
B = The set of polyhedra
C = The set of cones

Surfaces

26. What do we call surfaces in $A \cap B$?
27. What do we call surfaces in $B \cap C$?

Give an example of a surface in each set below.
28. $A \cap B'$ **30.** $B \cap (A \cup C)'$
29. $C \cap B'$ **31.** $(A \cup B \cup C)'$

Slopes of Lines

You can measure steepness two ways. One way is to find the angle of inclination above the horizontal using a protractor. The other is to calculate a number called the slope.

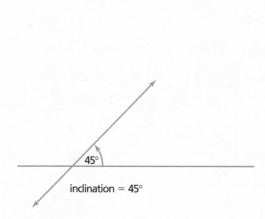

inclination = 45°

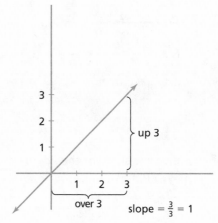

up 3

over 3

slope = $\frac{3}{3}$ = 1

The **slope of a line** is a ratio obtained from two points of a line by dividing the difference in y coordinates by the difference in x coordinates.

$$m = \frac{y_2 - y_1}{x_2 - x_1}$$

A shorter definition of the slope of a line is "rise over run" or "the vertical change divided by the horizontal change." Look at the graph of the line.

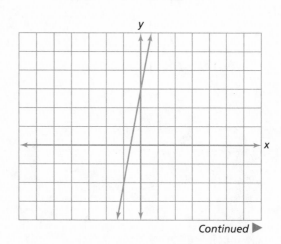

Continued ▶

First look for points of a line where it is easy to identify coordinates. You should see that the line passes through (0, 3) and (−1, −2). Then find the slope of this line by determining the vertical change divided by the horizontal change: $\frac{y_2 - y_1}{x_2 - x_1}$. Since this line passes through (0, 3) and (−1, −2), the slope is $m = \frac{-2 - 3}{-1 - 0} = \frac{-5}{-1} = 5$. You can determine the slope of any line by using this method.

▶ Exercises

Give the slope of the line passing through
1. (0, 0) and (3, −2).
2. (−2, 3) and (2, 4).
3. (6, −2) and (5, 3).

Give the slope of each line shown.

4.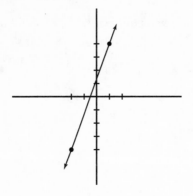

5.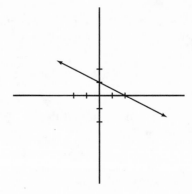

4.3 Angle Bisectors

How many angles are in the diagram below?

Notice that each ray is a side of two of the three angles. However, the large angle shares interior points with each of the smaller angles. The small angles do not share interior points and $\overrightarrow{NK}$ is a side of each. $\angle MNK$ and $\angle PNK$ are called adjacent angles.

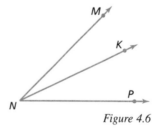

Figure 4.6

Definition

Adjacent angles are two coplanar angles that have a common side and a common vertex but no common interior points.

The Angle Addition Postulate describes angles much as betweenness describes points. The definition of *betweenness* says that B is between A and C if $AB + BC = AC$.

Postulate 4.3

Angle Addition Postulate. If K lies in the interior of $\angle MNP$, then $m\angle MNP = m\angle MNK + m\angle KNP$.

You can see that $\angle MNK$ and $\angle KNP$ are adjacent angles. So, the postulate shows you how to find the angle measure of a larger angle from related adjacent angles.

EXAMPLE 1   Find $m\angle XYZ$ if $m\angle XYT = 25°$ and $m\angle TYZ = 15°$.

Answer   According to the Angle
Addition Postulate,
$m\angle XYZ = m\angle XYT + m\angle TYZ$
$m\angle XYZ = 25 + 15$
$m\angle XYZ = 40°$

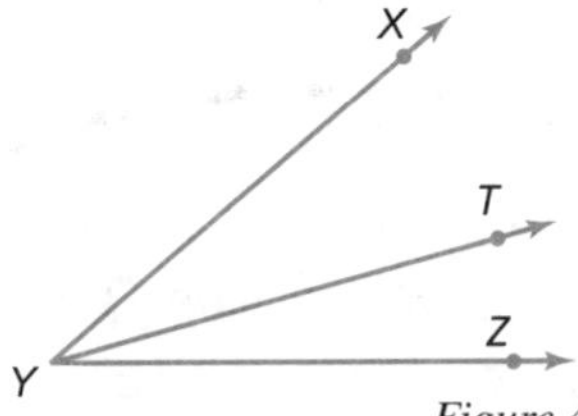

Figure 4.7

EXAMPLE 2   Find $m\angle DBC$ if $m\angle ABC = 90°$
and $m\angle ABD = 70°$.

Answer   According to the Angle
Addition Postulate,
$m\angle ABC = m\angle ABD + m\angle DBC$
$90 = 70 + m\angle DBC$
$20° = m\angle DBC$

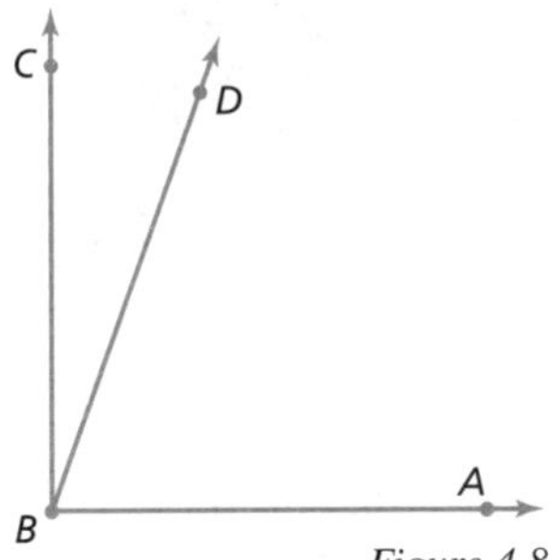

Figure 4.8

You know that distance is to points as degree measure is to angles. This analogy goes further. In this lesson, you have seen that betweenness is like the Angle Addition Postulate. Now you will see a similarity between midpoints and angle bisectors.

Definition

An **angle bisector** is a ray that (except for its origin) is in the interior of an angle and forms congruent adjacent angles.

If $\overrightarrow{XT}$ is the angle bisector, then
$\angle YXT \cong \angle ZXT$.

Notice that $\frac{1}{2} m\angle YXZ = m\angle YXT$.

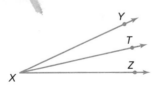

Figure 4.9

What happens if a straight angle is bisected as in figure 4.10?

Since the measure of any straight angle is 180, its bisector divides it into two right angles. When lines intersect to form right angles, the lines are called *perpendicular* lines.

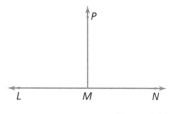

Figure 4.10

Thomas Jefferson's home at Monticello near Charlottesville, Virginia, uses an angle bisector to divide the front facade into two equal halves.

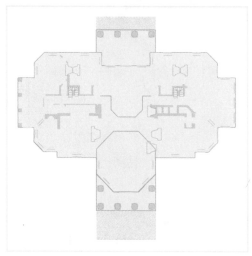

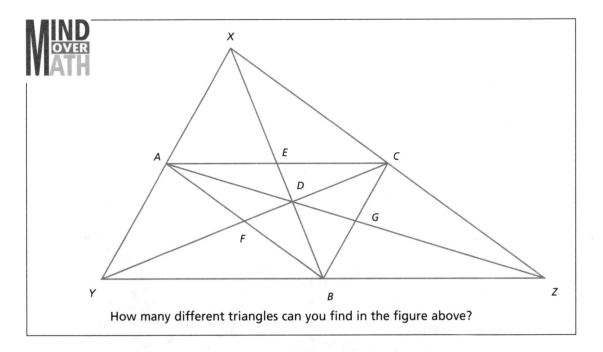

How many different triangles can you find in the figure above?

Perpendicular lines are lines that intersect to form right angles.

The symbol for perpendicular is ⊥. In figure 4.11, $\overleftrightarrow{PR} \perp \overleftrightarrow{QS}$ (read "line *PR* is perpendicular to line *QS*").

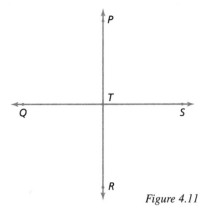

Figure 4.11

Four right angles are formed when two perpendicular lines intersect. Can you name the right angles in figure 4.11?

▶ A. Exercises

Use the diagram for exercises 1-10. Find the measures using a protractor.

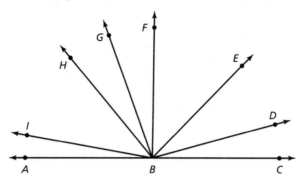

1. $m\angle FBE$	6. $m\angle IBD$
2. $m\angle HBE$	7. $m\angle GBE$
3. $m\angle DBC$	8. $m\angle ABH$
4. $m\angle IBH$	9. $m\angle ABE$
5. $m\angle FBC$	10. $m\angle ABD$

Use the diagram for exercises 11-14.

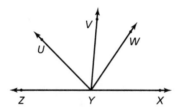

11. Find $m\angle UYX$ if $m\angle UYW = 75°$ and $m\angle WYX = 35°$.
12. Find $m\angle UYW$ if $m\angle UYV = 58°$ and $m\angle VYW = 29°$.

▶ B. Exercises

13. Find $m\angle UYV$ if $m\angle UYW = 85°$ and $m\angle VYW = 15°$.
14. Find $m\angle ZYW$ if $m\angle WYX = 45°$.

Use the following figure for exercises 15-18.

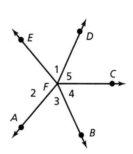

15. If $\overrightarrow{FD}$ is the bisector of $\angle EFC$, what is true about $\angle 1$ and $\angle 5$?
16. If $m\angle 3 = m\angle 4$, what is true about $\overrightarrow{FB}$?
17. If $\overrightarrow{FC}$ bisects $\angle DFB$ and $m\angle DFB = 92°$, what is $m\angle 5$?
18. If $\overrightarrow{FB}$ bisects $\angle AFC$ and $m\angle 3 = 54°$, what is $m\angle AFC$?

Answer exercises 19-20 based on the figure shown.

19. Name an angle whose measure equals 103.

20. Name three right angles and the perpendicular lines forming each of them.

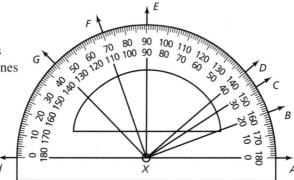

Use the figure for exercises 21-24. Rewrite each sum or difference using a single angle.

21. $m\angle NRM + m\angle MRQ$

22. $m\angle ORP + m\angle PRQ$

23. $m\angle LRP - m\angle LRQ$

24. $m\angle ORL - m\angle MRL$

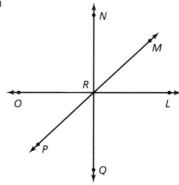

▶ **C. Exercises**

25. Draw a diagonal of a square. What is the measure of the angle between the diagonal and the side of the square? Will it always be the same? Why?

■ **Cumulative Review**

In each problem, identify the sets into which the first figure divides the second. Justify your answer with a postulate or theorem.

26. plane, space

27. point, line

28. polygon, plane

29. line, plane

30. polyhedron, space

4.4 Angle Properties

Look at figure 4.12. Notice that ∠*ABC* is an acute angle, ∠*CBD* is an obtuse angle, and both angles have the same vertex and a common side.

The Pacific Design Center in Los Angeles incorporates angles so that the smaller lower floors support even larger upper floors.

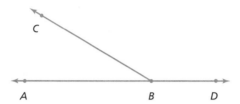

Figure 4.12

Definition

A **linear pair** is a pair of adjacent angles whose noncommon sides form a straight angle (are opposite rays).

∠*LMO* and ∠*LMN* are adjacent angles. If *m*∠*OMN* = 180°, then these two angles form a linear pair.

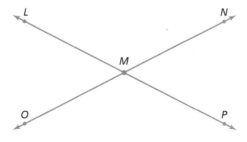

Figure 4.13

Another special pair of angles commonly seen in geometry are vertical angles. Vertical angles form two pairs of opposite rays. In figure 4.13 ∠*NMP* and ∠*LMO* are vertical angles.

Vertical angles are angles adjacent to the same angle and forming linear pairs with it.

Can you name another pair of vertical angles in figure 4.13?

Here are some other important pairs of angles.

Definitions

Two angles are **complementary** if the sum of their angle measures is 90°.

Two angles are **supplementary** if the sum of their angle measures is 180°.

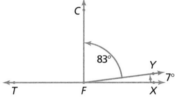

Figure 4.14

Since $m\angle CFY = 83°$ and $m\angle YFX = 7°$, $\angle CFY$ and $\angle YFX$ are complementary angles. Notice that together they form a right angle. If $m\angle TFY = 173°$, then $\angle TFY$ and $\angle YFX$ are supplementary angles. Notice that they form a linear pair.

Remember that sketches are freehand. Sketch the figure above without rulers or protractors. Now measure your angles. Are they exactly 83° and 7°? This shows you that the labels on a sketch are essential. Since measurements on a sketch are not reliable, only the labels provide exact values.

Since many diagrams are not intended to be accurate, whenever values are provided or labeled, you should use them regardless of the way the diagram looks. Unless you are told to use a protractor or make a construction, you should always treat diagrams as sketches. The sketches will help you understand which points are collinear and therefore which pairs of angles are supplementary or vertical. The following theorems can also help you calculate measures of related angles from given information.

You will justify these theorems in the exercises (the last is in the Chapter Review).

Theorem 4.1

All right angles are congruent.

Theorem 4.2

If two angles are adjacent and supplementary, then they form a linear pair.

Theorem 4.3

Angles that form a linear pair are supplementary.

Theorem 4.4

If one angle of a linear pair is a right angle, then the other angle is also a right angle.

Theorem 4.5

Vertical Angle Theorem. **Vertical angles are congruent.**

Theorem 4.6

Congruent supplementary angles are right angles.

Theorem 4.7

Angle Bisector Theorem. **If $\overrightarrow{AB}$ bisects $\angle CAD$, then $m\angle CAB = \frac{1}{2} m\angle CAD$.**

▶ A. Exercises

Use the figure for exercises 1-5. Do not use a protractor.

1. Find $m\angle IKH$.
2. Find $m\angle HKG$.
3. Name two pairs of congruent angles.
4. What kind of angles are $\angle GKJ$ and $\angle HKI$?
5. What is the measure of the complement of $\angle JKI$? the supplement?

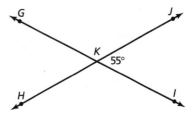

Use the figure below for exercises 6-10. $m\angle AGF = 40°$; $m\angle BGC = 50°$; $m\angle AGE = 90°$; $m\angle EGD = 90°$

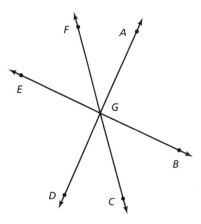

6. Name two pairs of adjacent angles.
7. Name two pairs of supplementary angles.
8. Name two pairs of complementary angles.
9. What is $m\angle FGE$?
10. What is $m\angle BGD$?
11. Name an angle in the diagram above for each of the four kinds of angles.

You should now be able to give the reasons for steps in proofs. Remember that every step requires a definition, postulate, or previous theorem to explain why it is correct.

Theorem 4.1
All right angles are congruent.

Given: $\angle A$ and $\angle B$ are right angles
Prove: $\angle A \cong \angle B$

STATEMENTS	REASONS
$\angle A$ and $\angle B$ are right angles	Given
12. $m\angle A = 90°$ $m\angle B = 90°$	12.
13. $m\angle A = m\angle B$	13.
14. $\angle A \cong \angle B$	14.

Answer exercises 15-17 to understand why the following theorem is true.

Theorem 4.2

If two angles are adjacent and supplementary, then they form a linear pair.

Given: ∠AXB and ∠BXC are adjacent and supplementary
Prove: ∠AXB and ∠BXC form a linear pair (∠AXC is straight)

15. What can you conclude from the supplementary angles?
16. What can you conclude from the Angle Addition Postulate?
17. What can you conclude from the transitive property?

▶ B. Exercises

Give the reason for each step in the proofs below.

Theorem 4.3

Angles that form a linear pair are supplementary.

Given: ∠PAB and ∠BAQ form a linear pair
Prove: ∠PAB and ∠BAQ are supplementary

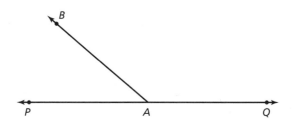

STATEMENTS	REASONS
∠PAB and ∠BAQ form a linear pair	Given
18. ∠PAB and ∠BAQ are adjacent and ∠PAQ is straight	**18.**
19. $m\angle PAB + m\angle BAQ = m\angle PAQ$	**19.**
20. $m\angle PAQ = 180°$	**20.**
21. $m\angle PAB + m\angle BAQ = 180°$	**21.**
22. ∠PAB and ∠BAQ are supplementary	**22.**

Answer exercises 23-24 to understand why the following theorem is true.

Theorem 4.4

If one angle of a linear pair is a right angle, then the other angle is also a right angle.

Given: ∠ABC and ∠CBD form a linear pair, and ∠ABC is a right angle

Prove: ∠CBD is a right angle

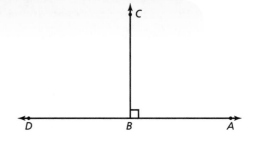

23. How do you know that the angles are supplementary?
24. Using the definitions of supplementary and right angles, write and solve an equation for m∠CBD.

Theorem 4.5

Vertical Angle Theorem. **Vertical angles are congruent.**

Given: ∠AXB and ∠CXD are vertical angles

Prove: ∠AXB ≅ ∠CXD

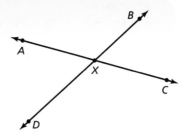

STATEMENTS	REASONS
25. ∠AXB and ∠CXD are vertical angles	**25.**
26. ∠BXC forms a linear pair with both angles above	**26.**
27. ∠BXC is a supplement to both angles above	**27.**
28. $m\angle AXB + m\angle BXC = 180$; $m\angle CXD + m\angle BXC = 180$	**28.**
29. $m\angle AXB + m\angle BXC = m\angle CXD + m\angle BXC$	**29.**
30. $m\angle AXB = m\angle CXD$	**30.**
31. ∠AXB ≅ ∠CXD	**31.**

Theorem 4.6

Congruent supplements are right angles.

Given: ∠A and ∠B are supplementary, and ∠A ≅ ∠B

Prove: ∠A and ∠B are right angles

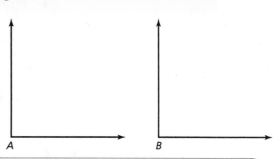

STATEMENTS	REASONS
32. ∠A ≅ ∠B, ∠A and ∠B are supplements	**32.**
33. $m∠A = m∠B$	**33.**
34. $m∠A + m∠B = 180°$	**34.**
35. $m∠A + m∠A = 180°$	**35.**
36. $2m∠A = 180°$	**36.**
37. $m∠A = 90°$	**37.**
38. $m∠B = 90°$	**38.**
39. ∠A and ∠B are right angles	**39.**

▶ **C. Exercises**

40. Discover a relationship between angle bisectors for the two angles of a linear pair. Justify your answer.

▪ Cumulative Review

Review properties of equality and inequality (Sections 3.1, 4.1). What would each property of inequality below be?

41. Addition property of ≤

42. Multiplication property of ≤

43. Reflexive property of ≤

44. Transitive property of ≤

45. Why is ≤ not an equivalence relation?

GEOMETRY AND ARCHITECTURE

Whose work is likely to last longer on this earth than any other's? Since the most obvious remains of ancient civilizations are buildings, architects probably make the most lasting contribution. The Egyptian pyramids are the last survivors of the seven wonders of the ancient world. The Greek Parthenon still stands on the acropolis in Athens. You can still walk into the Colosseum in Rome. What's the point? If the Lord tarries and the world lasts for another thousand years, people will probably learn about us from our buildings. Even the written records they find will probably have survived because they were protected by the ruins of buildings.

St. Paul's Cathedral in London, designed by Sir Christopher Wren, displays symmetry.

What does an architect need to know to do his job well? Many things are involved, of course. But one of the most basic skills he needs is an understanding of geometry. Why? He is trying to do two things. First, he wants to design a building that will be strong enough to last—or at least strong enough not to collapse on the people inside. Second, he wants this building to be pleasing to look at—he wants it to have aesthetic appeal. Geometric principles help him reach both of these goals.

By using geometry, the architect can be sure that his structures will be strong enough to support their own weight. For example, simple crossbeam supports illustrate the geometric principle of the plane. Suppose you were to build a square by fastening four wooden beams together at the corners. This structure would not be strong; it would tend to collapse. But if you nail a fifth beam from one corner to the opposite corner (that is, construct a diagonal), the square will be sturdier. Why? Because you have connected three points directly to one another. You have defined a plane by using three points. This is the same principle that makes a tripod more stable than a four-legged table.

Another geometric structure that helps architects strengthen their buildings is the arch. Before the arch was discovered, builders used a simple crosspiece atop two uprights. The builders of Stonehenge used this design. This kind of archway is only as strong as the crosspiece. When the crosspiece cannot hold any more weight, the whole structure collapses. The Romans discovered how the arch could hold more weight. They replaced the crosspiece with a rounded line of fitted stones. During the Middle Ages, designers developed a pointed arch that was even stronger.

The architect wants his buildings to be appealing as well as strong. Here geometry proves especially helpful. Designers have long known that certain ratios and shapes are pleasing to the eye. The Greeks often used such ratios in their designs. The Parthenon, for example, is designed around squares and rectangles of a certain shape.

Another pleasing shape is the circle. This simple shape occurs often in building design. Thomas Jefferson used it, for example, when he designed a building at the University of Virginia. It is also a basic element in the United States Capitol.

Architects make considerable use of the geometric principle of symmetry. Many buildings are beautiful simply because each side reflects the other. The Duomo in Milan, Italy, illustrated here, is a good example of this. Note that the right side of the building is a mirror image of the left side.

The greatest buildings in the world combine all these principles, and many others, in one solidly built, aesthetically pleasing structure. One of the best examples is St. Paul's Cathedral in London. From above it looks like a cross, symmetrically laid out. From the front it is also symmetrical. The ratio of its height to its width is aesthetically pleasing. Throughout the structure, arches and flying buttresses strengthen and beautify it. It is one of the world's best illustrations of man's use of God-given talents to design a functional work of art.

> *By using geometry, the architect can be sure that his structures will be strong enough to support their own weight.*

The ornate design of the facade of the Duomo Cathedral in Milan, Italy, shows perfect symmetry.

4.5 Polygons: Triangles and Quadrilaterals

Just as triangle and quadrilateral describe types of polygons, there are also terms that describe types of triangles and quadrilaterals. Most of these should be familiar, but the definitions will be important in later proofs.

Triangles can be classified by their angles.

Definitions

An **acute triangle** is a triangle with three acute angles.

A **right triangle** is a triangle with a right angle.

An **obtuse triangle** is a triangle with an obtuse angle.

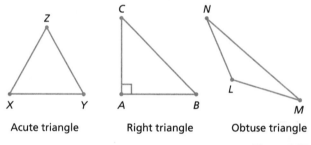

Acute triangle   Right triangle   Obtuse triangle

Figure 4.15

Can a triangle be both acute and obtuse?

In right triangle *ABC* (△*ABC*) the side opposite the right angle is called the *hypotenuse* of the triangle. The other two sides are called the *legs* of the triangle. So the hypotenuse of △*ABC* is $\overline{BC}$, and the legs are $\overline{AC}$ and $\overline{AB}$.

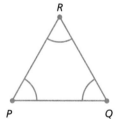

Figure 4.16

Since you learned equiangular and equilateral polygons in Chapter 2, you should recognize △*PQR* as an *equiangular triangle.* Notice the arc marks on the angles. These mean that the angles are congruent. Triangles can also be classified by the lengths of the sides.

A **scalene triangle** is a triangle with no congruent sides.

An **isosceles triangle** is a triangle with at least two congruent sides.

An **equilateral triangle** is a triangle with three congruent sides.

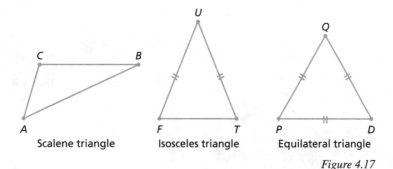

| Scalene triangle | Isosceles triangle | Equilateral triangle |

Figure 4.17

Notice the slash marks on the sides of △*FTU* and △*PDQ*. These slash marks indicate the congruent sides. The two congruent sides of an isosceles triangle have a special name, *legs*. Thus the legs of △*FTU* are $\overline{FU}$ and $\overline{UT}$. The other side of an isosceles triangle is called the *base*. The base in △*FTU* is $\overline{FT}$. The angles formed by the base and the legs are called *base angles,* and the angle formed by the two legs is called the *vertex angle*.

To classify quadrilaterals, you must know about opposite and consecutive sides and angles. The *opposite sides* of a quadrilateral are segments that have no points in common. *Consecutive sides* are two sides that intersect. *Opposite angles* of a quadrilateral are angles whose vertices are not the endpoints of the same side, whereas *consecutive angles* are angles whose vertices are end-points of the same side.

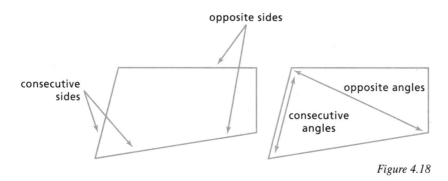

Figure 4.18

The Taj Mahal near Agra, India, incorporates rectangles in both the front facade and the floor plan.

Definitions

A **trapezoid** is a quadrilateral with a pair of parallel opposite sides.

A **parallelogram** is a quadrilateral with two pairs of parallel opposite sides.

A **rectangle** is a parallelogram with four right angles.

A **rhombus** is a parallelogram with four congruent sides.

A **square** is a rectangle with four congruent sides (or a rhombus with four congruent angles).

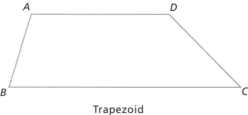

Trapezoid

Figure 4.19

Notice that trapezoids must have at least one pair of parallel opposite sides. The parallel sides are called *bases* of the trapezoid; these are $\overline{AD}$ and $\overline{BC}$ in figure 4.19. The other sides are called *legs* of the trapezoid. If the legs are congruent, it is an *isosceles trapezoid.* The four types of trapezoids are shown in figure 4.20.

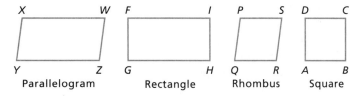

Parallelogram   Rectangle   Rhombus   Square

Figure 4.20

A Venn diagram in which the universal set is the set of quadrilaterals can show the relationships of the set of quadrilaterals.

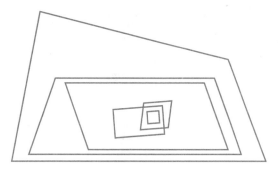

Figure 4.21

▶ A. Exercises

Classify each triangle by angle measure and by side lengths.

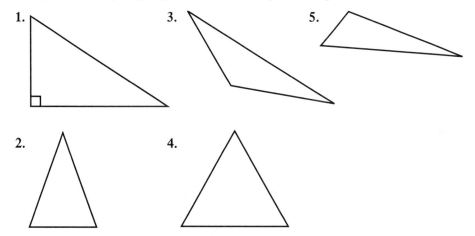

Identify two sets of

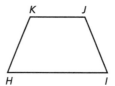

6. opposite sides in quadrilateral *PQRS*.
7. opposite angles in quadrilateral *HIJK*.
8. consecutive angles in quadrilateral *HIJK*.
9. consecutive sides in quadrilateral *PQRS*.
10. bases and legs in quadrilateral *HIJK*.

Answer exercises 11-15 based on the two triangles below.

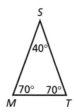

11. Name the legs and base of △*MTS*.
12. Name the vertex angle and base angles of △*MTS*.
13. What do you notice about the base angles? The sum of the angle measures?
14. Name the hypotenuse and legs of △*XYZ*.
15. Express the relationship between the lengths of the legs and the hypotenuse of the right triangle. What is the name for this relationship?

▶ B. Exercises

Give another name for each.
16. equilateral parallelogram
17. equiangular quadrilateral
18. equiangular triangle
19. regular quadrilateral
20. trapezoid with congruent bases

Sketch a diagram of a parallelogram.
21. Do any sides of the parallelogram appear to be congruent? the diagonals? any angles?

Express the given quantity using the values in the diagram.

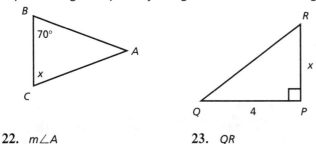

22. $m\angle A$ 23. QR

▶ C. Exercises

Write an equation and solve to find the measures of the angles.

24. A triangle is an isosceles right triangle.
25. The vertex angle of an isosceles triangle is ten degrees less than 3 times the base angles.

▶ Dominion Thru Math

26. Three houses are located at the vertices of a scalene triangle. A well is to be located so that the amount of pipe connecting the well to each house will be a minimum. Draw a triangle and label the vertices (houses) *A*, *B*, and *C*. Construct an equilateral triangle on the outside of each side of the triangle by using the side as radius and swinging an arc from each vertex of that side. Label the new vertex of the triangle on *AB* as *C'*, on side *BC* as *A'*, and on side *AC* as *B'*. Connect *A* to *A'*, *B* to *B'*, and *C* to *C'*. The point *W* where these segments meet is the location of the well that uses the minimum amount of pipe.
27. With a protractor, measure $\angle AWC$, $\angle BWC$, and $\angle AWB$.

▋ Cumulative Review

Express each quantity.

28. One angle of a linear pair has measure *x*. Express the other angle's measure.
29. If $x + 23$ is the measure of one angle, find its complement.

Write an equation and solve.

30. One angle measures 46° more than its complement. Find the measures.
31. Find the measures of two supplementary angles if one is 15° more than twice the other.
32. Suppose that *C-P-D*. If $CD = 152$ inches and *PC* is 4 inches more than 3 times *PD*, find the lengths.

4.6 Angle Constructions

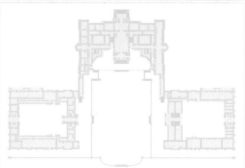

Blenheim Palace, Oxfordshire, England, incorporates many angles.

In the last chapter you learned to construct a bisector of a segment. The bisector of construction 3 was a *perpendicular bisector* since the bisector was perpendicular to the original segment.

Just as you construct segment bisectors, you can also construct angle bisectors. Construction 4 shows you how to bisect angles.

onstruction 4

Bisect an angle.

Given: ∠ABC
Construct: The angle bisector of ∠ABC

1. Place the point of the compass at *B* and mark an arc on both sides (rays) of the angle.

2. Move the point of the compass to the point of intersection between an arc and side. Mark an arc in the interior of the angle. Repeat this process at the other arc and side intersection. The interior arcs intersect at a point *D*.

3. Draw $\overrightarrow{BD}$ to form the angle bisector.

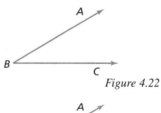

Figure 4.22

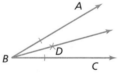

Figure 4.23

In Chapter 6 you will prove that this construction works. You can also construct a copy of an angle.

Copy an angle.

Given: ∠ABC
Construct: An angle congruent to ∠ABC

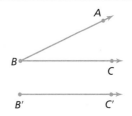

1. Draw a ray, $\overrightarrow{B'C'}$, with a straightedge.

2. Place the point of the compass on the given angle at *B* and construct an arc that intersects both sides of the given angle.

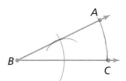

3. Without changing the compass, place its point on *B'* and construct an arc that corresponds to the arc on the given angle.

4. Adjust the compass to measure the length between the two intersection points of the arc and the sides of the given angle.

5. Using this measurement, place the point of the compass at the intersection of the arc and the constructed ray. Draw a short arc that intersects the other arc. Label the point of intersection of the arcs *A'*.

6. Connect *B'* with *A'* to form $\overrightarrow{B'A'}$ and thus ∠A'B'C', which is congruent to ∠ABC.

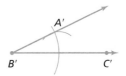

Figure 4.24

Notice how neat and orderly these constructions are. Your constructions should be just as neat. God expects us to do all things in a neat, orderly manner. God created the universe in a neat and orderly manner. If He had not, scientists would not be able to predict planetary motions, seasons, tides, chemical reactions, and other phenomena in nature. Psalm 19:1-2 states that the very heavens themselves declare God's glory.

We can be confident that the Lord's orderliness will be maintained (Gen. 8:22). You should be striving to develop characteristics that reflect the character of God. Construction exercises give you practice in developing your character of orderliness and neatness.

▶ A. Exercises

Construct the following.
1. An angle congruent to ∠PQR
2. An angle congruent to ∠STU
3. An angle congruent to ∠XYZ

Construct angles with measures as follows.

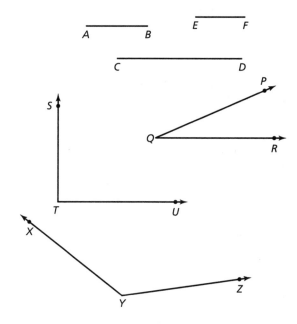

4. $\frac{1}{2}m\angle XYZ$

5. $\frac{1}{2}m\angle PQR$

6. $2m\angle PQR$

7. $m\angle PQR + m\angle STU$

Construct the following.
8. The angle bisector of ∠STU
9. The angle bisector of either angle formed by the angle bisector of ∠STU
10. What is the relation of the angles constructed in exercises 2 and 9?

▶ B. Exercises

Construct angles having the given measures.
11. $m\angle XYZ - m\angle PQR$

12. $3m\angle PQR$

13. $m\angle XYZ + m\angle PQR - m\angle STU$

14. $m\angle PQR + \frac{1}{2}m\angle XYZ$

15. Construct the bisector of the angle with measure equal to $m\angle STU - m\angle PQR$.

Construct each triangle.
16. Make one angle congruent to ∠PQR and adjacent sides congruent to $\overline{AB}$ and $\overline{CD}$.
17. Make one angle congruent to ∠STU and adjacent sides congruent to $\overline{AB}$ and $\overline{EF}$.
18. Make one angle congruent to ∠XYZ and adjacent sides congruent to $\overline{CD}$ and $\overline{EF}$.
19. Make the adjacent sides congruent to $\overline{AB}$ and $\overline{CD}$ and an angle of measure $\frac{1}{2}m\angle PQR$.

▶ C. Exercises

20. Construct a scalene triangle with one angle congruent to ∠STU. Bisect all three angles.

■ Cumulative Review

Construct the following.
21. A circle with $\overline{AB}$ as a diameter

A ———————— B

22. Perpendicular bisector of $\overline{EF}$

C ———————— D

23. A segment of length $AB + \frac{1}{2}EF$

E ———————— F

24. A rectangle with length equal to CD and width equal to $\frac{1}{2}CD$
25. A right triangle with legs of length AB and CD

Geometry and Scripture

Architecture and Right Angles

Now you know about angles and you can draw right angles. Architects must also draw many angles, including right angles, as they prepare blueprints for buildings and archways.

1. What groups prepared timber and stones for the temple in I Kings 5:18 (note verses 5-6)?

Gebal, also called Byblos, is a port on the Mediterranean Sea north of Sidon. The men of Gebal earned a reputation as architects because of their skill in cutting stones for buildings. The term "stonesquarers" (I Kings 5:18) refers to the men of this city.

Of course stonesquarers needed to cut square stones. Right angles were important so that the stones would fit together. Right angles were important in many constructions in the Bible.

> **HIGHER PLANE:** List all Bible references that use forms of the word "square" (including "squared" and "foursquare").

2. Name the "square" item described in each of the verses below.

Exodus 39:8-9

Ezekiel 40:47

Ezekiel 41:21

Ezekiel 43:16

3. Are the "square" items always cubes?

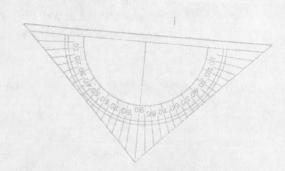

Match each object to its surface shape.

4. Revelation 21:16 (New Jerusalem)
5. Exodus 25:10 (ark of the covenant)
6. Jeremiah 52:21 (temple pillars)
7. Jeremiah 52:23 (hanging pomegranates)

A. Sphere
B. Cube
C. Cylinder
D. Prism

Noah's ark was one of the greatest construction projects in history. Turn to Genesis 6.

8. Give the dimensions of the ark (verse 15).
9. Convert the dimensions to feet (see p. 113). Was the ark larger than a football field?
10. What kind of quadrilateral served as the floor of the ark? What kind of angles were needed?
11. How high was each story (verses 15-16)?

Arches in architecture are also mentioned in the Bible. All fourteen references occur in the same chapter.

12. What chapter mentions arches?

God also builds. We know that Jesus is even now preparing a place for us (John 14:1-2).

In fact, Abraham looked forward to seeing that city whose builder and maker is God (Heb. 11:10-16). Good architects use geometry carefully and precisely. God even recorded the measurements of the heavenly city. Notice that the theme verse also refers to right angles.

Line upon Line

AND THE CITY lieth foursquare, and the length is as large as the breadth: and he measured the city with the reed, twelve thousand furlongs. The length and the breadth and the height of it are equal. ≥≥

REVELATION 21:16

Measure each angle below. Give the angle measure and state whether the angle is acute, obtuse, straight, or right.

1.

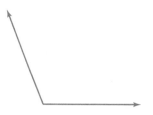

4.

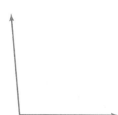

2.

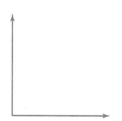

5.

3.

6.

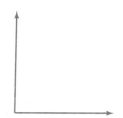

7. Draw a linear pair of angles that are congruent adjacent angles.

Use the figure for exercises 8-9.
  8. Name two pairs of vertical angles.
  9. Name three pairs of congruent angles.

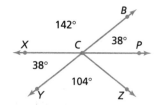

Answer these questions.
  10. What is the measure of the complement of $\angle STL$ if $m\angle STL = 49°$?
  11. What is the measure of the supplement of $\angle XYZ$ if $m\angle XYZ = 112°$?
  12. Name five types of quadrilaterals and explain how they relate to each other.

Use the figure for exercises 13-14. Rewrite each sum using a single angle.

13. $m\angle LMN + m\angle NMP$

14. $m\angle NMP - m\angle OMP$

Given that $x = 3 + y$ and that $-2y + 3 < -1$, provide reasons for each step in the following solution for x.

STATEMENTS	REASONS
15. $x = 3 + y$ and $-2y + 3 < -1$	15.
16. $x > y$	16.
17. $-2y < -4$	17.
18. $y > 2$	18.
19. $x > 2$	19.

20. Sketch an angle congruent to $\angle ABC$.
21. Draw an angle congruent to $\angle ABC$.
22. Construct an angle congruent to $\angle ABC$.
23. Construct the angle bisector of $\angle ABC$.

Prove Theorem 4.7, the Angle Bisector Theorem: The measure of a portion of a bisected angle is one-half the measure of the original angle.

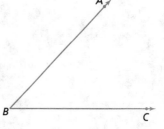

STATEMENTS	REASONS
24. $\overrightarrow{AD}$ bisects $\angle BAC$	24.
25. $\angle BAD \cong \angle DAC$	25.
26. $m\angle BAD = m\angle DAC$	26.
27. $m\angle BAC = m\angle BAD + m\angle DAC$	27.
28. $m\angle BAC = m\angle BAD + m\angle BAD$	28.
29. $m\angle BAC = 2m\angle BAD$	29.
30. $\frac{1}{2}m\angle BAC = m\angle BAD$	30.

31. Explain the mathematical significance of Revelation 21:16.

Here lies the heart of modern technology. Such computer chips control our high-tech industry. The microprocessor, for instance, uses only the power of a night-light but can perform a million calculations per second.

One chip is like making a map of a city on the head of a pin. One-quarter inch of one chip contains one million electronic components. Looking at a schematic drawing of a chip is like looking at a huge metropolis from the air.

The logic and memory of the computers that use these chips are astounding. Think of all the things they run today: pacemakers, cars, radios, gas pumps, space shuttles, and on and on. With all its marvelous powers of memory and calculation, the computer has greatly benefited modern society.

Can the human mind compete? Absolutely! Unlike a computer chip, you can prove theorems. However, even if computers someday generate proofs, the human mind will always surpass the mechanical mind of the computer in several ways. The human mind provided the wisdom to design and make the first computers. Furthermore, it alone can discern right from wrong in moral issues. But the most marvelous power given to the human mind, the ability to know God and to act upon that knowledge to glorify Him, is beyond the reach of any mechanical thing. Christians alone have that great privilege and awesome responsibility.

After this chapter you should be able to

1. use charts and diagrams as tools in reasoning.

2. identify negation (~) and the four connectives $\wedge$, $\vee$, $\rightarrow$, $\leftrightarrow$.

3. symbolize statements and evaluate them with truth tables.

4. classify inductive arguments.

5. apply five types of deductive arguments.

6. distinguish between valid and sound arguments.

7. recognize common fallacies.

8. perform and prove several constructions.

5.1 Introduction to Reasoning

Just what is this process that we call reasoning? Is reasoning really important to you as a high school student? Will you ever need to reason on the job in the future? Do you really know how to reason effectively?

Definition

Reasoning is the step-by-step process that begins with a known fact or assumption and builds to a conclusion in an orderly, concise way. This is also called logical thinking.

Reasoning, then, is a mental process. You may say, "I already know how to think. I've been doing it for years." This is a true statement. You have been able to think for years, but is your thinking logical?

One of the goals of geometry is to help you learn how to reason effectively. This aspect of geometry is different from any math class you have previously had. At times you may even wonder if you are still in math class. But be assured that if you can learn to reason effectively, you will use these skills not only in math class but throughout your life.

Sherlock Holmes, the famous detective in the stories by Sir Arthur Conan Doyle, was a master at using logic to solve complicated cases.

God created man as a thinking being. This fact shows the greatness of God and shows us our responsibility to develop what God has given us to His honor and glory. God expects us to understand what we are doing. Proverbs 4:7 tells the Christian to strive for understanding. This means that he is to try to understand biblical principles, grammatical structures, historical precepts, literary forms, scientific phenomena, and mathematical facts. It is not good enough to recite the facts. We must understand why the facts are true and be able to apply our understanding to new problems we face.

To understand, you must be able to reason effectively. God emphasizes the importance of reasoning. "Come now, and let us reason together, saith the Lord: though your sins be as scarlet, they shall be as white as snow; though they be red like crimson, they shall be as wool" (Isa. 1:18). Every Christian should reason well enough to gain principles from the Bible and to apply these principles to his life. The Christian needs to have a keen reasoning ability not only for his own use but also for dealing with unsaved individuals. Paul tried to deal with Felix in Act 24:24-25. After Paul faithfully proclaimed the Word of God, the Bible says that Felix trembled, undoubtedly because of the conviction of the Holy Spirit. Every Christian should learn to reason and proclaim the gospel message to the unsaved world around him (Matt. 28:19-20).

This chapter and those that follow will help you gain reasoning skills. Take these chapters seriously and apply yourself to learning the proper methods of effective reasoning. The formal logic introduced in this chapter will help you improve your reasoning.

The study of formal logic dates back to the days of Aristotle, who first classified patterns of reasoning. George Boole, an English mathematician, transferred Aristotle's work into symbolic notation. His symbols greatly facilitated the study of logic. Alfred Whitehead and Bertrand Russell invented an entire symbolic logic system in 1913 and published it in their work *Principia Mathematica.*

A table will often enable you to solve logic problems. Study how the table is used to solve the next problem.

EXAMPLE Bob, Jim, Phil, and Brent are part of the crew of a commercial tugboat. Their jobs, not in any particular order, are captain, first mate, cook, and mechanic. The mechanic is the first mate's younger brother. The captain has no relatives on the crew, but Brent does. The first mate is not the cook's uncle, and the cook is not the mechanic's uncle, but Bob is Phil's uncle. What job does each man have, and how are the men related to each other?

Answer Read the given information carefully. Construct a table and place all given information in the table under the proper column and row. Start with the easiest facts.

	Captain	First Mate (brother to mechanic)	Cook	Mechanic (brother to first mate)
Bob (Phil's uncle)	X	X	–	C
Jim	C	–	–	–
Phil (Bob's nephew)	X	X	C	X
Brent	X	C	–	

1. Since the captain has no relatives on the crew, he cannot be Bob or Phil. Cross out Bob and Phil with an X under the captain column.
2. Brent is related to at least one of the crew members, which means that he cannot be the captain. Cross out Brent as captain.
3. Now the only man left to be captain is Jim. Mark C for correct in that position. You can also write a dash in the rest of Jim's row because he cannot fill any of the other three jobs. You can use an X instead of the dash if you prefer, but the dash can remind you that you ruled those out after making an identification. You used the easiest information first and already have one man's job identified!
4. Now some tentative guesses will be necessary, but with the captain identified, the options are reduced. Suppose Phil is one of the brothers (mechanic or first mate). Since Bob is

Continued ▶

Phil's uncle, he cannot also be Phil's brother—which would eliminate him from being mechanic or first mate. Since the cook is not supposed to be the uncle, Bob would not be the cook either. No job is left for Bob, so our guess that Phil is one of the brothers must be wrong. Therefore we can cross out Phil from the jobs of mechanic and first mate. The only job left for Phil is cook. Put a C there and use dashes to show that no one else is the cook.

5. Phil's uncle Bob cannot be first mate (first mate not cook's uncle). When you X out Bob as first mate, the only job left for Bob is mechanic, and the only man left for first mate is Brent. Put a C in each position.

6. Finally, we were asked about the relationships. These are now clear: Bob and Brent are brothers, and Phil is Bob's nephew.

The example above is one of the harder logic problems. In many problems, you will find the answer simply by putting the given information in the table. If you cannot answer the question after charting the clues, use the trial-and-error method, similar to step 4 in the example.

The exercises in this section are designed to make you think and reason effectively. See if you can find the correct conclusions and write down the reasoning process that you used to come to your conclusions.

▶ A. Exercises

Lance, Terri, Jen, and Kelly prefer each a different type of nut. Lance detests almonds, and Kelly likes cashews. Jen cares neither for almonds nor walnuts.

1. Who likes pecans?
2. What does Lance like?

Dean, Kevin, and Frank have each a different hobby: hang-gliding, rappelling, and skydiving. Each also has a different favorite team sport: basketball, hockey, and soccer. Frank likes soccer, the skydiver likes hockey, and Kevin is the hang glider.

3. Make a table of men and hobbies and label the given information. Do not solve.
4. Make a chart of men and team sports and label the given information. Do not solve.
5. Make a table of hobbies and team sports and label the given information. Do not solve.

6. Give the hobby and sport of each man. Complete one of the tables in exercises 3-5.

7. Seven boys are sitting in a circle, and each one has a marble. Mark and Jim have marbles of the same color, and David and John both have yellow marbles. Paul and Bill both have blue marbles. Al is the only boy with a unique color of marble. No two boys with the same color of marble sit together. Jim is on the right side of Paul and on the left side of David. Arrange the boys in the proper order.

▶ B. Exercises

Fill in the Venn diagrams to answer each question below.

8. A survey of 40 shoppers who buy cereal showed that 32 buy Crunchy Flakes and 25 buy Vitabran. How many buy both?

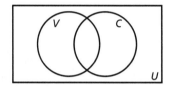

9. Of a sample of 77 students, 60 were single and 40 were women. If 35 of the women were single, how many men were married?

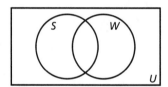

10. Of a sample of fifty mothers, thirty enjoy gardening, twenty-five collect antiques, and twenty-one knit. Only seven of the fifty participate in all three activities, but all fifty enjoy at least one. Five of the gardeners knit but don't collect antiques. Thirteen mothers both garden and collect antiques. How many knit but neither garden nor collect antiques?

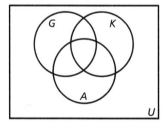

11. Five friends are all pilots: Mr. Johnson, Mr. Bruckner, Dr. Thompson, Mr. Landis, and Lord Christopher. Each pilot has one daughter but named his plane after a friend's daughter. Mr. Landis's plane is the *Carla Jean;*

Dr. Thompson's, the *Rene*. Elizabeth Johnson's father owns the *Lindsey Kay*. Mr. Bruckner owns the *Mary Lynn*, which is named after Mr. Landis's daughter. Carla's father used Lord Christopher's daughter's name for his plane. Who is Lindsey Kay's father?

12. Mike has a rowboat for crossing the river. The rowboat can hold only 200 pounds without sinking. If Mike has his two little brothers with him, Jerry and David, who weigh 100 and 80 pounds respectively, and Mike weighs 160 pounds, how can all three boys get across the river?

13. Toggleland has two groups of residents. The Toggles can never tell the truth; they always lie. The Krinks, on the other hand, are always honest and always tell the truth. A stranger visits Toggleland and meets three of its residents. The stranger asks the first resident if he is a Toggle, and the resident tells his answer to the second resident. The second resident then tells the stranger that the first resident said he was not a Toggle. After this the third resident says that the first resident was really a Toggle. How many of the residents are Toggles? (*Hint:* Use the trial-and-error method.)

▶ C. Exercises

14. Beth, Maria, and Danielle have different favorite colors, participate in different sports, cook different favorite recipes, and play different musical instruments. Red is someone's favorite color, and the flute is one of the instruments. Beth dislikes chili and swimming. Maria dislikes soccer and violins. Danielle dislikes yellow and basketball. The violinist cooks lasagna, the pianist cooks chow mein, and the swimmer dislikes purple. Maria dislikes yellow, but the girl who likes yellow cannot play the violin. The girl who cooks chili cannot swim. Decide each girl's favorite color, sport, recipe, and instrument.

A publishing company has hired six translators to translate their books into Portuguese, Russian, Swahili, Chinese, French, and Arabic. By coincidence the last name of each translator begins with a letter that starts the name of one of the languages. Each translator can efficiently translate two of the six languages, but no two of them can write the same two languages. None of these translators write the language that starts with the first letter of his last name. Can you name the two languages that Mr. A. can write? Here are a few hints. Mr. S. can write Russian and Portuguese. One of the translators can write both Russian and Chinese. Mr. F. and Mr. R. can write the four languages that do not start with the letters of their last names. The translators whose last names begin with the same letters as the languages that Mr. R. can translate can both write French. Neither of the translators who write Portuguese can write Chinese.

$A = \{4, 5, 6, 7, 8\}$ $B = \{3, 5, 7, 9\}$ $C = \{1, 2\}$ $U = \{1, 2, 3, \ldots, 9\}$.

Find the following.

15. A' **16.** $A \cap B$ **17.** $C \cap B$

True/False

18. $A \cup B = C'$ **20.** $2 \subseteq C$

19. $7 \in A$ **21.** $\{3, 4\} \subseteq A \cup B$

5.2 Statements and Quantifiers

1. George Washington was the first president of the United States.

2. $2 \cdot 3 = 9$

3. $x + 5 = 23$

4. The Bible is the Word of God.

5. $8 + 7 = 15$

Reasoning and logic are concerned with many different kinds of sentences like the five given above. A statement is a special kind of sentence.

True or false: This nuclear power plant is a cylinder.

Definition

A **statement** is a sentence that is either true or false, but not both.

According to this definition, which of the five sentences above are statements? Sentences 1, 4, and 5 are true statements, sentence 2 is a false statement, and sentence 3 is not a statement. You may wonder why sentence 3 is not a statement. We do not know whether statement 3 is true or false until the value of x is assigned.

Mathematical reasoning and true conclusions are built on a series of true statements. In mathematical logic, statements are usually symbolized by a letter such as *p* or *q*. A statement can also be negated. For example,

> *Gold is a precious metal.*

This is obviously a true statement. The *negation* of this statement is

> *Gold is not a precious metal.*

This second statement has the opposite truth value from the first, so the second statement is false. In symbolic logic, if we call the first statement *p*, then the negation of statement *p* is symbolized by ~*p*, read "not *p*." A simple rule that you should remember is that if *p* is true, then ~*p* is false, and if *p* is false, then ~*p* is true.

In logical reasoning you will often see a statement in the following form.

> *All men are sinners.*

The words *all* and *every* are represented by the upside down A ($\forall$). If *p* represents "Men are sinners," then in symbolic logic the above statement could be symbolized $\forall p$. *All* is called the universal quantifier.

Another type of quantifier is the existential quantifier ($\exists$), which implies "one or more." Here is an example of a statement using an existential quantifier.

> *p*: Boys like to play baseball.

> $\exists p$: There exists a boy who likes to play baseball.

Here are some examples of how to negate a quantifier.

p	~*p*
All flowers are pretty.	Some flowers are not pretty.
There exists a student who gets straight A's.	No students get straight A's.
Some pies are not cherry pies.	All pies are cherry pies.
No square is a circle.	There exist squares that are circles.

► A. Exercises

Decide which of the following sentences are statements and tell whether each statement is true or false.

1. The Bible is true.
2. There are only fifty-five seconds in every minute.
3. Is Kelly a basketball player?
4. $2 + 9 = 11$
5. $x + 8 = 12$

6. $x + 5 = 5 + x$

7. A skill used in volleyball is dribbling the ball.

8. There is a literal hell.

9. Hello!

10. There are only 364 days in a year.

11. $x + 6 = 18$

12. Phoenix is the capital of Arizona.

▶ B. Exercises

Write the following symbolizations in words given:

p: Ben Franklin was an inventor.

q: Prime numbers are divisible by 2.

r: Men are bald.

s: Students should study mathematics.

13. $\forall r$

14. $\sim p$

15. $\sim q$

16. $\exists s$

17. $\forall s$

18. $\exists q$

19. $\sim \exists r$

20. $\sim \forall r$

21. Tell which of the statements in exercises 16-20 are true.

22. Negate *p*: "Obedience is better than sacrifice." Which is true, *p* or $\sim p$?

▶ C. Exercises

Negate each statement.

23. Some cats have green eyes.

24. All roses are red.

25. There is a number *x* so that $x + 2 = 9$ ($\exists x, x + 2 = 9$).

■ Cumulative Review

True/False

26. All obtuse triangles are scalene.

27. A rhombus is a trapezoid.

28. Some triangles are not convex.

29. No cones are polyhedra.

30. Some half-planes intersect.

5.3 Truth Values and Connectives

Now that you know what a statement is and how to symbolize a statement and its negation, we will begin an in-depth study of the truth value of these statements. What is truth? A dictionary might define *truth* to be a reality or actuality. But a Christian knows that God is truth. Deuteronomy 32:4 says that God is "a God of truth." In John 14:6 Jesus says, "I am the way, the truth, and the life: no man cometh unto the Father, but by me."

Many times you will have to determine the truth value of a statement. To reason to a correct conclusion, you must build your argument on true statements. You can decide whether a statement is true by comparing it with Scripture or with some proven fact that you have learned at a previous time in your life.

Truth tables provide a summary of the truth values for statements. Truth tables use *T* for true and *F* for false. The simple truth table below shows that statement *p* has only two possible values. Each row then shows the resulting value of ~*p* for the given value of *p*.

p	~*p*
T	F
F	T

Simple statements can be combined with *connectives* to express more complicated ideas. The two connectives in this section are *conjunction* and *disjunction*. Conjunctions connect statements using the word *and*. Disjunctions coordinate statements using the word *or*. You should remember conjunctions and disjunctions from algebra when you solved absolute value inequalities. These connectives can also be used to symbolize other words or phrases with a similar meaning. The coordinating conjunctions and punctuation you learned in grammar are symbolized logically by these connectives.

Definitions

A **conjunction** is a statement in which two statements, *p* and *q*, are connected by *and*. The notation for the conjunction "*p* and *q*" is denoted *p* ∧ *q*.

A **disjunction** is a statement in which two statements, *p* and *q*, are connected by *or*. The notation for the disjunction "*p* or *q*" is denoted *p* ∨ *q*.

The study of truth tables for disjunctions and conjunctions is interesting. For example, if *p*: Mars is a planet, and *q*: The moon is green cheese, are two statements, then is the disjunction true or false? Because of the word *or*, "Mars is a planet, or the moon is green cheese" is true. A disjunction is true if either of the statements is true.

A truth table for disjunction is given here. Notice that with two variables there are four possible combinations of truth values. So the table has four rows. Also notice that the example involving Mars and the moon corresponds to the second row.

p	*q*	*p* ∨ *q*
T	T	T
T	F	T
F	T	T
F	F	F

Here is the truth table for conjunction.

p	q	$p \wedge q$
T	T	T
T	F	F
F	T	F
F	F	F

Notice that both statements of a conjunction must be true in order for the conjunction to be true.

EXAMPLE 1   Is this statement true or false?
"Mars is a planet, and the moon is green cheese."

Answer   False. Because of the word *and*, it is a conjunction. Since *p* is true and *q* is false, the second line of the truth table shows us that the conjunction is false.

You can also make truth tables for more complicated expressions. First, decide how many combinations of true or false there should be. If there are *n* statements ($p_1, p_2, \ldots, p_n$), then there are 2^n rows in the truth table. List these combinations of true and false in an orderly fashion, then fill in the connectives using the rules you have learned for $\wedge$, $\vee$, and ~.

EXAMPLE 2   Give a truth table for $(p \wedge r) \vee (q \wedge \sim r)$.

Answer   Since there are three letters (*p*, *q*, and *r*), there should be 2^3 or 8 rows in the table. The patterns shown in the *p*, *q*, and *r* columns organize the eight combinations conveniently. The parentheses show that the two conjunctions must be done first.

---

p	q	r	$(p \wedge r)$	$\sim r$	$(q \wedge \sim r)$	$(p \wedge r) \vee (q \wedge \sim r)$
T	T	T	T	F	F	T
T	T	F	F	T	T	T
T	F	T	T	F	F	T
T	F	F	F	T	F	F
F	T	T	F	F	F	F
F	T	F	F	T	T	T
F	F	T	F	F	F	F
F	F	F	F	T	F	F

To get the $p \wedge r$ column, we look for the rows in which p and r are both true (first and third). These two are true, the rest false. The $\sim r$ column is just the negation of the r column. For $q \wedge \sim r$, look down the q and $\sim r$ columns for a row where both are true (second and sixth). The last column is a disjunction, so look for the rows where both parts are false. Those four spots will be false and the rest true.

▶ A. Exercises

Tell whether the following statements are disjunctions or conjunctions and then tell whether they are true or false.

1. A spider is not an insect, or a porpoise is not a fish.
2. Los Angeles is the capital of California, or Sacramento is the capital of California.
3. A stop sign is red, and a yield sign is blue.
4. $12 - 5 = 9$ or $12 = 6$
5. Lettuce is orange, or mustard is pink.
6. $6 + 3 = 9$ and $12 - 8 = 4$
7. A man will go to heaven, or he will go to hell.
8. December has thirty-one days, and September has thirty days.
9. $9 \geq 9$ and $2 < 5$
10. Either man evolved from animals by chance, or man evolved from animals through God's providence.

The Supreme Court of the United States of America must seek truth to determine justice in legal matters.

▶ B. Exercises

Tell whether the following statements are disjunctions or conjunctions and then tell whether they are true or false.

11. Orange is made from green and yellow, while lavender is made from red and blue.
12. Snow is falling outside, and it is 83° outside.
13. Five plus three is eight, but five times three is fifteen.
14. Venus is a star; the sun is a planet.

Make truth tables for each compound statement below. Be sure to list all possible combinations of *T* and *F* in an orderly manner.

15. $p \wedge \sim p$
16. $(m \vee n) \wedge m$
17. $(a \wedge b) \vee \sim a$
18. $(p \wedge q) \vee r$
19. $p \wedge (q \vee r)$

▶ C. Exercises

20. Compare exercises 18 and 19. What do you conclude?

▶ Dominion Thru Math

21. Your friend Juan claims that saying "It is not the case that you did not pass the course or study" is the same as saying "You passed the course and did not study." Use a truth table and logic connectives to assess whether Juan is right or not.

22. Jack explained to his dad, "The ballgame was not played but it did not rain." His dad replied, "It must not be true that the ballgame was played and it rained." Logically, did they say the same thing?

◼ Cumulative Review

True/False
23. A square is a rhombus, and a square is a rectangle.
24. A prism is a cone, and a prism is a cylinder.
25. A polygon is both a simple curve and a closed curve.
26. A tetrahedron has four edges, or it has four vertices.
27. An octahedron has eight edges, or it has eight vertices.

BOETHIUS

Boethius was the foremost scholar of his day. He knew astronomy, languages, theology, and politics as well as geometry and logic. His textbooks preserved Greek learning for over 800 years during the Middle Ages.

Toward the close of the ancient Roman Empire (ca. A.D. 475), Boethius was born at Rome. His Christian family had been influential for over a century in the capital. Boethius had the privilege of growing up in a Christian home and becoming a Christian early.

As a Christian, Boethius took his education seriously. There were seven main subjects in ancient education, and this wealthy family provided education in all seven liberal arts for their son. The first three subjects, the trivium, were considered basic: grammar, rhetoric (debate), and logic. After Boethius learned these basics, he applied his skills to learning the four advanced studies, the quadrivium. In fact, he mastered the four advanced studies so well that he wrote textbooks on them. He drew on Ptolemy and Nichomachus in writing his *Arithmetic, Music,* and *Astronomy* texts. His *Geometry* text summarized Euclid.

Boethius also left a good testimony as a politician. God apparently helped him to stand for right in a wicked nation. He displayed integrity and generosity, reformed the money system, and denounced officials who embezzled funds from the provinces.

Boethius translated as much Greek learning into Latin as he could. He wrote commentaries on Aristotle and independent works on logic. He was the most important logician of ancient Rome. He used his knowledge of logic for God by defending the doctrine of the Trinity and refuting the Arians, who taught that Jesus was not God.

Unfortunately for Boethius the emperor Theodoric, who was Arian, and the politicians Boethius exposed were against him. Boethius was imprisoned, but he used his imprisonment wisely. His most famous book, *The Consolation of Philosophy,* was written while he was facing death. He argued for moral responsibility based on the Greek philosophy of Aristotle and Plato, which his society accepted. This book was translated into English by King Alfred (9th century) and Chaucer (14th century). It was again translated by Elizabeth I in the 16th century, a millennium after he wrote it!

In 524 Boethius became a martyr. The legend goes that he was beaten to death without a trial in the baptistry of the church. His death signaled the end of ancient mathematics, but for another thousand years society felt the influence of his scholarship. Boethius provided the salt and light needed by his generation.

He used his knowledge of logic for God by defendng the doctrine of the Trinity.

5.4 Conditional Statements

Besides the coordinating connectives, there are two other connectives. These connectives make one statement contingent upon another statement.

The most basic contingent relationship is the conditional statement. Conditional statements are essential to reasoning and argumentation. Here is an example of one.

> If a man is unsaved, then the man will spend eternity in hell.

> Let p: The man is unsaved.
> q: The man will spend eternity in hell.

Actions, like conditional statements, have consequences.

Definition

A **conditional statement** is a statement of the form "If p, then q," where p and q are statements. The notation for this conditional statement is $p \rightarrow q$.

Notice that the p represents the "if" statement, called the *hypothesis*, and the q represents the "then" statement, called the *conclusion*. You should learn how to write conditional statements in if-then form.

EXAMPLE 1   Write the following statements in if-then form.
  a. There are no clouds in the sky, so it is not raining.
  b. School will be canceled if a blizzard hits.

Answer   **a.** If there are no clouds in the sky, then it is not raining.
  b. If a blizzard hits, then school will be canceled.

How do you determine the truth of a conditional statement? Look at the truth table for a conditional.

p	q	$p \rightarrow q$
T	T	T
T	F	F
F	T	T
F	F	T

The first two lines in the table are easily understood. When the hypothesis is true, the conclusion determines the truth of the conditional statement.

The following two statements about a kangaroo named Roo illustrate the first two lines of the truth table. The first statement is true and the second false.

> If Roo is a kangaroo, then Roo has a pouch.
> If Roo is a kangaroo, then Roo has wings.

Now the last two lines of the truth table may seem puzzling to you. Suppose you said, "If it rains, then the creek will flood." You did not lie, even if it doesn't rain because you said "if." Your claim was contingent upon rain. If it doesn't rain, there is no claim and no one can call you a liar. Without rain your statement has to be considered true whether the creek stays low or floods.

EXAMPLE 2   Which are true?
If horses had wings, then horses could fly.
If ocean water is grade-A milk, then ocean water is a nourishing beverage.
If whales walk, then $4 + 1 = 5$.

Answer   All three are true since each hypothesis is false. The symbolic forms are F → F, F → F, and F → T.

Occasionally you will see a conditional that uses the word *implies* instead of the words *if-then*. A conditional that is always true is sometimes called an *implication*. The example below states a theorem in both forms.

> If two lines in a plane do not intersect, then they are parallel.
> The fact that two lines in a plane do not intersect implies that they are parallel.

In this example you could interchange the hypothesis and conclusion to state, "If two lines are parallel, then they do not intersect." Conditional statements that can be expressed in both directions are called *biconditional statements*. To form a biconditional, we combine the two conditionals with a conjunction as in the following example.

The sun will shine tomorrow if and only if there are no clouds in the sky.

This biconditional means two statements: "If the sun shines tomorrow, then there are no clouds in the sky" and "If there are no clouds in the sky, then the sun will shine tomorrow." Besides the simplified wording "if and only if," a special symbol, $p \leftrightarrow q$, expresses biconditionals. Notice that the two statements are essentially equivalent, so either could substitute for the other.

Definition

A **biconditional statement** is a statement of the form "*p* if and only if *q*" (symbolized by $p \leftrightarrow q$), which means $p \rightarrow q$ and $q \rightarrow p$.

The next truth table gives the truth values of a biconditional statement. You can justify the truth values in this table by using those for a conditional and a conjunction. For statements *p* and *q* to be equivalent, they must have the same truth values.

p	q	$p \leftrightarrow q$
T	T	T
T	F	F
F	T	F
F	F	T

Just as it is useful to be able to rephrase an article in your own words, it is sometimes useful to express connectives in terms of other connectives. The following theorem allows us to change a conditional to a disjunction.

Theorem 5.1
The conditional $p \rightarrow q$ is equivalent to the disjunction ~*p* or *q*.

p	q	$p \rightarrow q$	$\sim p$	$\sim p \vee q$	$(p \rightarrow q) \leftrightarrow (\sim p \vee q)$
T	T	T	F	T	T
T	F	F	F	F	T
F	T	T	T	T	T
F	F	T	T	T	T

Because the truth values for $p \rightarrow q$ and $\sim p$ or q are the same, they are equivalent. Theorem 5.1 means that any disjunction can be changed to a conditional and vice versa. Equivalent expressions are important because they permit you to make substitutions, just as equal quantities in algebra may replace each other.

EXAMPLE 3	Change the following conditional statement to a disjunction. If a child disobeys, then he will be disciplined.
Answer	Using the theorem, $p \rightarrow q$ is the same as $\sim p$ or q. Notice that the simpler wording (obeys) is preferred for $\sim p$ (not disobeying). A child obeys, or the child will be disciplined.

Now consider three statements that can be formed from a conditional statement.

Definitions

The **converse** of a conditional statement is obtained by switching the hypothesis and conclusion. The converse of $p \rightarrow q$ is $q \rightarrow p$.

The **inverse** of a conditional statement is obtained by negating both the hypothesis and conclusion. The inverse of $p \rightarrow q$ is $\sim p \rightarrow \sim q$.

The **contrapositive** of a conditional statement is obtained by switching and negating the hypothesis and conclusion. The contrapositive of $p \rightarrow q$ is $\sim q \rightarrow \sim p$.

EXAMPLE 4		Write the converse, inverse, and contrapositive of the implication below.
	Implication	If we have a blizzard, then school will be canceled.
Answer	*Converse*	If school is canceled, then we had a blizzard.
	Inverse	If we do not have a blizzard, then school will not be canceled.
	Contrapositive	If school is not canceled, then we did not have a blizzard.

Of these three, only the contrapositive is equivalent to the original. See exercise 25.

> **Theorem 5.2**
>
> **Contrapositive Rule. A conditional statement is equivalent to its contrapositive. In other words, $p \rightarrow q$ is equivalent to $\sim q \rightarrow \sim p$.**

▶ A. Exercises

Write the following statements in if-then form.
1. When I study my geometry, I get good grades.
2. A square is not a triangle.
3. Two intersecting lines are not parallel.
4. I will go sledding when it snows.
5. The flowers bloom because the sun shines.

State whether the conditionals are true or false.
6. If you are saved, then you will go to heaven.
7. If a figure is a triangle, then it has four sides.
8. If $5x = 20$, then $x = 5$.
9. If our pig is a clean animal, then we will keep it in the house.
10. If all men have sinned, then all men need a Savior.
11. The roads will be slick if and only if there is ice on the roads.
12. A quadrilateral is a trapezoid if and only if it has a pair of parallel opposite sides.
13. Give the contrapositive of the statement "if a thief comes, then he will steal something."

Consider this conditional statement.

   If a quadrilateral is a rectangle, then the quadrilateral is a trapezoid.

14. Write the converse.
15. Write the inverse.

▶ B. Exercises

16. Give the converse, inverse, and contrapositive of "if two angles are vertical angles, then they are congruent."

Write each biconditional as two conditionals.
17. You will get an A in geometry if and only if you study hard.
18. You will go to heaven if and only if you are saved.

Change each conditional to a disjunction.
19. If $x = 10$, then $x + 6 = 16$.
20. If candy has sugar in it, then it is sweet.
21. If the stoplight is green, then you can go.

22. If lines are parallel, then the lines never intersect.
23. What would the inverse of the converse of a statement be?

▶ C. Exercises

Make truth tables to prove that each pair is equivalent.
24. $p \lor q$ and $q \lor p$. What could you name this property?
25. $p \to q$ and $\sim q \to \sim p$.

▶ Dominion Thru Math

Given that a and b are true statements while c and d are false statements, determine the truth value of each of the following statements. These variables could represent complicated statements or simple statements such as a) ripe apples are red, b) ripe bananas are yellow, c) chocolate is poisonous, and d) donuts are fat free.
26. $\sim b \land [(a \to \sim b) \leftrightarrow (\sim c \land d)]$
27. $\{[a \to (b \to c)] \to d\} \leftrightarrow (\sim d \to \sim a)$
28. $\{(\sim a \lor c) \land [\sim b \to (c \land \sim d)]\} \to [(a \land \sim b) \leftrightarrow (c \lor \sim d)]$
29. Give a single conditional that logically follows from the following propositions.

   If two lines are perpendicular, then they intersect.
   If two lines are non-coplanar, then they do not intersect.
   If two lines are coplanar, then they are not skew.

▮ Cumulative Review

True/False
30. If a triangle is isosceles, then it is equilateral.
31. Theorem 1.3 could be worded "if two lines intersect, then there is a plane containing them; and if two lines intersect, then there is at most one plane containing them."

How many sides or faces does each figure have?
32. heptagon
33. icosahedron
34. decagon

5.5 Proofs

Paul says, "prove all things; hold fast that which is good" (I Thess. 5:21). This verse tells Christians to test everything against Scripture. We should use good reasoning to make sure that what we believe is built on the sure foundation of the Bible.

Definition

A **proof** is a system of reasoning or argument to convince a person of the truth of a statement.

Now you will learn various procedures for proving statements—mathematical or otherwise. Before a proof can be meaningful to you, you must question the truth of the statement. You must not accept every statement that you hear as truth without some solid evidence that it is true.

Compare the two main types of reasoning.

Definitions

Inductive reasoning is an argument to establish that a statement is probably true.

Deductive reasoning is an argument to establish that a statement is absolutely certain.

Classify the arguments below as inductive or deductive. Try to decide before you read the answer.

EXAMPLE 1	All squares are rectangles and all rectangles are polygons. Therefore all squares are polygons.
Answer	The conclusion is definitely true if the premises are. The argument is deductive.

EXAMPLE 2	A meteorologist studied weather patterns for the last ten years in his city. Based on the trends, he predicts a high of 25°F in January.
Answer	The conclusion may be probable but it is not certain. The argument is inductive.

The inductive argument in example 2 appeals to a pattern or a trend. An argument such as this that relies on statistics is an *appeal to tendency*. Statistics is an important branch of mathematics but is very different from geometry. In geometry, you will use only deductive proof so that the truth of the theorems is certain. Proofs of theorems will help you understand math as a system and appreciate God's precise and orderly creation. They should also help you see the importance of good reasoning and make you alert to deceptions and faulty reasoning.

In a reasoned argument, the supporting statements are called *premises*. The statement that they support is the *conclusion*. If the connection between premises and conclusion is logically correct, it is a valid argument. True premises and valid reasoning result in a true conclusion. Such arguments are sound.

Definitions

An argument is **valid** if the reasoning proceeds logically from the premises to the conclusion.

An argument is **sound** if it is valid and the premises are true.

EXAMPLE 3	Analyze this argument. Is it deductive or inductive? Identify the premises and conclusion. Is the conclusion valid? Is it sound? **1.** If a man is saved, then he is a Christian. **2.** If a man is a Christian, then he should display the fruit of the Spirit. *Therefore* **3.** If a man is saved, then he should display the fruit of the Spirit (Gal. 5:22-23).
Answer	The argument is deductive. The first two statements are the premises. Statement 3 is the conclusion. It is valid since the reasoning is good. It is sound since the premises are also true.

Symbolic logic can help you analyze complicated arguments. For instance, in the above example let

> p: A man is saved.
>
> q: A man is a Christian.
>
> r: A man should display the fruit of the Spirit.

The premises $p \rightarrow q$ and $q \rightarrow r$ certainly guarantee the conclusion $p \rightarrow r$. In this argument we can be certain that the premises are true because the Bible says they are true.

In summary, truth deals with statements, while validity deals with reasoning. Soundness requires both. Classify each argument as to validity and soundness.

EXAMPLE 4　　**1.** All of the animals at the San Diego Zoo are zebras.
2. All zebras eat marshmallows.
Therefore
3. All of the animals at the San Diego Zoo eat marshmallows.

Answer　　Since both premises are false, the argument is not sound. Check the reasoning. This is a valid argument that results in a false conclusion.

You can show that the conclusion is false by giving a *counterexample*. A counterexample is a specific instance that fits the hypothesis but not the conclusion. In the conclusion above, a lion would be a good counterexample. A lion is an animal at the San Diego Zoo (fits hypothesis), but lions do not eat marshmallows (contradicts conclusion).

Although counterexamples can disprove a statement, one must be careful when he cannot find a counterexample. *Lack of counterexample* is evidence that a statement may be true, but it is only inductive evidence. You may think of a counterexample tomorrow even though you could not think of one today.

Because we will focus on deductive proof, only two inductive types have been discussed: Appeals to tendency because of their statistical premises, and lack of counterexample because of the use of counterexamples in mathematics. The other four types of appeals involve appeals to authority, experience, analogy, and utility.

In the study of reasoning you must be aware that validity and truth are two different things. An argument that is valid can result in a false conclusion, and an argument that is invalid can result in a true conclusion. You must be able to distinguish valid arguments from true conclusions. If a conclusion of an argument is false but the reasoning is valid, then the argument is not sound and you automatically know that at least one of the premises is false.

EXAMPLE 5
1. Pigs are dirty animals.
2. Cows provide milk.
Therefore
3. Rural areas have a low population density.

Answer The conclusion in this example is true, and the two premises are true, but the argument is invalid. There is no logical connection between the premises and the conclusion. Since it is invalid, it is also unsound.

This is a good example of getting a true conclusion for the wrong reasons. When we use poor reasoning in an argument, we commit a fallacy. Every invalid argument contains a fallacy, and every unsound argument must either contain a logical fallacy or a false premise.

The word *fallacy* is derived from the Latin word *fallere*, which means "to deceive." Second Corinthians 11:14 warns that "Satan himself is transformed into an angel of light," so the Christian must beware of deceivers (I Cor. 6:9; 15:33; Gal. 6:7). Strive to avoid fallacies in math and in your everyday life so that you do not deceive others or be deceived yourself.

Several fallacies are common in mathematical proofs. *Hasty generalization* occurs when you jump to a conclusion based on insufficient evidence. Be sure to justify each step carefully. Likewise, a *circular argument*, also called "begging the question," occurs when you assume what you are trying to prove. Never use the conclusion that you are trying to prove as a reason in the steps of the proof. You cannot assume the truth of the conclusion before you prove it.

The fallacy of *accident* occurs when you misapply a general principle to a situation for which it was not intended. For example, measurement is a useful principle, but it is not intended to apply to a sketch. The size of an angle may be "accidental," and you must not assume that $\angle ABC$ in the diagram below is a right angle just because it looks like one. You cannot call it a right angle unless that fact is part of the given (often shown by a right angle symbol on the diagram).

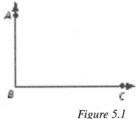

Figure 5.1

The following table summarizes the relation between truth, validity, and soundness. In this table, V = valid and S = sound, I = invalid and U = unsound. Notice that a sound argument requires both true premises and valid reasoning.

Premises	Reasoning	Argument
T	V	S
T	I	U
F	V	U
F	I	U

▶ A. Exercises

In each problem state whether the argument is valid or invalid. Also state whether the argument is sound.

1. Horses are purple.
   Apache is a horse.
   Apache is purple.

2. Moses thought he was inadequate for the work of God.
   Moses was used by God in a mighty way.
   A man can be used by God in a mighty way even if he feels inadequate.

3. If a car's gray iron combines chemically with oxygen, then rust forms.
   If a car is rusty, then it has a brownish red iron oxide on it.
   If oxygen and gray iron combine chemically, then a brownish red iron oxide forms.

4. If a man is an engineer, then he earns a large salary.
   Bill is an architect.
   Bill earns a large salary.

5. All men are mortal.
   Socrates is a man.
   Socrates is mortal.

6. All tulips are plants.
   All daisies are plants.
   All tulips are daisies.

7. A penguin is a bird.
   All birds have feathers.
   Penguins have feathers.

8. All poodles are dogs.
   All poodles are mammals.
   All dogs are mammals.

9. All poodles are fish.
   All fish are mammals.
   All poodles are mammals.

10. A hole is a collection of groundhogs.
    The 21 groundhogs belong to 7 holes, but every groundhog belongs to exactly 2 holes. Therefore, each hole has 6 groundhogs.

▶ B. Exercises

Match each argument to the letter of the description that best describes it.

   A. deductive, valid
   B. deductive, invalid
   C. inductive, appeal to tendency
   D. inductive, appeal to lack of counterexample
   E. inductive, appeal to authority

11. Jill heard Larry's father say that if he did not pass his exams, he would be grounded. The next week Jill found out that Larry was grounded, and she assumed that Larry must have failed his exams.

12. Dr. Nelson Glueck of Hebrew Union College argued from Deuteronomy 8:7-9 that copper deposits must exist in Israel. (He found them in 1957, and a profitable copper-mining operation resulted.)

13. In the trial of a civil suit, no individuals who were questioned testified unfavorably concerning the reliability of a certain witness. The testimony of the witness was therefore accepted fully by the judge.

14. By surveying thousands of individuals throughout the U.S., the Gallup Poll predicted the winner of the campaign for president.

15. All flags are cows and all cows are radios. Therefore all flags are radios.

Classify each argument as sound or unsound. If it is unsound, state why.

16. All planets are stars.
    All stars are created by God.
    Therefore, all planets are created by God.

17. No dogs are cats.
    Some dogs are poodles.
    Therefore, no poodles are cats.

18. If a man drinks poison, then he will die.
    Abraham Lincoln died.
    Therefore, Abraham Lincoln drank poison.

Classify the following arguments as inductive or deductive. Identify any fallacies committed.

19. People who take things without the owner's permission are thieves. Policemen confiscate drugs without the owner's permission. Therefore policemen are thieves.

20. The atomic bomb was developed from Einstein's work on the special theory of relativity. Since the bomb worked, some argue that the theory must be true.

21. A television repairman narrowed down the problem in a malfunctioning television set to two possibilities—the video detector or the horizontal oscillator. Through further tests he learned that the problem was not in the horizontal oscillator, so he replaced the video detector.

22. Joe claims that man is basically good because of evolution. The "survival of the fittest" has weeded out destructive evils that hinder survival. Human evolution must be occurring because man's goodness enables him to make choices for his own betterment.

Give a counterexample to disprove each statement.

23. If a quadrilateral has two pairs of parallel opposite sides, it is a rectangle.

24. Subtraction is associative.

▶ C. Exercises

Give a counterexample to disprove each statement.

25. If $A \vee B$ is true and if $B \vee C$ is true, then $A \vee C$ is true.

26. $a \rightarrow b$ therefore $b \rightarrow a$

■ Cumulative Review

Draw each using rulers and protractors.

27. A one inch by two inch rectangle

28. A 146° angle

29. An isosceles right triangle

30. The midpoints of two parallel chords of a circle

31. A convex heptagon

5.6 Deductive Proof

In I Corinthians 15:13-20, Paul used a form of deductive reasoning to conclude that there is a resurrection from the dead. Deductive reasoning is the type of reasoning that you will use most often in geometry and other math classes. Since deduction is used so frequently, this section is devoted to different types of deductive reasoning. Deductive reasoning can be split into two major sections: proof by direct argument and proof by contradiction. Some of the kinds of direct argument proofs are given below.

1. *Law of Deduction*
2. *Modus ponens*
3. *Modus tollens*
4. *Transitivity*

The judge must evaluate the validity and soundness of arguments made by the lawyer based on the testimony of each witness.

We will study only these four types of deductive arguments, although there are others. The Law of Deduction is perhaps the most important.

The **Law of Deduction** is a method of deductive proof with the following symbolic form.

p (assumed)	q_n (statements known to be true)
q_1	r (deduced from statements above)
q_2	$p \rightarrow r$ (conclusion)

Assume hypothesis p. Use p together with any number of true statements $q_1, q_2, q_3, \ldots q_n$ to deduce r. This shows that the working hypothesis p results in conclusion r, so we know that $p \rightarrow r$.

EXAMPLE 1 *Prove:* If T is the midpoint of $\overline{SV}$, then $ST = \frac{1}{2}SV$.

Answer

p: T is the midpoint of $\overline{SV}$	Assumed
q_1: $ST = TV$ and S-T-V	Definition of midpoint
q_2: $ST + TV = SV$	Definition of between
q_3: $ST + ST = SV$	Substitution
q_4: $2ST = SV$	Distributive property
r: $ST = \frac{1}{2}SV$	Multiplication property of equality
$p \rightarrow r$: If T is the midpoint of $\overline{SV}$, then $ST = \frac{1}{2}SV$	Law of Deduction

---

If you understand this example of the Law of Deduction, you should be able to understand the proofs in the rest of the book. To prove the conditional $p \rightarrow r$, you must assume that p is true and show that r is also true. This shows "if p, then r," or in symbols $p \rightarrow r$.

Modus ponens is a method of deductive proof with the following symbolic form.

Premise 1: $p \rightarrow q$
Premise 2: p
Conclusion: q

EXAMPLE 2   Classify this argument.
>   ***Premise 1:***   If man is sinful, then he must die (Rom. 6:23).
>   ***Premise 2:***   Man is sinful (Rom. 3:23).
>   ***Conclusion:***   Man must die (second death; Rev. 21:8).

Answer   This argument has the form of modus ponens and is therefore valid. Since it is based on true biblical premises, it is also sound.

It is wonderful that the Bible also gives us the remedy for the fearsome conclusion found above. The remedy is found in Romans 10:9-13.

>   ***Premise 1:***   If a man will confess the Lord Jesus Christ and believe on Him in his heart, then he shall be saved.
>   ***Premise 2:***   A man asks Jesus into his heart.
>   ***Conclusion:***   This man will be saved.

You use modus ponens so often that you do so without thinking about it. Modus ponens is the rule that lets you place your own name in the arguments above. Likewise, every time you substitute numbers in a formula, you are relying on modus ponens. In fact, modus ponens is the rule that lets you apply principles to specific situations.

For instance, you used modus ponens even in example 1. The multiplication property of equality says that if $b = c$, then $ab = ac$ for any real number a. Of course, these letters did not occur in example 1. Your substitution of distances for these letters (in statement r) is an application of modus ponens. Since almost every step of every proof is an application of some previous definition, postulate, or theorem, you will use modus ponens frequently without realizing it.

The form of deductive argument Paul used in I Corinthians 15:13, 20 is called *modus tollens.*

>   ***Premise 1:***   "If there be no resurrection of the dead, then is Christ not risen" (I Cor. 15:13).
>   ***Premise 2:***   "Now is Christ risen from the dead" (I Cor. 15:20).
>   ***Conclusion:***   There is a resurrection of the dead.

Definition

Modus tollens is a method of deductive proof with the following symbolic form.

>   ***Premise 1:*** $p \rightarrow q$
>   ***Premise 2:*** $\sim q$
>   ***Conclusion:*** $\sim p$

This method can be justified two ways. One way would be to combine the premises with "and" ($\wedge$) and show with truth tables that the conclusion follows from the premises: $[(p \to q) \wedge \sim q] \to \sim p$. This truth table would involve seven columns, and it would be easy to make a mistake on one of the 28 truth values. A shorter way is to derive it by steps using previous methods as reasons.

STATEMENTS	REASONS
1. $p \to q$; $\sim q$	**1.** Given
2. $\sim q \to \sim p$	**2.** Contrapositive Rule (Theorem 5.2)
3. $\sim p$	**3.** Modus ponens (premises: step 2 and $\sim q$)
4. $[p \to q$ and $\sim q] \to \sim p$	**4.** Law of Deduction (premises: steps 1 and 3)

Finally, *transitivity* should sound familiar. The transitive property of numbers is if $a = b$ and $b = c$, then $a = c$. The deductive proof called transitivity is similar to this. An example of a proof by transitivity is given here.

Premise 1: If a triangle is equilateral, then its angles have equal measure.

Premise 2: If the angles of a triangle have equal measure, then none of the angles are right angles.

Conclusion: If a triangle is equilateral, then none of its angles are right angles.

Definition

Transitivity is a method of deductive proof with the following symbolic form.

Premise 1: $p \to q$
Premise 2: $q \to r$
Conclusion: $p \to r$

You will prove transitivity in the exercises.

Be careful to use these four deductive proof methods correctly. Two fallacies result from common misuse of modus ponens and modus tollens. These fallacies are assuming the converse and assuming the inverse.

EXAMPLE 3   Classify the fallacy committed by this argument.
All dogs are mammals.
Fido is a mammal.
Therefore, Fido is a dog.

Answer $d \rightarrow m$ 1. Symbolize the argument.
   m ($\therefore$ means *therefore*)
   $\therefore d$
   Assuming the converse 2. This would be modus ponens
                             only if the first premise were
                             reversed to say $m \rightarrow d$. Reversing
                             these statements forms the con-
                             verse of the given premise, which
                             is not the same as the original.

Likewise, the argument below contains a fallacy, though it may at first remind
you of modus tollens. The name of this fallacy is assuming the inverse.

All dogs are mammals. $d \rightarrow m$
Fido is not a dog. $\sim d$
Therefore, Fido is not a mammal. $\therefore \sim m$

▶ A. Exercises

Suppose you want to prove the Vertical Angle Theorem using the Law of
Deduction: "If two angles are vertical angles, then they are congruent."
   1. What should you assume?
   2. What should you derive from the assumption?

Suppose you know that ∠1 and ∠2 are vertical angles.
   3. What would you conclude from the Vertical Angle Theorem?
   4. What rule of logic lets you draw the conclusion?

Suppose you know that ∠A and ∠B are not congruent.
   5. What could you conclude from the theorem?
   6. What rule of logic lets you draw the conclusion?
   7. If $A \rightarrow B$ and $B \rightarrow C$, then what?

Identify the fallacy in each of the following arguments.
   8. If smoke comes from a house, it must be on fire. No smoke comes from
      the house. Therefore, the house is not on fire.
   9. If the temperature is above 100°C, then water will boil. The water on the
      campfire was boiling. Thus, its temperature was above 100°C.

Prove transitivity by supplying the missing reasons.

STATEMENTS	REASONS
Assume p	Assume for Law of Deduction
$p \to q$	Given (premise 1)
$q \to r$	Given (premise 2)
10. q	10.
11. r	11.
12. $p \to r$	12.

▶ **B. Exercises**

Look at the following arguments and state the form of deductive reasoning used.

13. If I break my arm, then I will go to the hospital. I broke my arm. I will go to the hospital.

14. If it is warm in April, then the flowers will bloom in May. The flowers did not bloom in May. It was not warm in April.

15. If you study for your geometry test, then you will get a high score on the test. If you get a high score on the test, then you will get a good grade on your report card. If you study for your geometry test, then you will get a good grade on your report card.

16. Two lines intersect. Every line contains at least two points. Three distinct noncollinear points lie in exactly one plane. Therefore, two intersecting lines lie in exactly one plane.

17. A cat is a mammal. All mammals have four-chambered hearts. A cat has a four-chambered heart.

18. If a triangle has two congruent sides, then the triangle is an isosceles triangle. △ABC is not an isosceles triangle. △ABC does not have two congruent sides.

Give the symbolic logic form needed to prove each of the following.
19. modus ponens   20. transitivity

▶ **C. Exercises**

Construct truth tables for the rules of logic in exercises 21-22.
21. modus ponens
22. $(p \land q) \to p$

Prove that the arguments in exercises 23-24 are invalid using a truth table.
Circle a truth value that disproves each argument.

23. *Premise 1:* $p \rightarrow q$
    Premise 2: q
    Conclusion: p

24. *Premise 1:* $p \rightarrow q$
    Premise 2: $\sim p$
    Conclusion: $\sim q$

Cumulative Review

25. p: A quadrilateral is a square if and only if it is both a rectangle and a rhombus.

    Using s: "the quadrilateral is a square,"
    r: "the quadrilateral is a rectangle," and
    h: "the quadrilateral is a rhombus," which of the following symbolic expressions below best represents statement p?

    A. $s \rightarrow (r \wedge h)$
    B. $(s \rightarrow r) \wedge h$
    C. $s \leftrightarrow (r \wedge h)$
    D. $(s \leftrightarrow r) \wedge h$
    E. $(s \leftrightarrow r) \wedge (s \leftrightarrow h)$

For each step in the solution of the inequality, state the key property that justifies it.

Solve $5 - 2x \leq 8$.

STATEMENTS	REASONS
$-2x + 5 \leq 8$	Given
26. $-2x \leq 3$	**26.**
27. $\frac{-1}{2}(-2x) \geq \frac{-1}{2} \cdot 3$	**27.**
28. $\left(\frac{-1}{2} \cdot -2\right)x \geq \frac{-1}{2} \cdot 3$	**28.**
29. $1x \geq \frac{-3}{2}$	**29.**
30. $x \geq \frac{-3}{2}$	**30.**

Graphing Lines

In algebra you used the slope-intercept form of the line, $y = mx + b$, to graph lines. Review the process.

EXAMPLE Graph $y = \frac{-3}{4}x + 2$

Answer
1. Find the slope and y-intercept by comparing the equation to $y = mx + b$.

$m = \frac{-3}{4}$ and $b = 2$

2. Plot the y-intercept: $(0, 2)$.

3. m = vertical change over horizontal change. From the y-intercept, go down 3 (vertical change = -3) and over 4 (horizontal = $+4$).

4. Draw the line.

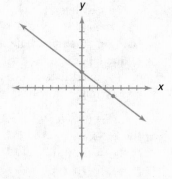

You can also find the equation of a line from the slope and the y-intercept. Sometimes you may need to find the slope from given points as in Chapter 4.

Can you prove that $(0, b)$ is the y-intercept? Since the y-intercept is on the y-axis, it must have an x-coordinate of zero. Substituting 0 into $y = mx + b$, you get $y = m \cdot 0 + b$ or $y = b$. Therefore $(0, b)$ is the y-intercept of the line.

Often a property is discovered inductively. On dozens of line graphs you may notice that $(0, b)$ is always the y-intercept. This appeal to tendency draws the correct conclusion, but only the deductive proof above guarantees that there are no exceptions.

▶ **Exercises**

Graph each line.
1. $y = \frac{2}{3}x - 1$
2. $y = -2x + 3$
3. $y = \frac{1}{3}x$

Find the equation of each line.
4. The slope is $\frac{1}{2}$ and the intercept is $(0, 3)$.
5. The line passes through $(2, 5)$ and $(0, -1)$.

5.7 Construction and Proof

In earlier chapters you learned how to construct segments and angles. Now you will see how to make three constructions by combining earlier constructions with careful reasoning.

Construction 6

Perpendicular through point

Given: $\overleftrightarrow{AB}$ and a point P

Construct: A line that passes through a given point P and is perpendicular to a given $\overleftrightarrow{AB}$

1. Place the point of the compass at point P and make arcs that intersect $\overleftrightarrow{AB}$ in two places. Call these points C and D.

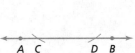

2. Move the point of the compass first to C and then to D, making intersecting arcs in the half-plane that does not contain P (or in either half plane if P is on the line). Call this point of intersection Q.

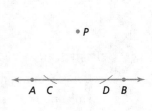

3. Draw $\overleftrightarrow{PQ}$ to form a line perpendicular to $\overleftrightarrow{AB}$ through P.

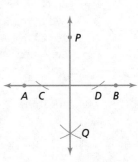

Figure 5.2

Why does this construction work?

STATEMENTS	REASONS
1. Step one identifies two points C and D equidistant from P	1. $\overline{PC}$ and $\overline{PD}$ are radii of the same circle (construction 1)
2. Starting from points C and D, point P and the opposite point Q are both equidistant	2. $\overline{CQ}$ and $\overline{DQ}$ have the same length (construction 2)
3. $\overleftrightarrow{PQ}$ is the perpendicular bisector of $\overline{CD}$	3. Construction 3
4. $\overleftrightarrow{PQ}$ is perpendicular to $\overleftrightarrow{AB}$	4. Since perpendicular bisectors are perpendicular and $AB = CD$

Construction 7

45° angle

Construct: 45° angle

1. Draw $\overleftrightarrow{AB}$ and construct a perpendicular line through B, thus forming four right angles.

2. Bisect one of the right angles, thus forming two 45° angles; $m\angle ABC = 45°$.

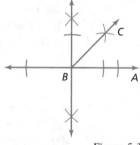

Figure 5.3

Construction 8

Equilateral triangle

Construct: An equilateral triangle

1. Mark off any segment, $\overline{AB}$. Place the point of the compass at each endpoint and, using length AB, mark intersecting arcs above $\overline{AB}$. Call the point where the arcs intersect point C.

2. Draw $\overline{AC}$ and $\overline{BC}$.

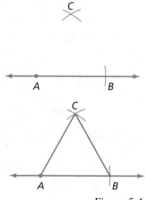

Figure 5.4

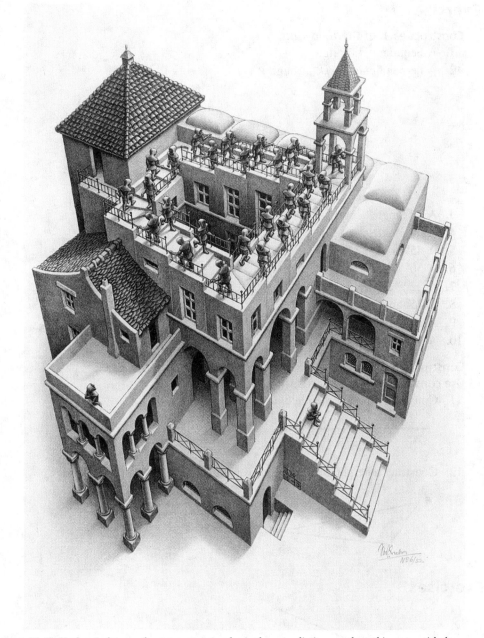

M. C. Escher is famous for art containing logical contradictions such as this one entitled Ascending and Descending.

Can you see why all of the angles in construction 8 measure 60°? You will prove this fact in Chapter 6, so you can include 60° in the set of angles you can construct.

▶ A. Exercises

Construct each of the following.
1. An equilateral triangle
2. A perpendicular to $\overline{LK}$ through P

3. A 90° angle
4. A 60° angle
5. A 45° angle
6. A 30° angle
7. A $22\frac{1}{2}$° angle
8. A 15° angle
9. A 135° angle
10. A 105° angle

Construction 6 works even when the given point is on the line.
Use construction 6 for exercise 11.
11. Construct a line perpendicular to $\overleftrightarrow{XY}$ through L.

12. Construct a perpendicular segment from S to $\overleftrightarrow{RT}$.

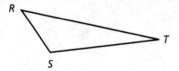

▶ B. Exercises

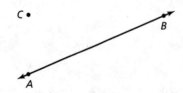

13. Construct a line parallel to $\overleftrightarrow{AB}$ through C.
14. Explain in words how you solved exercise 13.

Complete the proof of construction 8.

STATEMENTS	REASONS
15. $\overline{AC} \cong \overline{AB}$	15.
16. $\overline{AB} \cong \overline{BC}$	16.
17. $AC = AB$, $AB = BC$	17.
18. $AC = AB = BC$	18.

▶ C. Exercises

Complete the proof of construction 7.

STATEMENTS	REASONS
19.	19. Construction 6 (or 3)
20.	20. Construction 4
21.	21. Definition of right angle
22.	22. Definition of angle bisector

▶ Dominion Thru Math

23. Is the following a valid and sound argument? If golf balls are plums, then golf balls are a healthful snack. Golf balls are not plums; therefore, golf balls are not a healthful snack.

■ Cumulative Review

24. Make a Venn diagram of $A = \{1, 2\}$, $B = \{2, 3\}$, $C = \{1, 3\}$.
25. Draw an equilateral pentagon that is not regular.
26. Define *congruent*.
27. If $m\angle ABC$ is $57°$, find the measures of a supplement, a vertical angle, and a complement.
28. How many truth values are there? Name them.

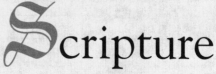

Geometry and Scripture

Statements

You have already seen that the Bible warns us against being deceived.

1. What does Colossians 2:8 warn us to do?

2. What does Exodus 23:7 command of us?

The Bible warns us against error and falsehoods. It also speaks about truth.

3. Give a reference that proves that Jesus Christ is the Truth.

4. What verses in Ephesians 4 command us to speak the truth?

There are also connectives in Scripture. Match each verse to the type of statement that is best illustrated. Use each answer once.

5. Proverbs 21:14
6. Luke 17:33
7. John 8:36
8. Mark 12:30
9. John 11:35
10. Ecclesiastes 12:6
11. Galatians 5:8

A. simple statement
B. negation
C. conjunction
D. disjunction
E. conditional statement
F. biconditional statement
G. not a statement

Read I John 5:12. This is an excellent example of a biconditional.

12. Write 12a as a conditional statement.

13. Write 12b as a conditional statement.

14. Write the conditional statement of 12b without using negations.

15. Write I John 5:12 as a biconditional (using "if and only if").

We can also find various types of proofs in the Bible. Decide whether each is deductive or inductive. Then also explain which form of deductive or inductive reasoning is used.

16. Romans 8:30
17. Romans 10:9-13
18. I Corinthians 9:9-10
19. I Corinthians 15:13-20

HIGHER PLANE: Decide whether these are deductive or inductive and explain which form is used.

Matthew 4:10

John 9:25, 31

Finally, counterexamples are also used in Scripture. Read II Peter 3:3-6.

20. What counterexample does Peter use to refute the claim of the scoffers?

Line upon Line

ᕼE THAT HATH THE SON hath life; and he that hath not the Son of God hath not life. ৶

I JOHN 5:12

1. Alice, Beth, Christine, and Denise each own a parakeet and a collie. Each of the girls' pets is named after one of the other girls. No two collies and no two parakeets have the same name. John, a friend of the girls, is trying to find the owner of the parakeet called Denise. John knows that Denise's parakeet and Christine's collie are both named after the owner of the collie Christine. The girl for whom Beth's collie is named owns the collie that is named for the owner of the parakeet Alice. Solve John's problem.

Decide which of the following are statements and determine whether each is true or false.

2. Is it cold outside?
3. A square has two nonparallel opposite sides.
4. Every rectangle is a parallelogram.
5. $3x = 9$

Let p: Girls have blond hair. Write the meaning of each symbolization.

6. $\forall p$
7. $\sim p$
8. $\exists p$

State whether the following statements are disjunctions or conjunctions and whether the statements are true or false.

9. Birds fly, or fish swim.
10. Blue is one of the three basic colors, and orange is one of the three basic colors.
11. $5 \cdot 7 = 25$ or $3 \cdot 1 = 5$
12. A square is a rhombus, and a square is a rectangle.

Change each statement to an if-then statement. Then change the conditional to a disjunction.

13. The pen writes because it has ink in it.
14. When the cornstalks are brown, the corn is ready to be picked.
15. Two parallel lines never intersect.

In each of the following problems, tell whether the argument is valid or invalid and state whether the conclusion is true or false.

16. Joe has a pet boa constrictor named Hugger. No snakes are pets. Therefore, Hugger is not a snake.

17. All lemons are sour.
    All grapefruits are sour.
    Therefore, all lemons are grapefruits.
18. Thirty-six inches make up three feet.
    Three feet make up a yard.
    So, thirty-six inches make up a yard.
19. A Firebird is an automobile.
    A Pontiac is an automobile.
    Thus, a Firebird is a Pontiac.
20. Give the converse, inverse, and contrapositive of this statement:
    If rhombi are trapezoids, then squares are trapezoids.

Read the following arguments and state the type of each deductive or inductive argument given.

21. If it rains, then the game will be rescheduled for tomorrow. The game was not rescheduled for tomorrow. Therefore, it did not rain.
22. Observations on the movements of Uranus in the 1800s led to the belief that a large planet was located in an orbit beyond Uranus. Calculations based on Newton's law of gravitation pinpointed the region of the sky where the eighth planet should be. When Neptune was found within one degree of the predicted location, the discovery was considered to be a major victory for Newton's law of gravitation.
23. Peter and Paul preached the gospel wherever they went; we as Christians should do likewise.
24. If Grandma makes cookies, then the children will eat them. If the children eat cookies, then they will not eat their supper. If Grandma makes cookies, then the children will not eat their supper.
25. If a person understands reasoning, then he is likely to succeed in geometry. You understand reasoning. Therefore, you are likely to succeed in geometry.
26. Make a truth table for the expression $(s \leftrightarrow t) \wedge \sim t$.
27. Make a truth table to prove that this argument form is invalid.
    Premise 1: $A \vee B$
    Premise 2: A
    Conclusion: B
28. List five common fallacies in geometry proofs.
29. Disprove this statement: If an organism lives underwater, then it is a fish.
30. Explain the mathematical significance of I John 5:12.

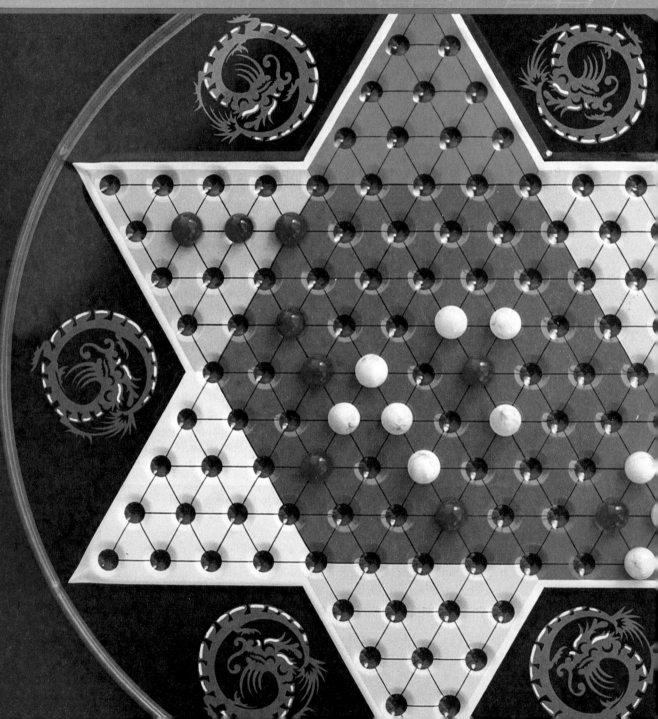

6 Congruence

Chinese checkers has little to do with either China or checkers. Chinese checkers actually originated in England and was called *Halma*, from the Greek word meaning "leap." Its rules correspond, to some degree, to the rules for checkers, but it uses colored marbles instead of checkers.

Like Chinese checkers, not everything is what it claims to be. The apostle John tells us to "try the spirits," or "test your teachers." No matter who proclaims principles from the Bible, the Christian must "search the Scriptures" to see whether those things are so. Remember that Luke praised the Bereans for checking what the apostle Paul taught. Such checking is the only way to avoid playing into the hands of the "many false prophets" who are "gone out into the world."

The board for Chinese checkers is a geometric masterpiece that illustrates congruent figures—that is, figures with the same size and shape. Besides congruent segments and congruent angles, you can find six large and colorful congruent triangles. You should also find numerous small congruent triangles as well as intermediate sizes. You can see how to prove polygons congruent by dividing them into triangles. Can you find congruent parallelograms, congruent trapezoids, and even congruent hexagons? You should even find several sizes of each!

After this chapter you should be able to

1. write formal proofs of theorems (after identifying the given information).

2. use definitions of congruent segments, angles, circles, triangles, and polygons.

3. identify transversals and the special angles formed.

4. express congruent figures correctly in symbols using corresponding parts.

5. distinguish proper criteria for congruence (SAS, ASA, SAA, SSS) from invalid criteria (SSA and AAA).

6.1 Congruent Segments

Snow fences on the plains, like this one near Ennis, Montana, incorporate congruent boards.

Proving theorems can be an exciting, creative skill to learn. To become successful at this new skill, you must attempt to prove new theorems and accept the challenge that each one presents. There is no set way of proving theorems. Each one presents some challenge or variation. However, as you prove more and more theorems, you will improve your ability to think. There are some general hints that you should keep in mind when you are attempting to prove a theorem.

1. Identify the premise (or given information).
2. Identify the conclusion you are trying to obtain.
3. Draw a picture to make sure you understand the theorem.
4. Write down any definitions, postulates, or previously proved theorems that relate to the theorem you are trying to prove.
5. Work backwards if necessary, starting with the conclusion.

The most important point is that you do not give up. You *must* have perseverance if you expect to succeed at proving theorems. You must also be willing to work at thinking. *Work* has negative connotations even among Christians, but the Bible says that we are to work at the tasks set before us (II Thess. 3:10-13). The thinking habits that you develop here will either aid or hinder you for the rest of your life. So do your best and do not give up.

Remember that to prove a theorem, you can use definitions, postulates, and previously proved theorems as true statements. Then use deductive reasoning to derive the desired conclusion.

First, we will prove that there is (at least) one plane with a certain property. We will prove that there is only one with this property in a later chapter when we discuss indirect proofs.

EXAMPLE 1 **Prove:** A line and point not on that line are contained in one and only one plane. (This is Theorem 1.2.)

Answer
1. Consider definitions and incidence postulates that you have learned.
2. Determine the premise and the conclusion.

   Premise: There is a line and a point not on that line.
   Conclusion: The line and point lie in exactly one plane.

3. Draw a picture to help you understand the premise and the conclusion.

4. *l* is the given line, and *K* is the point that is not on the line. You want to show that only one plane passes through both the line and the point. According to the Expansion Postulate, a line contains at least two points, so *l* contains at least two points: call them *X* and *Y*.

5. Now you see three non-collinear points, *X*, *Y*, and *K*. According to the Plane Postulate, three non-collinear points determine exactly one plane.
   ∴ *X*, *Y*, and *K* lie in one plane.

6. By the Flat Plane Postulate, since *X* and *Y* lie in a plane, the entire line *l* that contains them lies in the plane. Thus line *l* and point *K* lie in one plane.

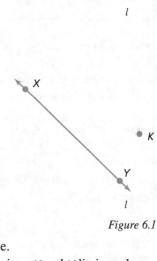

Figure 6.1

Notice that this proof follows the deductive type of reasoning called the Law of Deduction. Example 1 shows a valid, sound proof, but its form may not seem easy for you to follow. However, when you prove theorems, you should think in a manner similar to the way example 1 is written. The most common form for writing high school geometry proofs is called the two-column proof. Example 2 shows the same proof but in two-column form.

EXAMPLE 2 A line and a point not on that line are contained in one and only one plane.

Answer *Given:* a line *l*, a point *K* not on line *l*
 Prove: a plane *p* containing *l* and *K*

STATEMENTS	REASONS
1. There is a line, *l*, and a point, *K*, not on the line	1. Premise (Given)
2. Line *l* contains two points, *X* and *Y*	2. Expansion Postulate
3. *X*, *Y*, and *K* are noncollinear	3. Definition of noncollinear points
4. Points *X*, *Y*, and *K* determine exactly one plane *p*	4. Plane Postulate
5. Line *l* and point *K* are in plane *p*	5. Flat Plane Postulate

In the previous example, the Law of Deduction explains why the statement is proved when the "Prove:" part follows from the "Given:" part. Sometimes you will be asked to show the Law of Deduction step. You will know when you need to show this step because the "Prove:" part will be in "if-then" form.

Notice that Theorem 6.1 is a conditional statement, so the Law of Deduction will be needed for the last step in the proof. Notice also that properties of real numbers are often helpful in geometric proofs. After proving this theorem, you will have an opportunity to prove some theorems on your own. Study this proof carefully!

Theorem 6.1
Congruent Segment Bisector Theorem. **If two congruent segments are bisected, then the four resulting segments are congruent.**

First, draw a picture.

Next look for *p* in your conditional, *p* → *r*. Remember that *p* is your given and that *r* is the conclusion to be proved from *p*.

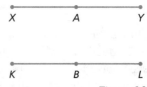
Figure 6.2

Given: Two congruent segments: $\overline{XY} \cong \overline{KL}$
Prove: If *A* and *B* are the midpoints of the congruent segments $\overline{XY}$ and $\overline{KL}$, then $\overline{AY} \cong \overline{XA} \cong \overline{KB} \cong \overline{BL}$

STATEMENTS	REASONS
1. $\overline{XY} \cong \overline{KL}$; A and B are midpoints	1. Given
2. $XY = KL$	2. Definition of congruent segments
3. $\frac{1}{2}XY = \frac{1}{2}KL$	3. Multiplication property of equality
4. $XA = \frac{1}{2}XY$; $KB = \frac{1}{2}KL$	4. Midpoint Theorem
5. $XA = KB$	5. Substitution (step 4 into 3)
6. $XA = AY$, $KB = BL$	6. Definition of midpoint
7. $\overline{XA} \cong \overline{KB}$; $\overline{XA} \cong \overline{AY}$; $\overline{KB} \cong \overline{BL}$	7. Definition of congruent segments
8. $\overline{AY} \cong \overline{XA} \cong \overline{KB} \cong \overline{BL}$	8. Transitive property of congruent segments
9. If A and B are the midpoints of the congruent segments $\overline{XY}$ and $\overline{KL}$, then $\overline{AY} \cong \overline{XA} \cong \overline{KB} \cong \overline{BL}$	9. Law of Deduction

There are a few pointers that you should be aware of before you officially start to prove theorems. The theorems you see proved in this book and by your teacher will look neat and polished. These proofs did not look this way the first time they were proved. Your proofs may not be neat the first time you reason through them either. This is all right; but when you write the proof for your homework or an exam, you must write it neatly and orderly to communicate the reasoning process that you used to prove the statement. Sometimes it helps to reason backwards from the conclusion; but when you write the proof, you must communicate it logically from the hypothesis (or premise) to the conclusion.

You will prove the next theorem in the exercises.

Theorem 6.2
Segment congruence is an equivalence relation.

Remember that equivalence relations have three properties. So the theorem above actually has three parts: the reflexive property of congruent segments, the symmetric property of congruent segments, and the transitive property of congruent segments.

Draw two quadrilaterals with perpendicular diagonals that are not rhombi.
Draw one convex (*kite*) and one concave (*deltoid*).

▶ **A. Exercises**

Write the reasons for the statements in each proof.

Given: A-X-B and X-B-Y

Prove: If AB = XY, then AX = BY

A   X   B   Y

STATEMENTS	REASONS
1. AB = XY, A-X-B, X-B-Y	1.
2. AX + XB = AB; XB + BY = XY	2.
3. AX + XB = XB + BY	3.
4. XB = XB	4.
5. AX = BY	5.

Prove: If two lines intersect, then there is a plane containing them (Theorem 1.3)

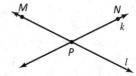

STATEMENTS	REASONS
6. Lines *l* and *k* intersect at point *P*	6.
7. Line *l* has at least one more point on it; call it *M*. Line *k* has at least one more point on it; call it *N*	7.
8. Points *M*, *N*, and *P* determine exactly one plane *p*	8.
9. Lines *l* and *k* lie in plane *p*	9.
10. If two lines intersect, then they lie in a plane	10.

Given: A-T-B and A-L-C
Prove: If $BT = CL$ and $TA = LA$,
then $\triangle ABC$ is an isosceles
triangle

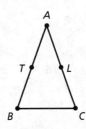

STATEMENTS	REASONS
11. $BT = CL$; $TA = LA$; A-T-B, A-L-C	11.
12. $BT + TA = CL + LA$	12.
13. $BT + TA = BA$; $CL + LA = CA$	13.
14. $BA = CA$	14.
15. $\overline{BA} \cong \overline{CA}$	15.
16. $\triangle ABC$ is isosceles	16.
17. If $BT = CL$ and $TA = LA$, then $\triangle ABC$ is an isosceles triangle	17.

▶ B. Exercises

18. Prove that two parallel lines are contained in one and only one plane.
19. *Given:* $LM = PQ$ in the figure at right
    Prove: $LP = MQ$

    $\overset{\bullet}{L}\quad\overset{\bullet}{M}\quad\overset{\bullet}{X}\quad\overset{\bullet}{P}\quad\overset{\bullet}{Q}$

20. *Given:* $\overline{AL} \cong \overline{MY}$; $\overline{MY} \cong \overline{PB}$
    Prove: $\overline{AL} \cong \overline{PB}$

21. *Given:* $\overline{AB} \cong \overline{CD}$
    Prove: $\overline{CD} \cong \overline{AB}$

22. Prove that every segment is congruent to itself.
23. What can you conclude from exercises 20, 21, and 22? What theorem have you just proved?
24. *Given:* $XR = PY$; $XQ = YQ$
    Prove: $\overline{QR} \cong \overline{PQ}$

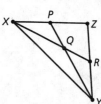

▶ C. Exercises

25. *Given:* $AB > AC$ where A, B, and C are noncollinear
    Prove: There is a point P on $\overrightarrow{AC}$ so that $AB = AC + CP$

State the first five postulates.
26. Expansion Postulate
27. Line Postulate
28. Plane Postulate
29. Flat Plane Postulate
30. Plane Intersection Postulate

6.2 Congruent Angles

The only way you can become proficient in the methods of proving theorems is to practice, practice, practice. In this section we will give three examples of proved theorems. Study these examples and think through the process of proof used for each example.

Skiers usually keep their legs together so that their legs make congruent angles with the skis.

EXAMPLE 1 *Prove:* All right angles are congruent.

    Answer Draw a picture of two
               arbitrary right angles.

               Now prove that $\angle B \cong \angle M$.

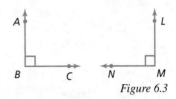

Figure 6.3

STATEMENTS	REASONS
1. $\angle B$ and $\angle M$ are right angles	1. Given
2. $m\angle B = 90°$; $m\angle M = 90°$	2. Definition of right angle
3. $m\angle B = m\angle M$	3. Transitive property of equality
4. $\angle B \cong \angle M$	4. Definition of congruent angles

Since $\angle B$ and $\angle M$ are arbitrary right angles,
the theorem is true for all right angles.

Do you recall doing this proof as an exercise in Chapter 4? It is Theorem 4.1.

Theorem 6.3
Supplements of congruent angles are congruent.

EXAMPLE 2 *Prove:* Theorem 6.3

    Answer Draw two congruent angles and their supplements.

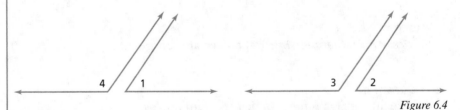

Figure 6.4

Assume that $\angle 1 \cong \angle 2$, $\angle 3$ is a supplement of $\angle 2$, and
$\angle 4$ is a supplement of $\angle 1$. Prove that $\angle 3 \cong \angle 4$.

Continued ▶

STATEMENTS	REASONS
1. $\angle 1 \cong \angle 2$; $\angle 1$ and $\angle 4$ are supplements; $\angle 2$ and $\angle 3$ are supplements	1. Given
2. $m\angle 1 = m\angle 2$	2. Definition of congruent angles
3. $m\angle 1 + m\angle 4 = 180$; $m\angle 2 + m\angle 3 = 180$	3. Definition of supplementary angles
4. $m\angle 1 + m\angle 4 = m\angle 2 + m\angle 3$	4. Transitive property of equality
5. $m\angle 4 = m\angle 3$	5. Addition property of equality
6. $\angle 4 \cong \angle 3$	6. Definition of congruent angles

Remember that you can use theorems that you have previously proved as valid reasons when proving a later theorem. Example 3 illustrates this. Before studying example 3, review Theorems 4.2 through 4.7.

Theorem 4.2
Adjacent supplementary angles form a linear pair.

Theorem 4.3
Linear pairs are supplementary.

Theorem 4.4
If one angle of a linear pair is a right angle, then so is the other.

Theorem 4.5
Vertical Angle Theorem. Vertical angles are congruent.

Theorem 4.6
Congruent supplementary angles are right angles.

Theorem 4.7
Angle Bisector Theorem. The measure of an angle is twice that of the angles formed by the angle bisector.

EXAMPLE 3 *Prove:* If $m\angle AXB = m\angle DXY$, then
$m\angle AXD = m\angle BXY$

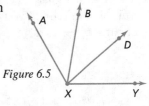

Figure 6.5

Answer

STATEMENTS	REASONS
1. $m\angle AXB = m\angle DXY$	1. Given
2. $m\angle AXB + m\angle BXD = m\angle BXD + m\angle DXY$	2. Addition property of equality
3. $m\angle AXB + m\angle BXD = m\angle AXD$; $m\angle BXD + m\angle DXY = m\angle BXY$	3. Angle Addition Postulate
4. $m\angle AXD = m\angle BXY$	4. Substitution (step 3 into 2)
5. If $m\angle AXB = m\angle DXY$, then $m\angle AXD = m\angle BXY$	5. Law of Deduction

The following important theorems are exercises.

Theorem 6.4
Complements of congruent angles are congruent (exercise 8).

Theorem 6.5
Angle congruence is an equivalence relation (exercises 10, 12–13).

Theorem 6.6
***Adjacent Angle Sum Theorem.* If two adjacent angles are congruent to another pair of adjacent angles, then the larger angles formed are congruent (exercise 15).**

Theorem 6.7
***Adjacent Angle Portion Theorem.* If two angles, one in each of two pairs of adjacent angles, are congruent, and the larger angles formed are also congruent, then the other two angles are congruent (exercise 16).**

Theorem 6.8
***Congruent Angle Bisector Theorem.* If two congruent angles are bisected, the four resulting angles are congruent (exercises 17).**

Theorem 6.8 guarantees that bisections of congruent angles are congruent. This should remind you of Theorem 6.1, which says that bisections of congruent segments are congruent.

▶ A. Exercises

Given: ∠4 ≅ ∠5 in the figure shown
Prove: ∠1 ≅ ∠7

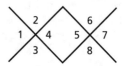

STATEMENTS	REASONS
1. ∠4 ≅ ∠5	1.
2. m∠4 = m∠5	2.
3. ∠1 ≅ ∠4; ∠5 ≅ ∠7	3.
4. m∠1 = m∠4; m∠5 = m∠7	4.
5. m∠1 = m∠7	5.
6. ∠1 ≅ ∠7	6.

7. *Given:* ∠AXB and ∠BXL are supplementary;
   ∠AXB ≅ ∠BXL
   Prove: $\overleftrightarrow{AL} \perp \overleftrightarrow{BT}$

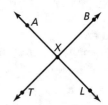

8. Complements of congruent angles are congruent. (Theorem 6.4)
9. *Given:* E is the midpoint of $\overline{AC}$ and $\overline{BD}$; $\overline{BE} \cong \overline{EC}$
   Prove: $\overline{AC} \cong \overline{BD}$
10. Transitive property of congruent angles
    Given: ∠A ≅ ∠B and ∠B ≅ ∠C
    Prove: ∠A ≅ ∠C
11. *Given:* $\overrightarrow{OZ}$ bisects ∠POY
    Prove: ∠QOX ≅ ∠POZ

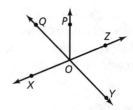

▶ B. Exercises

Prove each theorem.

12. Symmetric property of congruent angles

13. Reflexive property of congruent angles

14. If two angles are complementary and congruent, then each angle has a measure of 45°.

Use the following diagram for exercises 15-16.

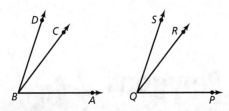

15. *Given:* $\angle ABC \cong \angle PQR$; $\angle DBC \cong \angle SQR$
    Prove: $\angle ABD \cong \angle PQS$

16. *Given:* $\angle ABC \cong \angle PQR$; $\angle ABD \cong \angle PQS$
    Prove: $\angle DBC \cong \angle SQR$

17. If two congruent angles are bisected, then the four resulting angles are congruent.

18. *Given:* $\angle ABC$ and $\angle XYZ$ are supplementary angles;
    $m\angle ABC = m\angle XYZ$
    Prove: $\angle ABC$ is a right angle

19. *Given:* $\angle 2$ and $\angle 3$ are supplements
    Prove: $m\angle 1 = m\angle 4$

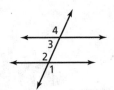

▶ C. Exercises

20. *Prove:*   If $m\angle ABC > m\angle DEF$, then there exists $\angle DEX$ adjacent to $\angle DEF$ so that $m\angle ABC = m\angle DEF + m\angle DEX$.

Diagram each theorem listed below.
21. All right angles are congruent. (Theorem 4.1)
22. If one angle of a linear pair is right, so is the other. (Theorem 4.3)
23. Adjacent supplementary angles form a linear pair. (Theorem 4.4)
24. Vertical Angle Theorem (Theorem 4.5)
25. Congruent supplementary angles are right angles. (Theorem 4.6)

6.3 Congruent Polygons and Circles

You have already studied congruent segments and congruent angles. Do you remember what *congruent* means? Congruent segments are two segments that have the same measure. You must remember that segments are not *equal* when they have the same measure; they are *congruent*. The symbol for congruence is ≅.

Fossils found by splitting a rock show two congruent halves. This fossil fern comes from Mazon Creek, Illinois.

In general, congruent figures are figures with the same size and shape. Notice how the definitions below guarantee that the figures have the same size and shape. The symbol ≅ is used for all congruent figures, not just for segments and angles.

The 18 wheels of a semi are congruent circles, and the dozens of nuts holding the wheels on are congruent hexagons.

Congruent circles are circles with congruent radii.

Congruent polygons are polygons that have three properties: **1)** same number of sides, **2)** corresponding sides are congruent, and **3)** corresponding angles are congruent.

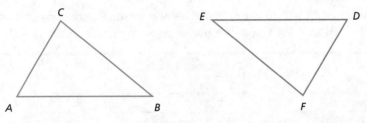

Figure 6.6

Are △*ABC* and △*DEF* congruent? Since both polygons are triangles, they have the same number of sides. If you measure each side of △*ABC* and compare those measurements to the measures of the sides of △*DEF*, you will see that *AC* = *DF*, *AB* = *DE*, and *BC* = *EF*. Since corresponding sides are congruent, you can mark congruent pairs with slash marks as shown in figure 6.7. Now compare the angle measures. Likewise, the angle markings show that ∠*A* ≅ ∠*D*, ∠*B* ≅ ∠*E*, and ∠*C* ≅ ∠*F*. Therefore, △*ABC* ≅ △*DEF* by the definition of congruent polygons, since all the corresponding parts are congruent.

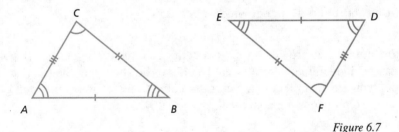

Figure 6.7

You must name congruent triangles by their corresponding parts. The correct one-to-one correspondence for the triangles in figure 6.6 pairs vertices *A*→*D*, *B*→*E*, and *C*→*F*. You can therefore write △*ABC* ≅ △*DEF* (or other corresponding orders such as △*BCA* ≅ △*EFD*). However, it would be incorrect to write △*ABC* ≅ △*EDF*, because this notation does not indicate the correct one-to-one correspondence of congruent parts.

Since triangles are the most important polygons, the definition for congruent triangles follows.

Congruent triangles are triangles in which corresponding angles and corresponding sides are congruent.

In other words, $\triangle ABC \cong \triangle DEF$ if and only if $\angle A \cong \angle D$, $\angle B \cong \angle E$, $\angle C \cong \angle F$, $\overline{AB} \cong \overline{DE}$, $\overline{AC} \cong \overline{DF}$, and $\overline{BC} \cong \overline{EF}$.

You have proved the following properties of congruence for segments and angles. In the exercises you will prove them for triangles and circles also.

Congruence Properties	
Property	Meaning
Reflexive	$X \cong X$
Symmetric	If $X \cong Y$, then $Y \cong X$
Transitive	If $X \cong Y$ and $Y \cong Z$, then $X \cong Z$

Do you remember the name for a relationship that is reflexive, symmetric, and transitive?

Theorem 6.9
Triangle congruence is an equivalence relation.

Theorem 6.10
Circle congruence is an equivalence relation.

Congruent triangles provide a way to identify congruent polygons. Any polygon can be divided into several triangles. If we can show that the triangles in one polygon are congruent to the corresponding triangles in another polygon, then the two polygons are congruent.

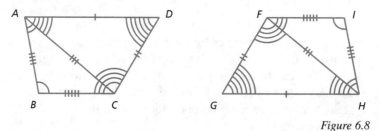

Figure 6.8

From the markings on these quadrilaterals, you can assume that $ABCD \cong HIFG$. Using the relation between triangle congruence and polygon congruence, you can prove that the properties of congruence apply to polygons too.

> **Theorem 6.11**
>
> **Polygon congruence is an equivalence relation.**

▶ A. Exercises

Write the correct triangle congruence statement for each pair. Pay particular attention to the markings on the diagrams.

1.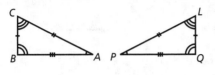

4.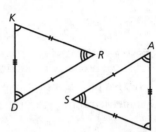

2.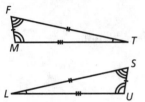

5.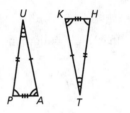

3.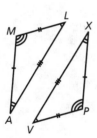

6.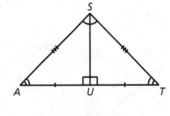

Name the congruent triangle using correct notation.

7. △TIS
8. △IST
9. △TSI
10. △SIT

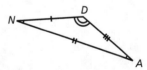

Name the congruent corresponding parts of the congruent triangles in exercises 11-13.

11. △QMN ≅ △LPS

13. △BFV ≅ △KEY

12.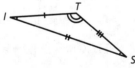

Use the figure for exercises 14-17.
14. Why are the angles at *B* congruent?
15. Why is *B* the midpoint of $\overline{CZ}$?
16. Name the congruent triangles.

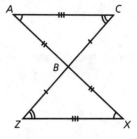

Prove each statement.
17. Transitive property of triangle congruence
18. Reflexive property of triangle congruence
19. Symmetric property of triangle congruence

► C. Exercises

20. Prove that circle congruence is an equivalence relation.

■ Cumulative Review

Match. Be as specific as possible.

21.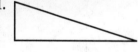

22.

23.

24.

A. Acute and equilateral

B. Acute and isosceles

C. Acute and scalene

D. Right and equilateral

E. Right and isosceles

F. Right and scalene

G. Obtuse and equilateral

H. Obtuse and isosceles

I. Obtuse and scalene

25. Which two choices above describe impossible triangles?

6.4 Parallel Lines and Transversals

Parallel lines have already been defined as two coplanar lines that do not intersect. There are many interesting facts about parallel lines and some special angles formed with them. You will study many of these properties in this chapter. Some of these properties are the bases for carpentry techniques.

Before you can study the theorems of this lesson, you need to know some basic definitions.

Definition

A **transversal** is a line that intersects two or more distinct coplanar lines in two or more distinct points.

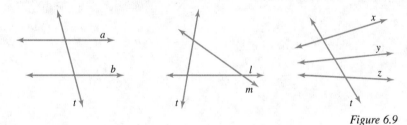

Figure 6.9

Notice that in figure 6.10, *t* is not a transversal. Since the definition requires at least two distinct points of intersection, the single point of concurrency does not satisfy the definition.

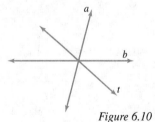

Figure 6.10

Look at the eight angles formed by transversal *t* in the next figure. Certain pairs of these angles have special names.

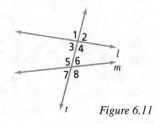

Figure 6.11

In this lesson you will see some relationships between the following special pairs of angles and parallel lines. Refer to figure 6.12.

Definitions

Alternate interior angles are angles such as ∠3 and ∠6, which are numbered on opposite sides of the transversal and between the other two lines.

Alternate exterior angles are angles such as ∠1 and ∠8; these angle numbers are on opposite sides of the transversal and outside the other two lines.

Corresponding angles are angles such as ∠2 and ∠6. These angle numbers are on the same side of the transversal and on the same side of their respective lines. ∠3 and ∠7 form another pair of corresponding angles.

Can you name another pair of each type?

Which of these roads in Washington, D.C., are transversals that cut across the grid of parallel and perpendicular streets?

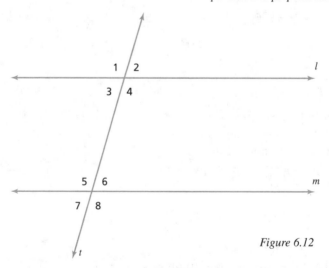

Figure 6.12

Use a protractor to measure ∠3 and ∠6. Now measure the other pair of alternate interior angles. Draw another pair of parallel lines with a more slanted transversal cutting them. Compare the measurements of the alternate interior angles. What seems to be true about the measures of alternate interior angles? This experimentation suggests the following postulate.

Postulate 6.1
Parallel Postulate. **Two lines intersected by a transversal are parallel if and only if the alternate interior angles are congruent.**

The two conditionals asserted by this biconditional are stated below. Can you state them before you look?

1. If two parallel lines are intersected by a transversal, then the alternate interior angles formed are congruent.
2. If two lines are intersected by a transversal in such a way that the alternate interior angles are congruent, then the two lines are parallel.

The form of the parallel postulate given above will be easier to work with than the historic form. Each form can be proved from the other, so both are correct. Because the historic form is very famous, you should be familiar with it.

Historic Parallel Postulate
Given a line and a point not on the line, there is exactly one line passing through the point that is parallel to the given line.

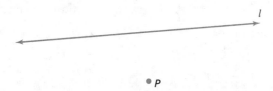

Figure 6.13

Can you sketch the line parallel to *l* through *P*?

Theorem 6.12
Alternate Exterior Angle Theorem. **Two lines intersected by a transversal are parallel if and only if the alternate exterior angles are congruent.**

Theorem 6.13
Corresponding Angle Theorem. **Two lines intersected by a transversal are parallel if and only if the corresponding angles are congruent.**

Remember that each biconditional requires two proofs, one for each conditional. Study the examples closely so that you can do the others in the exercises.

EXAMPLE 1 **Alternate Exterior Angle Theorem**—*Part One*

If two parallel lines are intersected by a transversal, then the alternate exterior angles are congruent.

Given: $a \parallel b$ and t is a transversal that forms the eight angles shown

Prove: $\angle 1 \cong \angle 8$

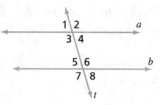

Figure 6.14

Answer

STATEMENTS	REASONS
1. $a \parallel b$; t is a transversal	**1.** Given
2. $\angle 4 \cong \angle 5$	**2.** Parallel Postulate
3. $\angle 1 \cong \angle 4$; $\angle 8 \cong \angle 5$	**3.** Vertical Angle Theorem
4. $m\angle 4 = m\angle 5$; $m\angle 1 = m\angle 4$; $m\angle 8 = m\angle 5$	**4.** Definition of congruent angles
5. $m\angle 1 = m\angle 8$	**5.** Substitution (see step 4)
6. $\angle 1 \cong \angle 8$	**6.** Definition of congruent angles
7. If two parallel lines are intersected by a transversal, then the alternate exterior angles formed are congruent	**7.** Law of Deduction

EXAMPLE 2 **Corresponding Angle Theorem**—*Part One*

If a transversal intersects two lines such that the corresponding angles are congruent, then the two lines are parallel.

Answer Draw a diagram to help visualize the theorem.

Given: $\angle 1 \cong \angle 2$

Prove: $a \parallel b$

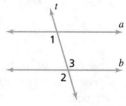

Figure 6.15

Continued ▶

STATEMENTS	REASONS
1. $\angle 1 \cong \angle 2$	1. Given
2. $\angle 2 \cong \angle 3$	2. Vertical Angle Theorem
3. $\angle 1 \cong \angle 3$	3. Transitive property of congruent angles
4. $a \parallel b$	4. Parallel Postulate
5. If $\angle 1 \cong \angle 2$, then $a \parallel b$	5. Law of Deduction

Part two of both theorems as well as the next theorem will be exercises. Study the proof in example 3 before attempting these exercises.

Theorem 6.14
If a transversal is perpendicular to one of two parallel lines, then it is perpendicular to the other also.

EXAMPLE 3

Theorem 6.15
If two coplanar lines are perpendicular to the same line, then they are parallel to each other.

Answer   *Given:* $p \perp t; q \perp t$
           Prove: $p \parallel q$

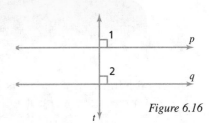

Figure 6.16

STATEMENTS	REASONS
1. $p \perp t; q \perp t$	1. Given
2. $\angle 1$ and $\angle 2$ are right angles	2. Perpendicular lines intersect to form right angles
3. $\angle 1 \cong \angle 2$	3. All right angles are congruent
4. $p \parallel q$	4. Corresponding Angle Theorem
5. If $p \perp t$ and $q \perp t$, then $p \parallel q$	5. Law of Deduction

▶ A. Exercises

Use the figure for exercises 1-10. Assume that $l \parallel m$.

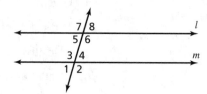

1. If $m\angle 4 = 42$, find $m\angle 5$.
2. If $m\angle 3 = 129$, find $m\angle 7$.
3. If $m\angle 1 = 60$, find $m\angle 2$.
4. If $m\angle 1 = 60$, find $m\angle 4$.
5. If $m\angle 7 = 110$, find $m\angle 2$.
6. If $m\angle 3 = 57$, find $m\angle 5$.
7. If $m\angle 2 = 120$, find $m\angle 6$.
8. If $m\angle 8 = 6x + 12$, find $m\angle 4$.
9. If $m\angle 3 = x^2 + 2x - 6$, find $m\angle 6$.
10. If $m\angle 1 = 3x + 9$, find $m\angle 3$.

Part one of the Corresponding Angle Theorem was proved in example 2. Prove part two by supplying the missing reasons.

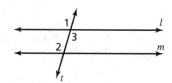

If a transversal intersects two parallel lines, then the corresponding angles formed are congruent.

Given: $l \parallel m$
Prove: $\angle 1 \cong \angle 2$

STATEMENTS	REASONS
11. $l \parallel m$	**11.**
12. $\angle 2 \cong \angle 3$	**12.**
13. $\angle 3 \cong \angle 1$	**13.**
14. $\angle 2 \cong \angle 1$	**14.**
15. If $l \parallel m$, then $\angle 1 \cong \angle 2$	**15.**

▶ B. Exercises

Use the same figure as exercises 1-10 for exercises 16-18.

16. Prove part two of the alternate exterior angle theorem. If a transversal intersects two lines so that the alternate exterior angles are congruent, then the lines are parallel.

17. *Prove:* If a transversal intersects two lines such that one pair of alternate interior angles is congruent, then the other pair of alternate interior angles is also congruent.

18. *Prove:* If a transversal intersects two lines and two corresponding angles are congruent, then two alternate interior angles are congruent.

Prove the following statements.

19. *Given:* t is a transversal; $\angle 2 \cong \angle 7$
   Prove: $\angle 4 \cong \angle 8$

20. *Given:* $p \parallel q$
   Prove: $\angle 4$ and $\angle 6$ are supplementary

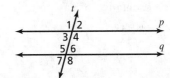

Use the figure to prove the following statements.

21. *Given:* $\angle 2 \cong \angle 6$; $d \parallel c$
   Prove: $\angle 2 \cong \angle 14$

22. *Given:* $\angle 9 \cong \angle 13$; $\angle 16 \cong \angle 5$
   Prove: $a \parallel b$; $c \parallel d$

23. *Given:* $a \parallel b$; $\angle 16 \cong \angle 4$
   Prove: $c \parallel d$

24. Prove Theorem 6.14.

If a transversal is perpendicular to one of two parallel lines, then it is perpendicular to the other one also.

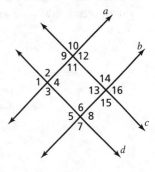

▶ C. Exercises

25. Suppose that line l is perpendicular to line m at point P and that lines m and n are parallel. If you are not given that l is a transversal (as in 24), how could you prove that l must be a transversal?

■ Cumulative Review

State the following postulates and theorems and sketch a picture for each.

26. Line Separation Postulate

27. Plane Separation Postulate

28. Jordan Curve Theorem

29. Midpoint Theorem

30. Angle Bisector Theorem

6 Analytic Geometry

Equations of Lines

By using the definition of slope, $m = \frac{y_2 - y_1}{x_2 - x_1}$, you can derive an important formula for finding equations of lines. Suppose (x_1, y_1) is a specific point on the line, and let (x, y) be representative of any point on the line.

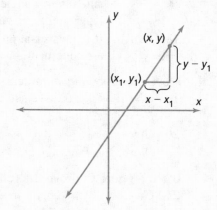

1. $\frac{y - y_1}{x - x_1} = m$ 1. Slope formula
2. $y - y_1 = m(x - x_1)$ 2. Multiplication property of equality

The formula is called the point slope formula since it uses the slope and one particular point.

EXAMPLE   Find the equation of the line passing through $(6, -1)$ and having a slope equal to $\frac{2}{3}$.

Answer $y - y_1 = m(x - x_1)$

$y - (-1) = \frac{2}{3}(x - 6)$

$y + 1 = \frac{2}{3}x - 4$

$y = \frac{2}{3}x - 5$

Given two points on the line, you can also obtain the equation of the line. First, calculate the slope using the definition. Then use either of the two points with the slope in the point-slope form. Slope is very useful for studying parallel lines. Do you know why?

▶ Exercises

Find the equation of the line passing through the given point(s).

1. $(5, -2)$ with slope 3
2. $(6, 3)$ with slope $\frac{-1}{2}$
3. $(4, 7)$ and $(2, 3)$
4. $(-1, 4)$ and $(2, -5)$
5. Give the y-intercept of $y - y_1 = m(x - x_1)$.

6.5 Angles of Polygons

Look at the triangles in figure 6.17.

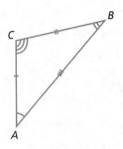 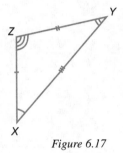

Figure 6.17

Road signs come in octagons, triangles, rectangles, and even this rhombus.

Is △*ABC* ≅ △*XYZ*? How do you know? The slash mark on $\overline{AC}$ and the one on $\overline{XZ}$ indicate that these two sides are congruent. What other sides are congruent? What angles are congruent? Since the corresponding parts are congruent, the two triangles are congruent by definition.

Since definitions are reversible, you also know that if two triangles are congruent, then their corresponding sides and angles are congruent. This is often used to prove that specific parts of triangles are congruent. To use this definition correctly, you must mark the corresponding parts correctly.

When doing proofs that involve triangles, you may often need to use *included sides* and *included angles*. In △*ABC* in figure 6.17, the included side between ∠*A* and ∠*B* is $\overline{AB}$. In △*XYZ* the included angle between sides $\overline{XZ}$ and $\overline{ZY}$ is ∠*Z*.

There are many interesting theorems about triangles. You will see just a few of them in this section.

The sum of the measures of the angles of any triangle is 180°.

Given: △ABC

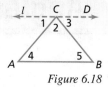

Figure 6.18

Draw a line that passes through C and is parallel to $\overleftrightarrow{AB}$. Make the new line dotted to show that it is not a part of the given information but is used in the proof. You may number the angles for convenience. *Note:* The dotted line is an *auxiliary figure* needed to prove the theorem. Auxiliary figures are valid to use if the figure drawn can really exist. This auxiliary line exists according to the Historic Parallel Postulate.

Prove: $m\angle 2 + m\angle 4 + m\angle 5 = 180°$

STATEMENTS	REASONS
1. △ABC; $l \parallel \overleftrightarrow{AB}$ through C	1. Given; auxiliary line
2. $m\angle 2 + m\angle 3 = m\angle ACD$	2. Angle Addition Postulate
3. $\angle 1$ and $\angle ACD$ are supplementary	3. Linear pairs are supplementary
4. $m\angle 1 + m\angle ACD = 180°$	4. Definition of supplementary
5. $m\angle 1 + m\angle 2 + m\angle 3 = 180°$	5. Substitution (step 2 into 4)
6. $\angle 1 \cong \angle 4$; $\angle 3 \cong \angle 5$	6. Parallel Postulate
7. $m\angle 1 = m\angle 4$; $m\angle 3 = m\angle 5$	7. Definition of congruent angles
8. $m\angle 4 + m\angle 2 + m\angle 5 = 180°$	8. Substitution (step 7 into 5)

By using this theorem, you will be able to find the measure of an angle of a triangle if you know the measure of the other two angles of the triangle.

Theorem 6.17

If two angles of one triangle are congruent to two angles of another triangle, then the third angles are also congruent.

Given: $\angle A \cong \angle X$; $\angle B \cong \angle Y$
Prove: $\angle C \cong \angle Z$

Draw and mark two triangles.

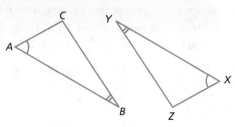

Figure 6.19

STATEMENTS	REASONS
1. $\angle A \cong \angle X$; $\angle B \cong \angle Y$	1. Given
2. $m\angle A + m\angle B + m\angle C = 180°$; $m\angle X + m\angle Y + m\angle Z = 180°$	2. The sum of the measures of the angles of a triangle is 180
3. $m\angle A + m\angle B + m\angle C = m\angle X + m\angle Y + m\angle Z$	3. Transitive property of equality
4. $m\angle A = m\angle X$; $m\angle B = m\angle Y$	4. Definition of congruent angles
5. $m\angle A + m\angle B + m\angle C = m\angle A + m\angle B + m\angle Z$	5. Substitution (step 4 into 3)
6. $m\angle C = m\angle Z$	6. Addition property of equality
7. $\angle C \cong \angle Z$	7. Definition of congruent angles
8. If $\angle A \cong \angle X$ and $\angle B \cong \angle Y$, then $\angle C \cong \angle Z$	8. Law of Deduction

Theorem 6.18
The acute angles of a right triangle are complementary (exercise 21).

You can also find the total measure of the angles of a polygon. First, subdivide the polygon into triangles. If you are careful to use only vertices of the original polygon, you will find the number of triangles is always the number shown on the table.

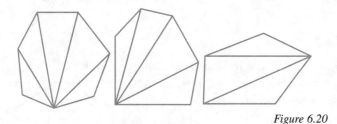

Figure 6.20

Number of Sides	Number of Triangles
4	2
5	3
6	4
7	5
8	6
9	7
10	8
n	$n - 2$

Since the sum of the measures of the angles of a triangle is 180°, simply multiply the number of triangles in the polygon by 180.

EXAMPLE What is the sum of the measures of the angles of a pentagon? If the pentagon is regular, find the measure of each angle.

Answer $3(180) = 540°$ 1. Find the sum of the angle measures.

$$\frac{540°}{5} = 108°$$ 2. Find the measure of one angle of a regular pentagon.

▶ A. Exercises

Use the figure for exercises 1-4.
1. Name the included angle between $\overline{AD}$ and $\overline{DC}$.
2. Name the included angle between $\overline{EA}$ and $\overline{EB}$.
3. Name the included side between $\angle DAE$ and $\angle AED$.
4. Name the included side between $\angle BCD$ and $\angle CDA$.

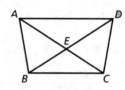

Give the sum of the measures of the angles in each convex polygon.
5. hexagon
6. decagon
7. n-gon
8. 100-sided polygon

Give the measure of an interior angle of each regular polygon.
9. heptagon
10. octagon
11. n-gon
12. 84-sided polygon

▶ B. Exercises

Use the figure to find the indicated measures.
13. $m\angle ABE = 78$; $m\angle BAE = 62$. Find $m\angle AEB$.
14. $m\angle DEC = 56$. Find $m\angle AED$.
15. $m\angle AED = 102$; $m\angle EDA = 49$. Find $m\angle DAE$.
16. $m\angle BEC = 82$. Find $m\angle AED$.
17. Find $m\angle DEC + m\angle ECD + m\angle CDE$.
18. $m\angle AEB = 40$; $m\angle ADE = 34$. Find $m\angle EAD$.
19. $m\angle ADC = 66$; $m\angle ADE = 28$; $m\angle AEB = 45$. Find $m\angle ECD$.
20. Find $m\angle ABC + m\angle BCD + m\angle CDA + m\angle DAB$.

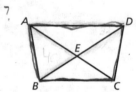

Prove the following statements.

21. Given any right triangle, its acute angles are complementary.

22. *Given:* ∠LMN ≅ ∠NOP
Prove: ∠NLM ≅ ∠NPO

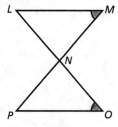

23. The angle measure of each angle of an equiangular triangle is 60°.

24. *Given:* ∠Y ≅ ∠Z; ∠YXW ≅ ∠ZXW
Prove: $\overleftrightarrow{XW} \perp \overleftrightarrow{YZ}$

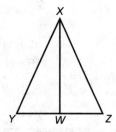

▶ C. Exercises

25. If ∠A and ∠B are trisected and
m∠C = 30 as shown, find m∠K and m∠R.

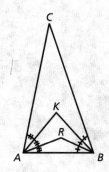

■ Cumulative Review

State each postulate.

26. Ruler Postulate

27. Protractor Postulate

28. Completeness Postulate

29. Continuity Postulate

30. Parallel Postulate

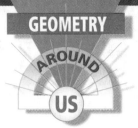
GEOMETRY AND SPORTS

Playing areas for many sports use geometry. A track, for example, is a circle that has been "stretched." The sides of the track are segments of parallel lines, but the ends are semicircles. A runner in the outside lane must run farther than one in the inside lane, since the circle around which he is running has a longer radius. How do we make the race fair for the runner in the outside lane? We "stagger" the starting lines: the outside runner starts slightly ahead of the others, and the closer to the inside lane the

These race cars look congruent.

runner is, the farther back he starts. Thus the advantage he gains from the curves makes up for the position from which he started. Olympic speed skaters use a slightly different technique. Only two skaters race on the track at the same time; every two laps, they trade lanes so that by the end of the race they have skated the same distance.

Other fields use congruent shapes. Can you identify the distinctive shapes of a football, baseball, or soccer field; a tennis, basketball, or racquetball court; a wrestling mat; a bowling alley; or a hockey rink? Regulation fields and courts must all be congruent, having the correct size with lines painted precisely. Distance is essential between yard lines in football, while the radius of the free-throw circle is important to basketball. Angles define proper base lines in baseball.

Athletic equipment also depends on geometry. Could you take a piece of wood and shape it into a baseball bat or hockey stick? Could you make a leather soccer ball, basketball, or football the right size so that it contains the proper amount of air when inflated? Have you ever tried making a baseball cover and core the precise size and weight required—with the exact number of stitches necessary? Could you make a puck or tennis ball of the proper size and weight? Geometric concepts of measurement, shape, and volume are important for making regulation equipment.

Playing areas for many sports use geometry.

The length of the "staggers" is dependent upon the radius of the semicircular ends of the track.

6.6 Congruence Postulates

So far, you must prove six congruences to prove that two triangles are congruent. Now you will learn faster methods.

Think of an angle measure and two segment lengths. Draw an angle with that measure and copy those lengths along the sides from the vertex. Complete the triangle by connecting the endpoints of the marked lengths. Notice that the two sides and included angle determined the size and shape of the whole triangle. Any triangles with those three measures are congruent. The postulate that summarizes this is called SAS, which abbreviates Side-Angle-Side.

The seven congruent triangles on the crown of the Statue of Liberty radiate liberty to all nations from New York harbor.

Postulate 6.2

SAS Congruence Postulate. **If two sides and an included angle of one triangle are congruent to the corresponding two sides and included angle of another triangle, then the two triangles are congruent.**

Figure 6.21 shows that △*LNO* ≅ △*XTU*.

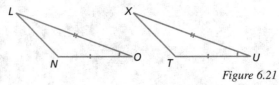

Figure 6.21

Figure 6.22 illustrates another postulate that can be used to show that two triangles are congruent. This postulate covers the case of two angles and the included side, so it is called ASA (Angle-Side-Angle).

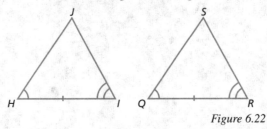

Figure 6.22

Postulate 6.3

ASA Congruence Postulate. **If two angles and an included side of one triangle are congruent to the corresponding two angles and included side of another triangle, then the two triangles are congruent.**

EXAMPLE 1   *Given:* $\overline{LM} \cong \overline{LO}$; $\angle MLN \cong \angle OLN$
                 Prove: $\triangle LMN \cong \triangle LON$

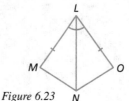

Figure 6.23

Answer

STATEMENTS	REASONS
1. $\overline{LM} \cong \overline{LO}$; $\angle MLN \cong \angle OLN$	1. Given
2. $\overline{LN} \cong \overline{LN}$	2. Reflexive property of congruent segments
3. $\triangle LMN \cong \triangle LON$	3. SAS

Now instead of proving six congruences, you prove only three. If you are asked to prove that two triangles are congruent, you must show that the conditions of one of these two postulates are met. In later sections you will see other ways of proving the congruence of two triangles.

Since $\triangle LMN$ and $\triangle LON$ are congruent, the definition of congruent triangles guarantees that the other three pairs are congruent, namely $\overline{MN} \cong \overline{ON}$; $\angle MNL \cong \angle ONL$; $\angle LMN \cong \angle LON$.

EXAMPLE 2   *Given:* $\overline{LN}$ bisects $\angle MLO$ and $\angle MNO$ (figure 6.23)
                 Prove: $\angle M \cong \angle O$

Answer   Notice that $\angle M$ and $\angle O$ are parts of triangles. This means you can prove the triangles congruent and then use the definition.

STATEMENTS	REASONS
1. $\overline{LN}$ bisects $\angle MLO$ and $\angle MNO$	1. Given
2. $\angle NLM \cong \angle NLO$; $\angle LNM \cong \angle LNO$	2. Definition of angle bisector
3. $\overline{LN} \cong \overline{LN}$	3. Reflexive property of congruent segments
4. $\triangle MLN \cong \triangle OLN$	4. ASA
5. $\angle M \cong \angle O$	5. Definition of congruent triangles

▶ A. Exercises

Look at the markings on the triangles in each exercise and state which postulate you would use to prove that the two triangles are congruent. If neither of the postulates would apply as indicated by the markings, write *neither*.

1.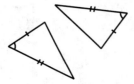

5.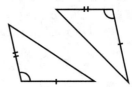

2.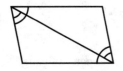

6.

3.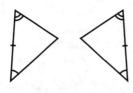

7.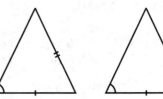

4.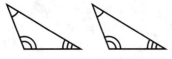

8.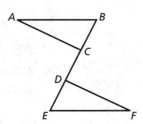

9. *Given:* $\overline{AC} \cong \overline{DF}$; $\overline{BC} \cong \overline{DE}$; $\angle EDF \cong \angle BCA$
   Prove: $\angle A \cong \angle F$

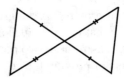

10. *Given:* $\overleftrightarrow{AD} \parallel \overleftrightarrow{BC}$; $\overleftrightarrow{AB} \parallel \overleftrightarrow{DC}$

    Prove: $\triangle ABD \cong \triangle CDB$

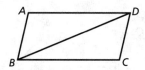

Prove the following statements.

11. *Given:* $\triangle XYZ$ and $\triangle LMN$ are equiangular; $\overline{LN} \cong \overline{YZ}$

    Prove: $\triangle XYZ \cong \triangle MNL$

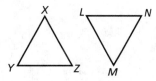

12. *Given:* V is the midpoint of $\overline{US}$; $\angle VUT \cong \angle VSK$; $\overline{TU} \cong \overline{KS}$

    Prove: $\triangle TUV \cong \triangle KSV$

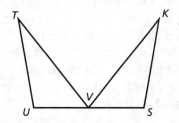

▶ B. Exercises

Use the diagram for the next two proofs.

13. *Given:* $\overleftrightarrow{AB} \parallel \overleftrightarrow{ED}$; $\overline{AC} \cong \overline{CD}$

    Prove: $\triangle CAB \cong \triangle CDE$

14. *Given:* C is the midpoint of $\overline{AD}$ and $\overline{BE}$

    Prove: $\triangle ACB \cong \triangle DCE$

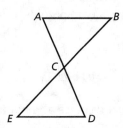

15. *Given:* $\overleftrightarrow{AB} \parallel \overleftrightarrow{EF}$; $\overline{AB} \cong \overline{EF}$; $\overline{BC} \cong \overline{DE}$

    Prove: $\overleftrightarrow{AC} \parallel \overleftrightarrow{DF}$

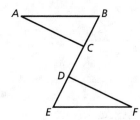

Use the diagram below for exercises 16-18.

16. *Prove:* If $\triangle ABD \cong \triangle DEA$, then $\angle CBD \cong \angle FEA$

17. *Given:* $\triangle ABD \cong \triangle DEA$; $\overline{EF} \cong \overline{BC}$

    Prove: $\triangle BCD \cong \triangle EFA$

18. *Given:* $\triangle ABD \cong \triangle DEA$; $\triangle ACD \cong \triangle DFA$

    Prove: $\triangle AFE \cong \triangle DCB$

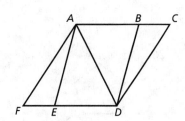

▶ C. Exercises

19. *Given:* $\vec{RT} \perp \vec{SQ}$;
   $\angle SRT \cong \angle STR$

   Prove: $\angle RQS \cong \angle TQS$

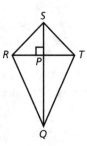

▶ Dominion Thru Math

A steel microwave tower is a square pyramid with the top cut off. It has four sloping corner members connected on all sides by isosceles trapezoids, including diagonals. The horizontal members of the trapezoids are 18 inches apart and start at the bottom as 8 feet. Each member above is one inch shorter, so the sides converge.

20. The tower has a 126-foot slant height. How wide is it at the top and how high does it reach?

21. Find the length of a diagonal brace for a trapezoid of base *b*.

22. Find the length for the bottom and top diagonal braces.

23. Identify any trapezoids or triangles that are congruent.

■ Cumulative Review

Express each distance as a number or algebraic expression.

24. Perimeter

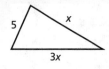

25. *AB*

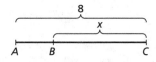

26. Perimeter

27. Circumference

28. Total length

29. *AC*

6.7 Congruent Triangles

Theorems about triangles and congruence get easier with practice. Only by perseverance will you become proficient at proving theorems. In Philippians 3:14 Paul gives Christians a principle for life: "I press toward the mark for the prize of the high calling of God in Christ Jesus." This verse teaches that we are to continue in the work that God has given us to do for Him. God has told every Christian to be a good testimony and to witness to others around him. As a Christian you must strive to finish every task that is set before you.

The metal parts of standard sizes used to make cranes guarantee that all the triangles are isosceles.

The next theorem provides another way to prove triangles congruent that will make some proofs easier. SAA abbreviates Side-Angle-Angle.

Theorem 6.19
***SAA Congruence Theorem.* If two angles of a triangle and a side opposite one of the two angles are congruent to the corresponding angles and side of another triangle, then the two triangles are congruent.**

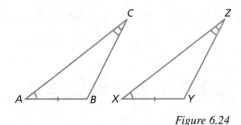

Figure 6.24

By the markings on the triangles in figure 6.24 and by the SAA Theorem, you can conclude that $\triangle ABC \cong \triangle XYZ$. Let us see whether we can prove it.

Given: $\angle A \cong \angle X$; $\angle C \cong \angle Z$; $\overline{AB} \cong \overline{XY}$
Prove: $\triangle ABC \cong \triangle XYZ$

STATEMENTS	REASONS
1. $\angle A \cong \angle X$; $\angle C \cong \angle Z$; $\overline{AB} \cong \overline{XY}$	1. Given
2. $\angle B \cong \angle Y$	2. Third angles are congruent
3. $\triangle ABC \cong \triangle XYZ$	3. ASA
4. If $\angle A \cong \angle X$, $\angle C \cong \angle Z$, and $\overline{AB} \cong \overline{XY}$, then $\triangle ABC \cong \triangle XYZ$	4. Law of Deduction

Study these sample proofs.

EXAMPLE 1   *Given:* M is the midpoint of $\overline{LN}$; $\overrightarrow{OM} \perp \overleftrightarrow{LN}$
Prove: $\triangle LMO \cong \triangle NMO$

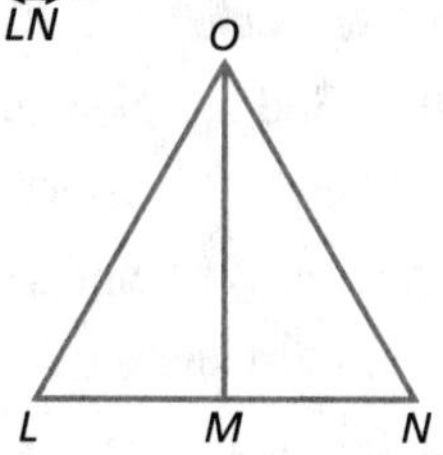

Answer

Figure 6.25

STATEMENTS	REASONS
1. M is the midpoint of $\overline{LN}$; $\overrightarrow{OM} \perp \overleftrightarrow{LN}$	1. Given
2. $LM = NM$	2. Definition of midpoint
3. $\overline{LM} \cong \overline{NM}$	3. Definition of congruent segments
4. $\overline{OM} \cong \overline{OM}$	4. Reflexive property of congruent segments
5. $\angle LMO$ and $\angle NMO$ are right angles	5. Definition of perpendicular
6. $\angle LMO \cong \angle NMO$	6. All right angles are congruent
7. $\triangle LMO \cong \triangle NMO$	7. SAS

You will often be required to prove a theorem that involves overlapping triangles. In such cases it may help you to draw separate diagrams for the triangles that you are proving congruent, mark them appropriately, and then complete the proof.

EXAMPLE 2   *Given:* $\angle A \cong \angle D$;
$\angle ABC \cong \angle DCB$

Prove: $\overline{AC} \cong \overline{DB}$

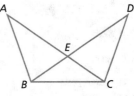

Figure 6.26

Answer   Separate the triangles and mark them according to the given information. Notice that $\overline{BC}$ has been drawn twice and can be marked the same.

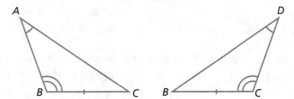

STATEMENTS	REASONS
1. $\angle A \cong \angle D$; $\angle ABC \cong \angle DCB$	1. Given
2. $\overline{BC} \cong \overline{BC}$	2. Reflexive property of congruent segments
3. $\triangle ABC \cong \triangle DCB$	3. SAA
4. $\overline{AC} \cong \overline{BD}$	4. Definition of congruent triangles

There are special kinds of triangles that have some interesting properties. You have already seen that each angle of an equiangular triangle measures 60° (see exercise 23, Section 6.5). Isosceles triangles are also interesting.

Theorem 6.20

Isosceles Triangle Theorem. **In an isosceles triangle the two base angles are congruent.**

Given: Isosceles $\triangle XYZ$; $\overline{XY} \cong \overline{XZ}$

Draw auxiliary $\overrightarrow{XW}$ to bisect $\angle YXZ$.

Prove: $\angle Y \cong \angle Z$

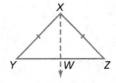

Figure 6.27

STATEMENTS	REASONS
1. $\overline{XY} \cong \overline{XZ}$	1. Given
2. $\overrightarrow{XW}$ bisects $\angle X$	2. Auxiliary ray
3. $\angle YXW \cong \angle ZXW$	3. Definition of angle bisector
4. $\overline{XW} \cong \overline{XW}$	4. Reflexive property of congruent segments
5. $\triangle XYW \cong \triangle XZW$	5. SAS
6. $\angle Y \cong \angle Z$	6. Definition of congruent triangles

The first theorem below is the converse of the Isosceles Triangle Theorem. You will prove both of the following theorems in the exercises.

Theorem 6.21
If two angles of a triangle are congruent, then the sides opposite those angles are congruent, and the triangle is an isosceles triangle.

Theorem 6.22
A triangle is equilateral if and only if it is equiangular.

▶ A. Exercises

Find the measure of each angle.

1.

2.

3.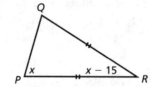

4.

Prove each statement.

5. *Given:* ∠CAB ≅ ∠CBA; ∠ADB ≅ ∠BEA
   Prove: ∠EAB ≅ ∠DBA

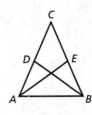

6. *Given:* $\overline{WY}$ bisects ∠XWZ and ∠ZYX
   Prove: △XWY ≅ △ZWY

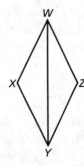

7. Prove Theorem 6.21.
8. Prove that equilateral triangles are equiangular. (part one of Theorem 6.22)
9. Prove that equiangular triangles are equilateral. (part two of Theorem 6.22)
10. Prove that equilateral triangles have three angles of sixty degrees each.

▶ B. Exercises

Use the illustration below for exercises 11-15.

11. *Given:* ∠1 ≅ ∠4; ∠2 ≅ ∠3
   Prove: △MPN ≅ △NOM
12. *Given:* Q is the midpoint of $\overline{PN}$ and $\overline{MO}$
   Prove: △PQM ≅ △OQN
13. *Given:* △MPN ≅ △NOM
   Prove: ∠1 ≅ ∠4
14. *Given:* ∠PMN ≅ ∠ONM; ∠MPN ≅ ∠NOM
   Prove: △LMO ≅ △LNP
15. *Given:* $\overline{LM}$ ≅ $\overline{LN}$ with midpoints P and O respectively
   Prove: △PQM ≅ △OQN

Prove the following statements.

16. *Given:* ∠DCE ≅ ∠B
   Prove: △ABC is an isosceles triangle

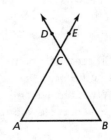

17. *Given:* ∠PQT ≅ ∠SRT
Prove: △TQR is isosceles

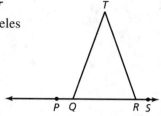

18. *Given:* UX ≅ UW;
∠VXW ≅ ∠YWX
Prove: YU ≅ VU

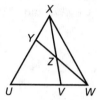

19. *Given:* BD ≅ CD; ∠1 ≅ ∠4
Prove: △ABC is an
isosceles triangle

20. *Given:* FE ≅ GE; IE ≅ HE
Prove: IH ∥ FG

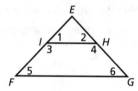

▶ C. Exercises

21. Explain in a paragraph why
△XYZ is isosceles if △PQR is
isosceles and X, Y, and Z are
midpoints.

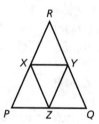

■ Cumulative Review

22. Review the proof of Theorem 6.20. What rule of logic enables us to
apply the reflexive property in step 4?

23. Name at least three of the equivalence relations you have studied so far.

24. *Given:* AD ∥ BC; ∠1 ≅ ∠3
Prove: ∠ABC ≅ ∠ACB

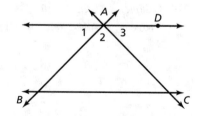

Let *P* represent a point and *l* and *m* lines.
∀ P ∧ ∀ l ∃ m so that m ∥ l ∧ P ∈ m

25. Write the symbolic statement as a
sentence. Do you recognize it?

6.8 Conditions for Congruent Triangles

Form a triangle using three straws of varying lengths. Next obtain three more straws identical in length to the first three. No matter how you form a triangle from the second set, it will be congruent to the first triangle. Since each straw represents a side of the triangle, this congruence illustrates the next theorem, SSS (Side-Side-Side).

Theorem 6.23

SSS Congruence Theorem. If each side of one triangle is congruent to the corresponding side of a second triangle, then the two triangles are congruent.

Given: $\overline{AB} \cong \overline{PQ}$; $\overline{BC} \cong \overline{QR}$; $\overline{AC} \cong \overline{PR}$

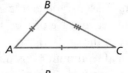
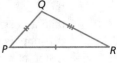

Draw: $\overrightarrow{AE}$ so that $\angle QPR \cong \angle CAE$; mark D on $\overrightarrow{AE}$ so that $\overline{AD} \cong \overline{AB}$; draw $\overline{CD}$

Prove: $\triangle ABC \cong \triangle PQR$

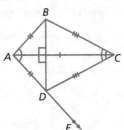
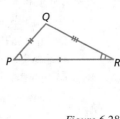

Figure 6.28

STATEMENTS	REASONS
1. $\overline{AB} \cong \overline{PQ}$, $\overline{BC} \cong \overline{QR}$, $\overline{AC} \cong \overline{PR}$	1. Given
2. Draw $\overrightarrow{AE}$ into the opposite half-plane (determined by $\overleftrightarrow{AC}$) from point B so that $\angle QPR \cong \angle CAE$	2. Continuity Postulate
3. Point D exists on $\overrightarrow{AE}$ so that $\overline{AD} \cong \overline{PQ}$	3. Completeness Postulate
4. $\overline{AD} \cong \overline{AB}$	4. Transitive property of congruent segments
5. $\triangle ADC \cong \triangle PQR$	5. SAS
6. $\overline{DC} \cong \overline{QR}$	6. Definition of congruent triangles
7. $\overline{DC} \cong \overline{BC}$	7. Transitive property of congruent segments

Continued ▶

8. Draw $\overleftrightarrow{BD}$	**8.** Line Postulate
9. △ABD and △BCD are isosceles	**9.** Definition of isosceles triangle (steps 4 and 7)
10. ∠ABD ≅ ∠ADB, ∠CBD ≅ ∠CDB	**10.** Isosceles Triangle Theorem
11. ∠ABC ≅ ∠ADC	**11.** Angle Addition Congruence Theorem
12. △ABC ≅ △ADC	**12.** SAS
13. △ABC ≅ △PQR	**13.** Transitive property of congruent triangles
14. If each side of one triangle is congruent to a corresponding side of a second triangle, then the two triangles are congruent	**14.** Law of Deduction

This is a difficult proof because of the auxiliary line in steps 2 and 3 and another one in step 8. You will not have to do a proof this complicated, but there are several lessons you can learn from it. First, SSS is a theorem rather than a postulate because we proved it. Second, placing auxiliary lines in useful spots is part of the art in discovering proofs. Third, you have not used the Completeness and Continuity Postulates very much, but that is because their main value is in justifying complicated auxiliary constructions like this one.

The following chart summarizes the ways to prove the congruence of two triangles. Use only these four methods. You cannot use SSA or AAA, and you will learn why in the exercises.

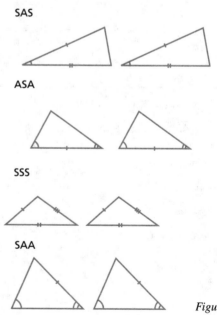

Figure 6.29

What theorem proves that the isosceles triangles in the boom of this crane are congruent?

EXAMPLE 1 *Given:* $\overleftrightarrow{ZX}$ bisects $\angle WZY$
and $\angle WXY$

Prove: $\overline{XY} \cong \overline{XW}$

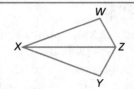

Figure 6.30

STATEMENTS	REASONS
1. $\overleftrightarrow{ZX}$ bisects $\angle WZY$ and $\angle WXY$	**1.** Given
2. $\angle WXZ \cong \angle YXZ$; $\angle WZX \cong \angle YZX$	**2.** Definition of angle bisector
3. $\overline{XZ} \cong \overline{XZ}$	**3.** Reflexive property of congruent segments
4. $\triangle XZW \cong \triangle XZY$	**4.** ASA
5. $\overline{XY} \cong \overline{XW}$	**5.** Definition of congruent triangles

EXAMPLE 2 *Given:* $\overline{ED} \cong \overline{BC}$, $\overline{EC} \cong \overline{BD}$
Prove: $\overline{AE} \cong \overline{AB}$

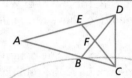

Figure 6.31

STATEMENTS	REASONS
1. $\overline{ED} \cong \overline{BC}$; $\overline{BD} \cong \overline{EC}$	**1.** Given
2. $\overline{DC} \cong \overline{DC}$	**2.** Reflexive property of congruent segments
3. $\triangle BCD \cong \triangle EDC$	**3.** SSS
4. $\angle EDC \cong \angle BCD$; $\angle BDC \cong \angle ECD$	**4.** Definition of congruent triangles
5. $\angle EDF \cong \angle BCF$	**5.** Adjacent Angle Portion Theorem
6. $\angle A \cong \angle A$	**6.** Reflexive property of congruent angles
7. $\triangle ABD \cong \triangle AEC$	**7.** SAA
8. $\overline{AE} \cong \overline{AB}$	**8.** Definition of congruent triangles

► A. Exercises

If △ABC ≅ △XYZ, find the following measures.

1. $m\angle B = 75$, find $m\angle Y$
2. $AC = 6$ inches, find XZ
3. $m\angle C = 60$, $m\angle A = 80$, find $m\angle Y$
4. $YZ = 18$ meters, find BC
5. How do you know that the answers you gave above are true?

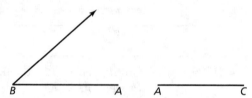

Consider the following six possible conditions for exercises 6-10.

SSS, SSA, SAS, SAA, ASA, and AAA

6. Which of the conditions can be used to prove two triangles congruent?
7. Sketch a triangle and then a larger triangle with angles congruent to the first triangle. Which condition above does this disprove?
8. Given $\overline{BA}$ and an angle, position a segment of length AC in two different ways to form △ABC. Which condition above does this disprove?

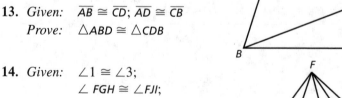

9. The AAS triangle congruence criterion is a synonym for one of the other criteria. Give the familiar name for AAS.
10. Whenever two angles and a side of one triangle are congruent to the corresponding angles and side of another triangle, the triangles are congruent. Why?

Prove the statements below.

11. *Given:* $\angle M \cong \angle P$; $\angle MNO \cong \angle PON$
    Prove: △MNO ≅ △PON

12. *Given:* $\angle I \cong \angle L$; $\overline{HJ} \cong \overline{KJ}$
    Prove: △HJI ≅ △KJL

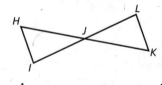

► B. Exercises

13. *Given:* $\overline{AB} \cong \overline{CD}$; $\overline{AD} \cong \overline{CB}$
    Prove: △ABD ≅ △CDB

14. *Given:* $\angle 1 \cong \angle 3$;
    $\angle FGH \cong \angle FJI$;
    $\overline{GH} \cong \overline{IJ}$
    Prove: △FGH ≅ △FJI

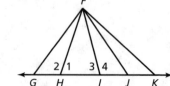

Use the diagram for exercises 15-16.

15. *Given:* $\overline{DC} \cong \overline{BA}$; $\angle BAC \cong \angle DCA$
   Prove: $\overleftrightarrow{AD} \parallel \overleftrightarrow{BC}$

16. *Given:* $\angle CAD \cong \angle ACB$; $\overline{AD} \cong \overline{CB}$
   Prove: $\angle D \cong \angle B$

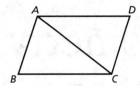

17. *Given:* $\angle 3 \cong \angle 4$; $\overline{HK} \cong \overline{JK}$
   Prove: $\overleftrightarrow{IK}$ bisects $\angle HIJ$

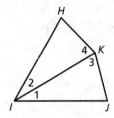

18. *Given:* $\overline{RS} \cong \overline{TS}$; $\overline{RU} \cong \overline{TU}$
   Prove: $m\angle RSU = 90°$

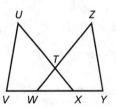

19. *Given:* $\overline{VW} \cong \overline{XY}$; $\overline{UV} \cong \overline{ZY}$; $\overline{UX} \cong \overline{ZW}$
   Prove: $\angle ZWV \cong \angle UXY$

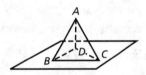

▶ C. Exercises

20. *Given:* $\overline{AC} \cong \overline{AB}$; $\overline{DB} \cong \overline{DC}$
   Prove: $\triangle ABD \cong \triangle ACD$

■ Cumulative Review

Match each statement to its type.

21. Two parallel lines are not concurrent.

22. A triangle is equilateral if and only if it is equiangular.

23. A square is a rectangle and a square is a rhombus.

24. If two lines are parallel, then corresponding angles are congruent.

25. $a < b$, $a = b$, or $a > b$ (where $a, b, \in \mathbb{R}$)

A. Conditional

B. Biconditional

C. Conjunction

D. Disjunction

E. Negation

Geometry and Scripture

Reasoning

After studying this chapter you should see why connectives, quantifiers, and valid arguments are so important in mathematics. Is this type of reasoning important to God?

In Chapter 5 you found connectives and arguments in the Bible. Now match each verse below to the quantifier that symbolizes it.

1. Matthew 7:21	**A.** ∀q
2. Isaiah 53:6a	**B.** ∃ r
3. I John 4:12a	**C.** ~∀ s
4. Proverbs 14:12a	**D.** ~∃ t

This shows us that a proper understanding of logic can help us understand Bible verses. There are many other proofs in the Bible for you to understand besides the ones you studied in Chapter 5. Here are three arguments. Classify each type and decide who was speaking.

5. Acts 11:15-17

6. Matthew 22:31-32

7. I Corinthians 6:15-18

Besides understanding the arguments in the Bible, reasoning enables us to follow the examples of Jesus and the apostles as you saw in the questions above. In fact, the Bible emphasizes repeatedly that Paul presented the truth in a reasonable way.

8. How many verses in Acts 17-19 include references to Paul arguing or reasoning when he presented the gospel? List the verses.

9. In which of them does it say that Paul *habitually* presented the gospel by reasoning from the Scripture?

We need to reason well to understand the arguments of the Bible, to follow the examples of the men of the Bible, and especially to obey the following commands of the Bible. Read the passages below and identify the commands of God. Notice how you will have to use good reasoning to obey them.

> Colossians 2:8-9
>
> Romans 16:17-18
>
> I Peter 3:15

10. Two of the commands above warn us not to be deceived by wrong reasoning. Which one tells us that we, like Paul, should also have reasonable answers for people when we share the gospel?

Finally, you studied congruence in this chapter. Congruent means having the same size and shape. Solomon beautified the temple with congruent figures.

HIGHER PLANE: In what verse do we read that Jesus is the "express image" of God?

Identify the congruent figures in each passage and give the verse that emphasizes the congruence of the figures.

11. I Kings 7:27-37

12. I Kings 6:23-30

Since congruent figures have identical characteristics, they are exact images of one another.

Line upon Line

AND PAUL, as his manner was, went in unto them, and three sabbath days reasoned with them out of the scriptures. ❧

ACTS 17:2

Chapter 6 Review

1. Draw a diagram and mark the corresponding parts if $\triangle HGI \cong \triangle LQP$.

2. Name the corresponding parts of the congruent triangles.

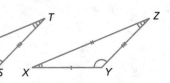

Give the measures of the indicated angles. Assume that $a \parallel b$.

3. $m\angle 3 = 115$, find $m\angle 6$
4. $m\angle 3 = 115$, find $m\angle 7$
5. $m\angle 3 = 115$, find $m\angle 2$
6. $m\angle 8 = 62$, find $m\angle 1$
7. $m\angle 6 = 120$, find $m\angle 5$
8. $m\angle 2 = 130$, find $m\angle 5$

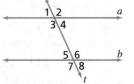

Use the following diagram for exercises 9-10.

9. *Given:* $l \parallel n; p \parallel q$
   Prove: $\angle 12 \cong \angle 5$

10. *Given:* $p \parallel q; \angle 2 \cong \angle 14$
    Prove: $l \parallel n$

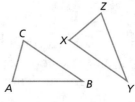

Prove the following statements.

11. *Given:* $\angle ABC \cong \angle XYZ$;
    $\angle ACB \cong \angle XZY; \overline{BC} \cong \overline{YZ}$
    Prove: $\triangle ABC \cong \triangle XYZ$

12. *Given:* T is the midpoint of $\overline{RV}$ and $\overline{US}$
    Prove: $\triangle RST \cong \triangle VUT$

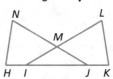

13. *Given:* $\overline{HN} \cong \overline{KL}; \angle H \cong \angle K; \overline{HI} \cong \overline{JK}$
    Prove: $\triangle HNJ \cong \triangle KLI$

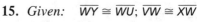

14. *Given:* $\odot P; \angle APB \cong \angle CPB$
    Prove: $\overline{AB} \cong \overline{CB}$

15. *Given:* $\overline{WY} \cong \overline{WU}; \overline{VW} \cong \overline{XW}$
    Prove: $\angle U \cong \angle Y$

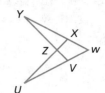

Give the measures of the indicated angles.

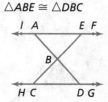

16. $m\angle 4 = 45$, $m\angle 11 = 68$, find $m\angle 5$
17. $m\angle 2 = 55$, find $m\angle 4$
18. $m\angle 4 = 70$, find $m\angle 3$
19. $m\angle 4 = 65$, $m\angle 11 = 55$, find $m\angle 6$
20. $m\angle 3 = 135$, $m\angle 11 = 87$, find $m\angle 5$
21. $m\angle A$ in a regular decagon
22. If *ABCDEFG* is a convex heptagon, find the sum of the angles.

23. *Given:* ⊙P
    Prove: $\angle PAC \cong \angle PCA$

24. *Given:* *B* is the midpoint of $\overline{AD}$;
    $\angle HCE \cong \angle FEC$
    Prove: $\triangle ABE \cong \triangle DBC$

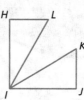

25. *Given:* $\overline{LI} \cong \overline{KI}$; $\angle HLI$ is complementary to $\angle LIH$;
    $\angle JKI$ is complementary to $\angle KIJ$;
    $\angle LIH \cong \angle KIJ$
    Prove: $\triangle LIH \cong \triangle KIJ$

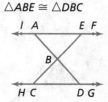

26. *Given:* $\angle BDC \cong \angle ECD$; $\angle BCD \cong \angle EDC$
    Prove: $\triangle BCD \cong \triangle EDC$

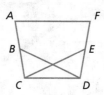

Use the diagram for exercises 27-28.

27. *Given:* $\angle VUP \cong \angle WUQ$; $\angle PUX \cong \angle QUX$;
    $\overline{VU} \cong \overline{WU}$
    Prove: $\triangle UVX \cong \triangle UWX$

28. *Given:* $\angle VUX \cong \angle WUX$; $\overrightarrow{UP}$ bisects $\angle VUX$;
    $\overrightarrow{UQ}$ bisects $\angle WUX$; $\angle QXU \cong \angle PXU$
    Prove: $\triangle PUX \cong \triangle QUX$

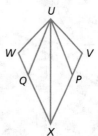

29. *Given:* *ABCDEF* is a regular hexagon and *P* is
    equidistant from the vertices
    Prove: All six triangles are congruent

30. Explain the mathematical significance of Acts 17:2.

7 Triangles and Quadrilaterals

The Dutch artist M. C. Escher, who drew this fascinating picture, was a lover of geometric patterns. He based many of his works on tessellations, geometric patterns that cover the plane so that no vacant spots remain. In the tessellation shown, he drew a grid of parallelograms and then embellished it with fish and birds. In other tessellations, he used squares for tadpoles, equilateral triangles for birds, and hexagons for lizards.

Escher also used other mathematical concepts. In some of his works, he included three-dimensional figures, such as the one-sided surface on page 70. He also delighted in impossible constructions and logical paradoxes (p. 199) to challenge the viewer's reasoning.

The Bible speaks of patterns too. Christians are to pattern their lives after the perfect attitudes and actions of Christ. However, the patterning of Christian lives after the life of Christ requires persistence in a spiritual struggle. So do not be discouraged by failures and flaws; the God who has begun a good work in you will continue it until the day of Jesus Christ. Only then will the pattern be complete.

After this chapter you should be able to

1. prove and apply right triangle congruence theorems.
2. find special points of concurrency for triangles.
3. prove and apply inequalities related to triangles.
4. prove and apply theorems about parallelograms.
5. construct congruent triangles, polygons, and parallel lines.
6. justify previous constructions.

7.1 Right Triangles

You can use the triangle congruence theorems from the last chapter to develop theorems specifically for right triangles. Let's briefly review the terminology needed to study these right triangles.

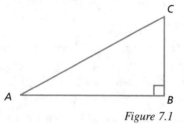

Figure 7.1

Why are the right triangle windows isosceles?

In right △*ABC*, ∠*B* is the right angle. The side opposite the right angle, $\overline{AC}$, is the *hypotenuse*. The other sides of the triangle, $\overline{AB}$ and $\overline{BC}$, are the *legs* of the right triangle.

The most important congruence theorem for right triangles is HL (Hypotenuse-Leg). However, its proof requires auxiliary lines. Once proved, it will provide an important way to prove two right triangles congruent.

Theorem 7.1

HL Congruence Theorem. If the hypotenuse and a leg of one right triangle are congruent to the hypotenuse and corresponding leg of another right triangle, then the two triangles are congruent.

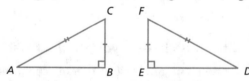

Figure 7.2

Given: $\overline{AC} \cong \overline{DF}$; $\overline{BC} \cong \overline{EF}$; ∠*B* and ∠*E* are right angles
Prove: △*ABC* ≅ △*DEF*

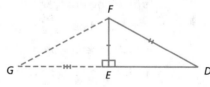

Figure 7.3

STATEMENTS	REASONS
1. $\overline{AC} \cong \overline{DF}$; $\overline{BC} \cong \overline{EF}$; $\angle B$ and $\angle E$ are right angles	1. Given
2. There is a point, G, on the ray opposite $\overrightarrow{ED}$ such that $\overline{EG} \cong \overline{AB}$; draw $\overline{FG}$	2. Auxiliary lines
3. $\angle FEG$ and $\angle FED$ form a linear pair	3. Definition of linear pair
4. $\angle FEG$ is a right angle	4. If one of the angles in a linear pair is a right angle, the other is also a right angle
5. $\angle B \cong \angle FEG$	5. All right angles are congruent
6. $\triangle ABC \cong \triangle GEF$	6. SAS
7. $\overline{AC} \cong \overline{FG}$	7. Definition of congruent triangles
8. $\overline{DF} \cong \overline{FG}$	8. Transitive property of congruent segments
9. $\triangle GDF$ is isosceles	9. Definition of isosceles triangle
10. $\angle FGE \cong \angle FDE$	10. Isosceles Triangle Theorem
11. $\triangle GEF \cong \triangle DEF$	11. SAA
12. $\triangle ABC \cong \triangle DEF$	12. Transitive property of congruent triangles

The other ways to prove right triangles congruent are LL (Leg-Leg), LA (Leg-Angle), and HA (Hypotenuse-Angle). These are easier to prove because they are special cases of theorems in the previous chapter. The only special feature is that the triangles are right, so the right angles are congruent. Study the proof of LL so that you will be able to do the others. HA is exercise 10. LA has two cases (exercises 11-12), depending on whether the leg is opposite or adjacent to the angle.

Theorem 7.2

LL Congruence Theorem. If the two legs of one right triangle are congruent to the two legs of another right triangle, then the two triangles are congruent.

Given: $\triangle HIJ$ and $\triangle MLN$ are right triangles; $\overline{HI} \cong \overline{ML}$; $\overline{JI} \cong \overline{NL}$

Prove: $\triangle HIJ \cong \triangle MLN$

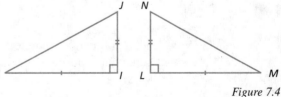

Figure 7.4

STATEMENTS	REASONS
1. $\triangle HIJ$ and $\triangle MLN$ are right triangles; $\overline{HI} \cong \overline{ML}$; $\overline{JI} \cong \overline{NL}$	1. Given
2. $\angle I$ and $\angle L$ are right angles	2. Definition of right triangle
3. $\angle I \cong \angle L$	3. All right angles are congruent
4. $\triangle HIJ \cong \triangle MLN$	4. SAS
5. If two legs of one right triangle are congruent to the two legs of another right triangle, then the triangles are congruent	5. Law of Deduction

Theorem 7.3

HA Congruence Theorem. **If the hypotenuse and an acute angle of one right triangle are congruent to the hypotenuse and corresponding acute angle of another right triangle, then the two triangles are congruent.**

Theorem 7.4

LA Congruence Theorem. **If a leg and one of the acute angles of a right triangle are congruent to the corresponding leg and acute angle of another right triangle, then the two triangles are congruent.**

▶ A. Exercises

1. State the four methods of proving that two right triangles are congruent. Draw a picture showing each case and mark the triangles accordingly.

Identify any triangle congruence theorem or postulate corresponding to each right triangle congruence theorem below. (Remember that the right angles are congruent.)

2. LL
3. LA, adjacent case
4. LA, opposite case
5. HL
6. HA
7. Which theorem for right triangle congruence does not correspond to another congruence postulate or theorem?

8. According to exercise 7, which condition is valid for right triangles that is not otherwise valid?

9. Use the diagram to state a triangle congruence. Which right triangle theorem justifies your statement?

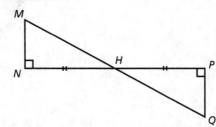

10. Prove HA.

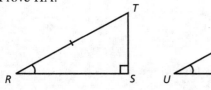

▶ B. Exercises

Use the same diagram as in exercise 10 for the proofs in exercises 11-12 (the two cases of the LA Congruence Theorem).

11. LA (opposite case)
    Given: △RST and △UVW are right triangles; $\overline{RS} \cong \overline{UV}$; $\angle T \cong \angle W$
    Prove: △RST ≅ △UVW

12. LA (adjacent case)
    Given: △RST and △UVW are right triangles; $\overline{RS} \cong \overline{UV}$; $\angle R \cong \angle U$
    Prove: △RST ≅ △UVW

Use the following diagram to prove exercise 13.

13. *Given:* $\angle P$ and $\angle Q$ are right angles;
    $\overline{PR} \cong \overline{QR}$
    Prove: $\overline{PT} \cong \overline{QT}$

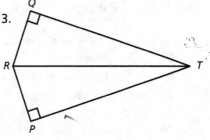

14. Use the diagram for this proof.
    Given: △EBC and △ADC are right triangles;
    $\overline{EB} \cong \overline{AD}$
    Prove: $\overline{DC} \cong \overline{BC}$

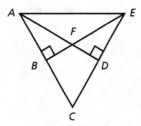

Use the following diagram to prove exercises 15-19.

15. *Given:* $\overline{WY} \perp \overline{XZ}$; $\angle X \cong \angle Z$
    Prove: $\triangle XYW \cong \triangle ZYW$

16. *Given:* $\angle XYW$ and $\angle ZYW$ are right angles;
    $\overline{XW} \cong \overline{ZW}$
    Prove: $\triangle XYW \cong \triangle ZYW$

17. *Given:* $\angle XYW \cong \angle ZYW$; $\overline{XY} \cong \overline{ZY}$
    Prove: $\triangle XYW \cong \triangle ZYW$

18. *Given:* $\triangle XYW$ and $\triangle ZYW$ are right triangles;
    $\angle XWY \cong \angle ZWY$
    Prove: Y is the midpoint of $\overline{XZ}$

19. *Given:* $\triangle XYW$ and $\triangle ZYW$ are right triangles; $\overline{XY} \cong \overline{ZY}$
    Prove: $\overline{XW} \cong \overline{ZW}$

▶ C. Exercises

Use the diagram above to prove the following.

20. *Given:* $\overleftrightarrow{WY}$ bisects $\angle XWZ$; $\overleftrightarrow{WY} \perp \overleftrightarrow{XZ}$
    Prove: $\triangle XZW$ is an isosceles triangle

▶ Dominion Thru Math

21. In miniature golf, the tee and the hole are the same distance from a wall. If you imagine the hole as if reflected in a mirror on the wall and aim at the image, the ball will bounce off the mirror into the hole. Which right triangle congruence theorem assures success by demonstrating congruent triangles?

■ Cumulative Review

Give the measure of the angle(s) formed by

22. two opposite rays.
23. perpendicular lines.
24. an equiangular triangle.
25. the bisector of a right angle.
26. Which symbol does not represent a set?
    $\angle ABC$, $\triangle ABC$, A-B-C, $\{A, B, C\}$

7.2 Perpendiculars and Bisectors

Besides right triangles, perpendicular lines are also used in several other ways. This section will present some important properties of perpendicular lines.

Each stained glass window of this Paris cathedral includes a bisector.

Theorem 7.5

Any point lies on the perpendicular bisector of a segment if and only if it is equidistant from the two endpoints.

Part One (*only if*)

Given: l is the perpendicular bisector of $\overline{AB}$ at D; C is a point on l

Draw: $\overline{AC}$ and $\overline{BC}$

Prove: $AC = BC$

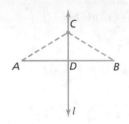

Figure 7.5

STATEMENTS	REASONS
1. $\overline{AB} \perp l$ at D; C is a point on l	1. Given
2. Draw $\overline{AC}$ and $\overline{CB}$	2. Auxiliary lines (Line Postulate)
3. $\angle CDA$ and $\angle CDB$ are right angles	3. Definition of perpendicular
4. $\triangle CDA$ and $\triangle CDB$ are right triangles	4. Definition of right triangle
5. $AD = DB$	5. Definition of segment bisector
6. $\overline{AD} \cong \overline{DB}$	6. Definition of congruent segments
7. $\overline{CD} \cong \overline{CD}$	7. Reflexive property of congruent segments

Continued ▶

8. $\triangle CDA \cong \triangle CDB$	8. LL (or SAS)
9. $\overline{AC} \cong \overline{BC}$	9. Definition of congruent triangles
10. $AC = BC$	10. Definition of congruent segments
11. If a point lies on the perpendicular bisector of a segment, then it is equidistant from the endpoints	11. Law of Deduction

Now study the proof of the other half of the biconditional (the *if* part).

Given: $\overline{AB}$ and point P so that $AP = BP$
Prove: P lies on the perpendicular bisector of $\overline{AB}$

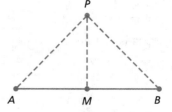

Figure 7.6

STATEMENTS	REASONS
1. $AP = BP$	1. Given
2. The midpoint M of $\overline{AB}$ exists	2. Completeness Postulate
3. $AM = MB$	3. Definition of midpoint
4. $\overline{AP} \cong \overline{BP}$; $\overline{AM} \cong \overline{BM}$	4. Definition of congruent segments
5. $\overline{MP} \cong \overline{MP}$	5. Reflexive property of congruent segments
6. $\triangle APM \cong \triangle BPM$	6. SSS
7. $\angle AMP \cong \angle BMP$	7. Definition of congruent triangles
8. $\angle AMP$ and $\angle BMP$ are right angles	8. Congruent angles that form a linear pair are right angles
9. $\overleftrightarrow{MP} \perp \overleftrightarrow{AB}$	9. Definition of perpendicular
10. $\overleftrightarrow{MP}$ bisects $\overline{AB}$	10. Definition of segment bisector (contains midpoint)
11. $\overleftrightarrow{MP}$ is the perpendicular bisector of $\overline{AB}$	11. Definition of perpendicular bisector
12. If P is equidistant from the endpoints of $\overline{AB}$, then it is on the perpendicular bisector of $\overline{AB}$	12. Law of Deduction

What happens if you draw the perpendicular bisectors of each side of a triangle? You should find that they are concurrent and that the distance from the point of concurrency to each vertex of the triangle is the same. Draw a circle through the vertices with the point of concurrency as the center. Because this circle circumscribes the triangle, the point of concurrency is called the *circumcenter*.

Theorem 7.6

Circumcenter Theorem. **The perpendicular bisectors of the sides of any triangle are concurrent at the circumcenter, which is equidistant from each vertex of the triangle.**

Given: △XYZ, where line *a* is the perpendicular bisector of $\overline{ZY}$, line *b* is the perpendicular bisector of $\overline{XY}$, and line *c* is the perpendicular bisector of $\overline{XZ}$

Prove: *a*, *b*, and *c* intersect at point *P*, and *P* is equidistant from each vertex

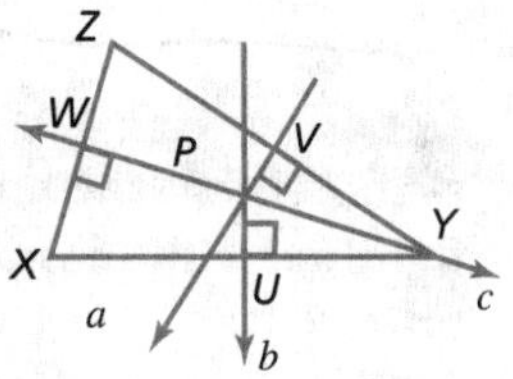

Figure 7.7

STATEMENTS	REASONS
1. *a*, *b*, and *c* are perpendicular bisectors to sides of △XYZ	1. Given
2. *a* and *b* intersect at *P*	2. Definition of parallel (coplanar lines *a* and *b* are not parallel and so must intersect at some point)
3. *PZ* = *PY*; *PY* = *PX*	3. Any point on a perpendicular bisector of a segment is equidistant from the endpoints
4. *PZ* = *PX*	4. Transitive property of equality
5. *P* is on the perpendicular bisector, *c*, of $\overline{XZ}$	5. Any point equidistant from the endpoints of a segment is on the perpendicular bisector of the segment

Now try bisecting the angles of a triangle. The point of concurrency is the *incenter*.

Theorem 7.7

Incenter Theorem. **The angle bisectors of the angles of a triangle are concurrent at the incenter, which is equidistant from the sides of the triangle.**

You can also draw the altitude or the medians of triangles to obtain points of concurrency. The points of concurrency in these cases are called the *orthocenter* and the *centroid* respectively.

Definitions

An **altitude of a triangle** is a segment that extends from a vertex and is perpendicular to the opposite side.

A **median of a triangle** is a segment extending from a vertex to the midpoint of the opposite side.

Theorem 7.8
Orthocenter Theorem. **The lines that contain the three altitudes are concurrent at the orthocenter.**

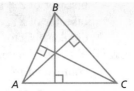

Figure 7.8

Theorem 7.9
Centroid Theorem. **The three medians of a triangle are concurrent at the centroid.**

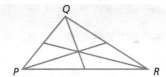

The word *concurrent* comes from a word that means "to run together." This word applies in different but related senses both to lines and to Christians. According to Hebrews 12:1 and I Corinthians 9:24-26, each Christian life is hotly contested by the enemy and requires great exertion and struggle to the finish. Christians run this race concurrently as a team. Lines also run together, but they do so at the same point.

▶ **A. Exercises**

Draw four obtuse triangles. Use one triangle for each of the next four exercises.
 1. Label the circumcenter C.
 2. Label the incenter I.
 3. Label the orthocenter O.
 4. Label the centroid D.

Define.

Reproduce each triangle below four times, then find the centroid, incenter, orthocenter, and circumcenter.

5.

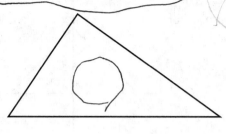

7.

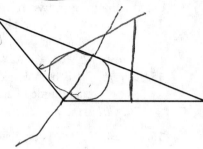

6.

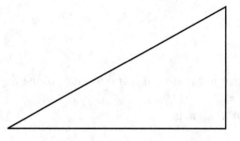

Use the diagram shown for exercises 8-10.

8. What is the name of point *P*? Draw a circle with a radius *PA*; what is the name of this circle?

9. If *PB* = 10 units, find *PA*.

10. If *PC* = 15 units, find *PA*.

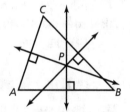

Use the diagram shown for exercises 11-13.

11. What is point *Q* called? Draw a circle with radius *QV*. What is the special name for this circle?

12. If *QV* = 26 units, find *QT*.

13. If *QT* = 8 units, find *QU*.

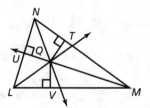

▶ B. Exercises

Use the diagram shown for exercises 14-15.

14. *Given:* $\overline{XY}$ is the perpendicular bisector of $\overline{MN}$ at point L

    Prove: $\triangle MLX \cong \triangle NLX$

15. *Given:* $\overline{XL}$ is both a median and an altitude of $\triangle MXN$

    Prove: $\triangle MLX \cong \triangle NLX$

16. *Given:* $\triangle ABC$ is equilateral; $\overline{BP}$ is a median

    Prove: $\angle ABP \cong \angle CBP$

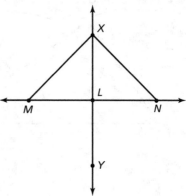

Prove the following theorems.

17. Any point on an angle bisector is equidistant from the sides of the angle. (*Hint:* Use congruent triangles.)

18. The median of an equilateral triangle is also an altitude.

▶ C. Exercises

19. For what kind of triangle are the incenter, orthocenter, centroid, and circumcenter the same? Prove your answer.

20. Prove that the medians from the base angles of an isosceles triangle are congruent.

▶ Dominion Thru Math

21. The trusses for the roof of a building are scalene right triangles sitting on the hypotenuses. The legs are 30 feet and 16 feet; overhangs on both sides are equal. How much overhang is on a 32-foot-wide building?

■ Cumulative Review

Given points A and B, consider the following:

$$\overline{AB}, \varnothing, \overleftrightarrow{AB}, AB, \overset{\circ}{AB}, \overrightarrow{AB}$$

22. Which symbol above is not a set?

23. Which set above is not a subset of any of the other sets?

24. Which set is a subset of all of the sets?

25. $\overrightarrow{AB}$ is a subset of which other sets above?

26. For which two of the sets is neither a subset of the other?

7.3 Exterior Angles and Inequalities

The *exterior* refers to the outside of something. Indeed, "man looketh on the outward appearance, but the Lord looketh on the heart" (I Sam.16:7). While only God can judge men justly, we are to evaluate professions of salvation by their exterior fruit (Matt. 7:15-20).

The *exterior angle* of a triangle is an angle on the outside. ∠ABD is an exterior angle of △BCD in figure 7.9. Draw the other five exterior angles of the triangle by extending different sides. Exterior angles have some special properties.

This room applies inequalities by incorporating rectangular windows of several unequal sizes.

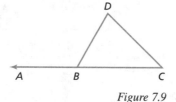

Figure 7.9

Definitions

An **exterior angle** of a triangle is an angle that forms a linear pair with one of the angles of the triangle.

The **remote interior angles** of an exterior angle are the two angles of the triangle that do not form a linear pair with a given exterior angle.

In figure 7.9, the remote interior angles of ∠ABD are ∠D and ∠C. Theorem 7.10 shows a relationship between an exterior angle and its two remote interior angles.

Theorem 7.10

Exterior Angle Theorem. The measure of an exterior angle of a triangle is equal to the sum of the measures of its two remote interior angles.

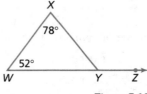

Figure 7.10

For example, the measure of ∠*XYZ* in figure 7.10 is 130 degrees. Do you understand why? It makes sense when you realize that $m\angle WYX$ is part of two 180-degree totals. The triangle and the linear pair suggest a proof for the theorem.

Given: △*HIJ* with exterior ∠*JHG*
Prove: $m\angle JHG = m\angle I + m\angle J$

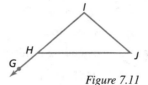

Figure 7.11

STATEMENTS	REASONS
1. △*HIJ* with exterior ∠*JHG*	1. Given
2. ∠*IHJ* and ∠*JHG* form a linear pair	2. Definition of exterior angle
3. ∠*IHJ* and ∠*JHG* are supplementary	3. Linear pairs are supplementary
4. $m\angle IHJ + m\angle JHG = 180$	4. Definition of supplementary
5. $m\angle IHJ + m\angle I + m\angle J = 180$	5. The measures of the angles of a triangle total 180
6. $m\angle IHJ + m\angle JHG = m\angle IHJ + m\angle I + m\angle J$	6. Transitive property of equality
7. $m\angle JHG = m\angle I + m\angle J$	7. Addition property of equality

The exterior angle theorem also shows that the exterior angle is larger than either remote interior angle. The proof of the inequality "larger than" will require properties of inequality. Since the rest of this chapter involves inequalities, review the properties of inequality closely. Assume *a*, *b*, and *c* are real numbers.

Inequality Facts	
Addition property	If $a < b$, then $a + c < b + c$
Multiplication property	If $a < b$ and $c > 0$, then $ac < bc$
	If $a < b$ and $c < 0$, then $ac > bc$
Transitive property	If $a < b$ and $b < c$, then $a < c$
Definition of greater than	If $a = b + c$ and $c > 0$, then $a > b$

Theorem 7.11

Exterior Angle Inequality. **The measure of an exterior angle of a triangle is greater than the measure of either remote interior angle.**

Given: $\triangle ABC$ and exterior $\angle DAB$

Prove: $m\angle DAB > m\angle B$ and
$m\angle DAB > m\angle C$

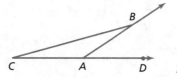

Figure 7.12

STATEMENTS	REASONS
1. $\triangle ABC$ and exterior angle $\angle DAB$	1. Given
2. $m\angle DAB = m\angle B + m\angle C$	2. Exterior Angle Theorem
3. $m\angle DAB > m\angle B$	3. Definition of greater than
4. $m\angle DAB > m\angle C$	4. Definition of greater than

▶ **A. Exercises**

Find the measures of the indicated angles.

1. $m\angle A$
   $m\angle BCA$
   $m\angle ACD$

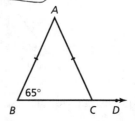

2. $m\angle ZXY$
   $m\angle WXY$

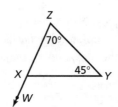

3. $m\angle KIH$
   $m\angle H$
   $m\angle K$

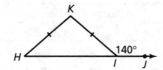

4. $m\angle S$
   $m\angle T$
   $m\angle TRS$
   $m\angle QRS$

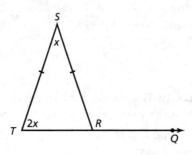

5. $m\angle ONL$
   $m\angle LNM$
   $m\angle L$
   $m\angle M$

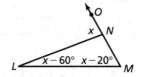

6. $m\angle SVU$
   $m\angle UVT$
   $m\angle T$
   $m\angle U$

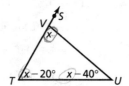

Refer to $\triangle ABC$; give the measure of the exterior angles at each vertex.

7. A
8. B
9. C

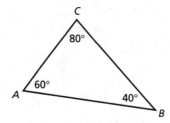

10. Do the largest interior and exterior angles of the triangle occur at the same vertex?

Answer each question.

11. If $\triangle ABC$ is equilateral, what is the measure of an exterior angle?

12. What are the measures of the acute angles in an isosceles right triangle?

▶ B. Exercises

13. If an exterior angle at the vertex of an isosceles triangle measures 78 degrees, what are the measures of the base angles?
14. What is the measure of an angle determined by a side and a diagonal of a square?
15. If *ABCD* is a rhombus and *C-D-E*, what is the relationship between $m\angle EDA$ and $m\angle DAC$?

Use theorems on exterior angles to prove exercises 16-20.
16. Angles adjacent to the hypotenuse of a right triangle are complementary.
17. A right triangle has two acute angles.
18. The exterior angle at the vertex of an isosceles triangle measures twice the measure of one of the base angles.
19. Prove that $m\angle ABE > m\angle CED$.

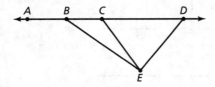

20. *Prove:* If *BCDE* is a parallelogram, then $m\angle ABC > m\angle FDE$.

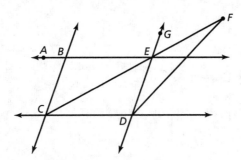

▶ C. Exercises

Exercises 21-23 develop the sum of the measures of the exterior angles for any convex polygon. Use only one exterior angle at each vertex. Find the sum of the measures for each polygon and justify your answer.
21. triangle
22. convex quadrilateral
23. *n*-gon

How does each diagram illustrate a method of proving two lines parallel? Identify the parallel lines in each case.

24.

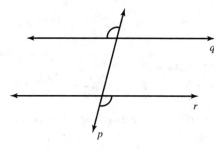

25.

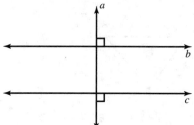

26.

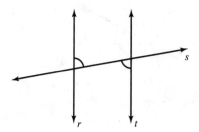

27.

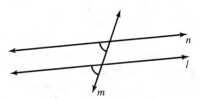

28.

GEOMETRY AND CARPENTRY

Miter saws, planes, and air nailers are tools for cutting, working, and joining timbers. A carpenter depends on these tools as well as others of his trade. He uses a framing square to test right angles between adjacent surfaces. He uses a bevel to check acute and obtuse angles. A gauge enables him to mark lines parallel to a surface.

Did you know that geometry is another essential tool of the carpenter? A carpenter faces a number of geometric problems, from laying out the foundation of a structure to cutting the proper angles for the rafters that support the roof. The foundation for a house usually consists of one or more rectangles that must be laid out perfectly. A carpenter cannot simply measure the lengths of the sides of the quadrilateral because that would guarantee only a parallelogram. Using theorems from geometry, he can either check that one of the angles is a right angle, or he can measure the lengths of the diagonals to see if they are the same.

A carpenter must also be able to read an architect's drawing. This is essential if he is going to build a house exactly the way the future owner wants it. If he makes

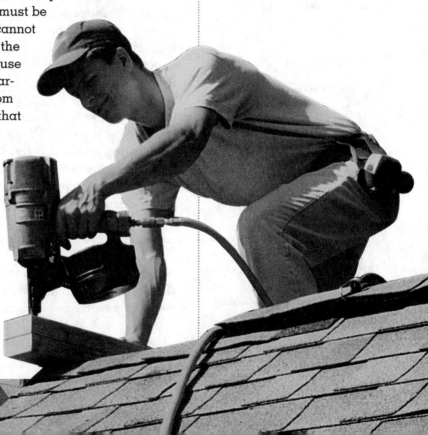

A carpenter nails boards for the roof with a pneumatic (air-driven) tool.

a utility room too small for the washer and dryer, it may be expensive to correct the mistake. If a door is too small or the stairs too narrow, it may be impossible to bring in the furniture. Reading these drawings requires a knowledge of proportion and ratio. The carpenter must also recognize errors in the drawings, if there are any.

When making door and window frames, the carpenter must be able to measure and construct right angles. When erecting the interior walls and their supports, he deals with perpendicular and parallel lines. He applies the theorem that in a plane, if two lines are perpendicular to the same line, then they are parallel. If an arch is needed, a carpenter must know whether it is to be a semicircle, pointed arch, parabola, or catenary curve and apply principles for building it.

Carpenters frame trapezoids, rectangles, and an octagon for windows as they construct this house.

Another geometric task that the carpenter must handle is constructing a flight of stairs. When handling such a job, he uses, perhaps without knowing its official name, the Pythagorean theorem. He uses a carpenter's square— a device that looks like a right triangle without its hypotenuse—for measuring and for marking each stair. After determining how many feet the stairs will rise vertically, he can divide by the maximum step height (riser) allowed by the local building code and determine the number of steps required.

Building codes also specify the minimum width of tread (where you place your foot). Multiplying this tread size by the number of steps, the carpenter can determine the horizontal distance required for the staircase. He then must determine a set ratio for each individual stair and use his carpenter's square to check each step.

Anyone who builds something, whether a house or a bookcase, uses the skills of the carpenter. The principles will serve you well as you apply them in the shop. Geometry is indeed one of a carpenter's most essential tools.

When handling such a job, he uses, perhaps without knowing its official name, the Pythagorean theorem.

7.4 Inequalities

The Exterior Angle Inequality is not the only important inequality for triangles. You will learn two more in this section. The first is a biconditional statement, but only one part will be proved.

> **Theorem 7.12**
> ***Longer Side Inequality.*** **One side of a triangle is longer than another side if and only if the measure of the angle opposite the longer side is greater than the measure of the angle opposite the shorter side.**

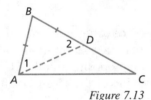

Figure 7.13

We will prove the *if* part of the Longer Side Inequality.

If $BC > AB$, then $m\angle A > m\angle C$.

Given: $\triangle ABC$; $BC > AB$

Construct: AD such that $B\text{-}D\text{-}C$ and $\overline{AB} \cong \overline{BD}$ (auxiliary line by Completeness Postulate)

Prove: $m\angle A > m\angle C$

Scalene right triangles flank the arched window of this home in Stillwater, Minnesota.

STATEMENTS	REASONS
1. $\triangle ABC$; $BC > AB$; $\overline{AB} \cong \overline{BD}$	1. Given
2. $\triangle ABD$ is an isosceles triangle	2. Definition of isosceles triangle
3. $\angle 1 \cong \angle 2$	3. Isosceles Triangle Theorem
4. $m \angle 1 = m \angle 2$	4. Definition of congruent angles
5. $m\angle CAD + m\angle 1 = m\angle CAB$	5. Angle Addition Postulate
6. $m\angle CAB > m\angle 1$	6. Definition of greater than
7. $m\angle CAB > m\angle 2$	7. Substitution (step 4 into 6)
8. $m\angle 2 > m\angle C$	8. Exterior Angle Inequality
9. $m\angle CAB > m\angle C$	9. Transitive property of inequality
10. If one side of a triangle is longer than another side, then the measure of the angle opposite the longer side is greater than the measure of the angle opposite the shorter side	10. Law of Deduction

Theorem 7.13

Hinge Theorem. Two triangles have two pairs of congruent sides. If the measure of the included angle of the first triangle is larger than the measure of the other included angle, then the opposite (third) side of the first triangle is longer than the opposite side of the second triangle.

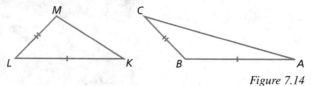

Figure 7.14

Given: $\triangle KLM$ and $\triangle ABC$ with $AB \cong KL$; $BC \cong LM$; and
$m\angle ABC > m\angle KLM$

Draw: $\overrightarrow{BD}$ so that $\triangle ABD \cong \triangle KLM$

Prove: $AC > KM$

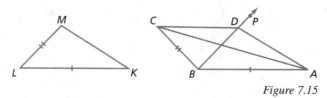

Figure 7.15

STATEMENTS	REASONS
1. $\triangle ABC$; $\triangle KLM$; $\overline{AB} \cong \overline{KL}$; $\overline{BC} \cong \overline{LM}$ and $m\angle ABC > m\angle KLM$	1. Given
2. $m\angle ABC = d + m\angle KLM$ ($d > 0$)	2. Definition of greater than
3. Draw $\overrightarrow{BP}$ with P on the same side of $\overleftrightarrow{AB}$ as C, so that $\angle ABP \cong \angle KLM$	3. Auxiliary line (Continuity Postulate)
4. Mark D on $\overrightarrow{BP}$ so that $\overline{BD} \cong \overline{LM}$	4. Completeness Postulate
5. $\triangle ABD \cong \triangle KLM$	5. SAS
6. $\overline{AD} \cong \overline{KM}$	6. Definition of congruent triangles
7. $AD = KM$	7. Definition of congruent segments
8. $\overline{BC} \cong \overline{BD}$	8. Transitive property of congruent segments (see steps 1 and 4)
9. $\triangle BCD$ is isosceles	9. Definition of isosceles triangle
10. $\angle BCD \cong \angle BDC$	10. Isosceles Triangle Theorem
11. $m\angle BCD = m\angle BDC$	11. Definition of congruent angles
12. $m\angle ADC = m\angle ADB + m\angle BDC$; $m\angle BCD = m\angle BCA + m\angle DCA$	12. Angle Addition Postulate
13. $m\angle ADC > m\angle BDC$; $m\angle BCD > m\angle DCA$	13. Definition of greater than
14. $m\angle ADC > m\angle BCD$	14. Substitution (step 11 into 13)
15. $m\angle ADC > m\angle DCA$	15. Transitive property of inequality
16. $AC > AD$	16. Longer side inequality
17. $AC > KM$	17. Substitution (step 7 into 16)

► **A. Exercises**

Give the order of sides from smallest to largest.

1.

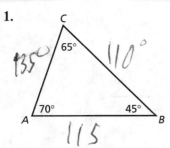

2.

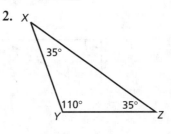

3.

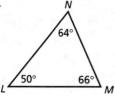

4.

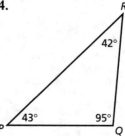

Give the order of angles from smallest to largest.

5.

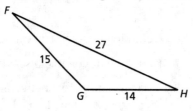

6.

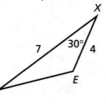

7.

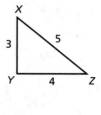

For each pair of triangles, compare an unlabeled pair of sides or angles.

8.

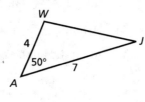

9.

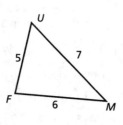
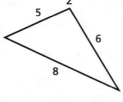

10.

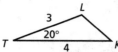
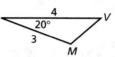

11.

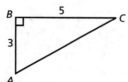
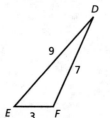

12.

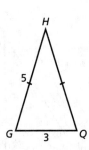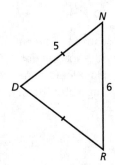

Prove the following statement.

13. In a right triangle the hypotenuse is the longest side.

▶ B. Exercises

Prove the following statement.

14. The shortest segment from a point to a line is a perpendicular segment.

15. *Given:* $AB > AC;\ AC > BC$
  Prove: $m\angle C > m\angle A$

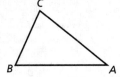

Prove exercises 16-17 using the following diagram and given information.

  Given: $\overline{MB}$ is the base of isosceles
       $\triangle BCM$ and M is the
       midpoint of $\overline{AB}$

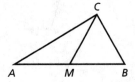

16. *Prove:* $AC > CM$

17. *Prove:* $m\angle B > m\angle A$

18. *Given:* $\overline{FG}$ is a median;
       $m\angle FGE > m\angle FGD$
   Prove: $m\angle D > m\angle E$

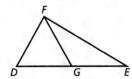

19. The longest side of an obtuse triangle is opposite the obtuse angle.

▶ C. Exercises

20. Prove that the centroid of an equilateral triangle divides the medians into segments of unequal lengths. Write your proof in paragraph form stating only the key steps.

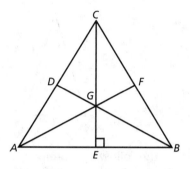

■ Cumulative Review

Give an algebraic expression or a numerical value for each indicated angle.

21. $m\angle PBA$ if $l \parallel m$

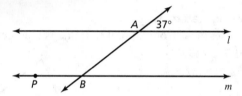

22. $m\angle A$

23. $m\angle ABC$

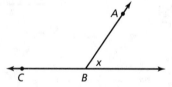

24. $m\angle R$

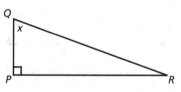

25. $m\angle A$

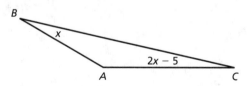

7.5 Triangle Inequality

In each of the three diagrams in figure 7.16, $BC = 8$. In diagram (a), $AB + AC = 8$. According to the definition of betweenness, the points are collinear and no triangle is formed.

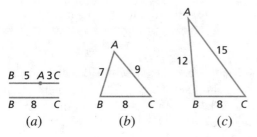

Figure 7.16

In diagrams (b) and (c), $AB + AC > 8$, and triangles are formed. This illustration suggests the Triangle Inequality Theorem.

The Chrysler Building in New York City incorporates triangles on its spire; it was the tallest skyscraper in the world for several months before the completion of the Empire State Building.

Theorem 7.14
Triangle Inequality. **The sum of the lengths of any two sides of a triangle is greater than the length of the third side.**

Given: $\triangle XYZ$ with no side longer than $\overline{XZ}$

Auxiliary line: $\overrightarrow{PY} \perp \overleftrightarrow{XZ}$ at point P

Prove: $XZ + YZ > XY$
$XZ + XY > YZ$
$XY + YZ > XZ$

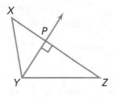

Figure 7.17

Trichotomy guarantees that you can list the three sides of a triangle in order of length from shortest to longest. (If two are the same length, either order will do.) Label side XZ so that there is no longer side. Since $XZ \geq XY$, it follows that $XZ + YZ > XY$ (distance $YZ > 0$). Similarly, $XZ \geq YZ$, so $XZ + XY > YZ$. These two inequalities were easy to prove, but the third is harder and requires the altitude from Y to $\overline{XZ}$ as an auxiliary line.

STATEMENTS	REASONS
1. $\triangle XYZ$ with $\overline{XZ}$ the longest side, altitude $\overline{YP}$	1. Given
2. $\overline{YP} \perp \overline{XZ}$	2. Definition of altitude
3. $\angle XPY$ and $\angle ZPY$ are right angles	3. Definition of perpendicular
4. $\triangle XPY$ and $\triangle ZPY$ are right triangles	4. Definition of right triangle
5. $XY > XP$, $YZ > PZ$	5. Hypotenuse is longest side
6. $XY + YZ > XP + PZ$	6. Addition property of inequality
7. $XP + PZ = XZ$	7. Definition of betweenness
8. $XY + YZ > XZ$	8. Substitution (step 7 into 6)

Now you can see at a glance that no triangle could exist with sides of lengths 2, 3, and 7. The diagram shows that the short sides are too short to intersect. The triangle inequality requires that $2 + 3 > 7$. Since this is false, the triangle inequality explains why no such triangle exists.

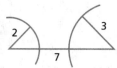

2   3

7

Figure 7.18

This principle applies in many practical situations as illustrated in the example.

EXAMPLE   In a business office the equipment must be arranged carefully for efficient use. Becky uses primarily the computer on her desk, the photocopier, and the fax machine. A triangular arrangement is desired to facilitate access between all three machines. The distance between her desk and the photocopier must be the shortest, and the total distance between the three items should not exceed 20 feet. The following table shows distances between the pieces of equipment. Which of these is possible and which is impossible? Why?

	Computer-Copier	Copier-Fax	Computer-Fax
1.	5 feet	7 feet	6 feet
2.	3 feet	6 feet	9 feet

Continued ▶

Answer *Layout 1* has a total distance between items of 18 feet, and the computer-copier distance is the shortest as desired. Every combination of two sides is greater than the third side, so layout 1 forms a legitimate triangular layout.

Layout 2 has a total distance of 18 feet, with the computer-copier distance being the shortest. However, the sum of the first two sides equals the third side. This combination is not triangular and therefore not efficient. Notice that the three items are collinear, so the photocopier is in the way of moving from the computer to the fax machine.

▶ A. Exercises

State whether triangles with the following side lengths exist. If not, state the reason.

	a	b	c	
1.	5	9	10	
2.	2	7	11	
3.	8	4	12	
4.	4	7	8	
5.	6	3	3	

AB, *BC*, and *AC* are given. Make a sketch and identify the type of triangle. If none is formed, explain why (impossible, collinear).

6. $AB = 4$, $BC = 7$, $AC = 5$

7. $AB = 3$, $BC = 5$, $AC = 2$

8. $AB = 1$, $BC = 1$, $AC = 1$

9. $AB = 5$, $BC = 3$, $AC = 1$

10. $AB = 2$, $BC = 5$, $AC = 2$

11. $AB = 5$, $BC = 3$, $AC = 3$

12. $AB = 5$, $BC = 5$, $AC = 10$

▶ B. Exercises

Refer to Becky's office in the example in this section. Are the following measurements possible? If not, explain why.

	Computer-Copier	Copier-Fax	Computer-Fax	
13.	5 feet	7 feet	8 feet	
14.	9 feet	4 feet	6 feet	
15.	6 feet	12 feet	10 feet	
16.	4 feet	10 feet	5 feet	
17.	5 feet	9 feet	6 feet	

18. Explain why the sum of the lengths of the legs of any right triangle is always greater than the length of the hypotenuse.

19. *Given:* △DEF with point *P* in the interior

    Prove: DF + FE > DP + PE

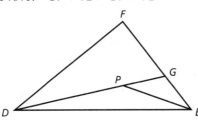

20. *Given:* $\overline{EC} \cong \overline{ED}$

    Prove: AB + AD > BC

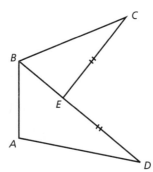

▶ **C. Exercises**

21. *Given:* △ABC with point *P* in the interior

    Prove: m∠APB > m∠ACB

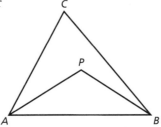

■ **Cumulative Review**

Supply the missing reasons needed to complete the following proof.

Given: $\overleftrightarrow{AD} \perp \overleftrightarrow{AB}$, $\overleftrightarrow{CB} \perp \overleftrightarrow{CD}$

Prove: ∠D ≅ ∠B

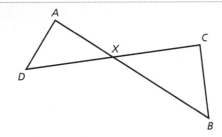

STATEMENTS	REASONS
22. $\overleftrightarrow{AD} \perp \overleftrightarrow{AB}$; $\overleftrightarrow{CB} \perp \overleftrightarrow{CD}$	22.
23. ∠A and ∠C are right angles	23.
24. ∠A ≅ ∠C	24.
25. ∠AXD ≅ ∠CXB	25.
26. ∠D ≅ ∠B	26.

7.6 Parallelograms

Remember that a parallelogram is a quadrilateral in which both pairs of opposite sides are parallel. Study the examples of proofs involving parallelograms. These important theorems will enable you to prove many other theorems.

The rectangles and squares of these windows are special parallelograms.

Theorem 7.15
The opposite sides of a parallelogram are congruent.

Given: Parallelogram *ABCD*
Prove: $\overline{AB} \cong \overline{CD}$ and $\overline{BC} \cong \overline{AD}$
Draw auxiliary diagonal $\overleftrightarrow{BD}$.

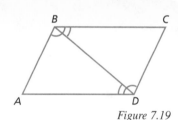

Figure 7.19

STATEMENTS	REASONS
1. *ABCD* is a parallelogram	1. Given
2. $\overleftrightarrow{AB} \parallel \overleftrightarrow{CD}$; $\overleftrightarrow{AD} \parallel \overleftrightarrow{BC}$	2. Definition of parallelogram
3. Draw $\overleftrightarrow{BD}$	3. Line Postulate
4. $\angle ABD \cong \angle CDB$; $\angle CBD \cong \angle ADB$	4. Parallel Postulate
5. $\overline{BD} \cong \overline{BD}$	5. Reflexive property of congruent segments
6. $\triangle ABD \cong \triangle CDB$	6. ASA
7. $\overline{AB} \cong \overline{CD}$; $\overline{AD} \cong \overline{BC}$	7. Definition of congruent triangles

The second example is a theorem on how to prove parallelograms congruent. Recall that you can prove polygons congruent by subdividing the polygons into corresponding triangles.

Theorem 7.16

SAS Congruence for Parallelograms

Given: Parallelograms *ABCD* and *PQRS*; $\overline{AB} \cong \overline{PQ}$; $\overline{AD} \cong \overline{PS}$; $\angle A \cong \angle P$
Prove: *ABCD* $\cong$ *PQRS*

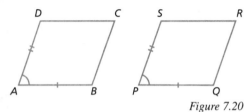

Figure 7.20

STATEMENTS	REASONS
1. Parallelograms *ABCD* and *PQRS*; $\overline{AB} \cong \overline{PQ}$; $\overline{AD} \cong \overline{PS}$; $\angle A \cong \angle P$	**1.** Given
2. Draw auxiliary lines: $\overleftrightarrow{BD}$ and $\overleftrightarrow{QS}$	**2.** Auxiliary lines
3. $\triangle ABD \cong \triangle PQS$	**3.** SAS
4. $\overline{BD} \cong \overline{QS}$	**4.** Definition of congruent triangles
5. $\overline{AB} \cong \overline{CD}$; $\overline{PQ} \cong \overline{RS}$; $\overline{AD} \cong \overline{BC}$; $\overline{PS} \cong \overline{QR}$	**5.** Opposite sides of a parallelogram are congruent
6. $\overline{BC} \cong \overline{QR}$; $\overline{CD} \cong \overline{RS}$	**6.** Transitive property of congruent segments (see steps 5, 1)
7. $\triangle BCD \cong \triangle QRS$	**7.** SSS
8. *ABCD* $\cong$ *PQRS*	**8.** Subdivision into corresponding congruent triangles

The following theorems concerning parallelograms will be proved as exercises.

Theorem 7.17
A quadrilateral is a parallelogram if and only if the diagonals bisect one another.

Theorem 7.18
Diagonals of a rectangle are congruent.

Theorem 7.19
The sum of the measures of the four angles of every convex quadri-lateral is 360°.

Theorem 7.20
Opposite angles of a parallelogram are congruent.

Theorem 7.21
Consecutive angles of a parallelogram are supplementary.

Theorem 7.22
If the opposite sides of a quadrilateral are congruent, then the quadri-lateral is a parallelogram.

Theorem 7.23
A quadrilateral with one pair of parallel sides that are congruent is a parallelogram.

▶ A. Exercises

Using parallelogram *ABCD*, find the measures of the indicated angles.

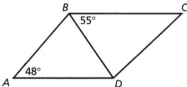

1. ∠*C*
2. ∠*ADC*
3. ∠*ABD*
4. The four angles of the parallelogram combined
5. An exterior angle of the parallelogram at angle *C*

Disprove the following statements. Remember that you disprove a statement by proving that it is false. This can be done with an illustration or a counterexample.

6. Adjacent angles form a linear pair.
7. Alternate interior angles are congruent.
8. Every pair of supplementary angles form a linear pair.
9. The acute angles of a triangle are complementary.
10. If two triangles have a pair of congruent angles, then the other pairs of angles are congruent.

▶ B. Exercises

11. Prove Theorem 7.17, part one: The diagonals of a parallelogram bisect each other.
12. Prove Theorem 7.17, part two: If the diagonals of *ABCD* bisect one another, then *ABCD* is a parallelogram.

Prove the following theorems.

13. Theorem 7.18: The diagonals of a rectangle are congruent.
14. Theorem 7.19: The sum of the measures of the four angles of every convex quadrilateral is 360.
15. Theorem 7.20: The opposite angles in a parallelogram are congruent.
16. Theorem 7.21: Consecutive angles in a parallelogram are supplementary.
17. Theorem 7.22: If opposite sides of a quadrilateral are congruent, then the quadrilateral is a parallelogram.
18. Theorem 7.23: If one pair of opposite sides of quadrilateral *ABCD* are both parallel and congruent, then *ABCD* is a parallelogram.

19. *Given:* *ABCD* and *DEFG* are parallelograms; $\overline{AG}$ and $\overline{EC}$ bisect each other at *D*

    Prove: *ABCD* ≅ *GFED*

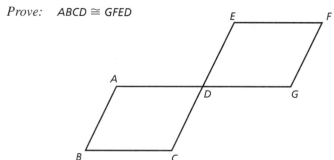

▶ C. Exercises

20. *Given:* Convex quadrilateral *ABCD* with ∠A ≅ ∠C and ∠ABC ≅ ∠D
   Prove: *ABCD* is a parallelogram

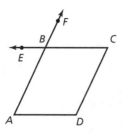

▶ Dominion Thru Math

Convex quadrilaterals are found in nature and man-made devices. Bisect the angles of each figure, extending the bisectors all the way across the figure. Analyze the resulting quadrilaterals and describe the results.

21. rectangle
22. parallelogram
23. isosceles trapezoid
24. Prove the results from exercise 22.

◼ Cumulative Review

Suppose two segments must be proved congruent. What reason could you use that involves the concept named?

25. distances
26. bisectors
27. one triangle
28. two triangles
29. a circle
30. a parallelogram

Midpoints

Study the proof of the midpoint formula. You will use some theorems from geometry in the proof, but you will also need to remember the formula for midpoints on number lines from algebra: $\frac{x_1 + x_2}{2}$

Finally, review the Cartesian plane. Vertical lines are parallel, horizontal lines are parallel, and vertical lines are perpendicular to horizontal lines. Also, points on a vertical line have the same x-coordinate, while points on a horizontal line have the same y-coordinate.

Midpoint Formula

If M is the midpoint of $\overline{AB}$ where $A(x_1, y_1)$ and $B(x_2, y_2)$, then $M\left(\frac{x_1 + x_2}{2}, \frac{y_1 + y_2}{2}\right)$.

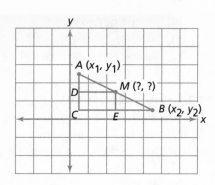

STATEMENTS	REASONS
1. M is the midpoint of $\overline{AB}$	1. Given
2. $\overline{AM} \cong \overline{MB}$	2. Definition of midpoint
3. Draw (parallel) vertical lines through A and M and (parallel) horizontal lines through B and M Note: $\angle ADM$ and $\angle MEB$ are right angles	3. Auxiliary lines
4. $C(x_1, y_2)$	4. Cartesian plane property of coordinates
5. Quadrilateral $MDCE$ is a parallelogram	5. Definition of parallelogram
6. $\overline{DC} \cong \overline{ME}, \overline{DM} \cong \overline{CE}$	6. Opposite sides of a parallelogram are congruent

Continued ▶

7. $\angle ADM \cong \angle MEB$	7. All right angles are congruent
8. $\angle A \cong \angle EMB$, $\angle B \cong \angle AMD$	8. Corresponding Angle Theorem
9. $\triangle AMD \cong \triangle MBE$	9. ASA
10. $\overline{AD} \cong \overline{ME}$, $\overline{DM} \cong \overline{EB}$	10. Definition of congruent triangles
11. $\overline{AD} \cong \overline{DC}$, $\overline{CE} \cong \overline{EB}$	11. Transitive property of congruent segments (steps 10, 6)
12. D is the midpoint of $\overline{AC}$; E is the midpoint of $\overline{BC}$	12. Definition of midpoint
13. $D\left(x_1, \dfrac{y_1 + y_2}{2}\right)$, $E\left(\dfrac{x_1 + x_2}{2}, y_2\right)$	13. Midpoint formula for number lines
14. $M\left(\dfrac{x_1 + x_2}{2}, \dfrac{y_1 + y_2}{2}\right)$	14. M has the same x-coordinate as E and the same y-coordinate as D

▶ Exercises

Find the coordinates of the midpoint between the two points.

1. $(4, 8)$ and $(2, -3)$
2. $(3, 5)$ and $(3, 9)$
3. $(-1, -4)$ and $(6, -2)$
4. Find the perimeter; then label and give the midpoints of each side.

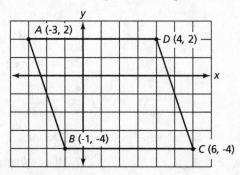

5. Use the Midpoint Formula to prove that the diagonals of a parallelogram bisect each other.

7.7 Polygon Constructions

These windows from Anderson Windows Company shows a full range of polygons.

Proverbs 4:1-7 emphasizes understanding the things you do. Christians must know the biblical basis for their actions, their beliefs, and the standards that they hold. Likewise, in mathematics you should know the reasons for every statement and application. You are now ready to understand the reasons for several more constructions. Construction 3 from Chapter 3 is proved below and you will prove others in the exercises.

 Construction 3

A line perpendicular to a given line through a given point of the line

Justification: Show that $\angle PXA$ is a right angle, and thus $\overrightarrow{PX} \perp \overleftrightarrow{AB}$

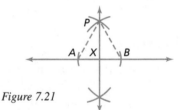

Figure 7.21

STATEMENTS	REASONS
1. $\overleftrightarrow{AB}$ contains point X	1. Given
2. $\overline{PA} \cong \overline{PB}$; $\overline{AX} \cong \overline{BX}$	2. Radii of congruent circles are congruent
3. $\overline{PX} \cong \overline{PX}$	3. Reflexive property of congruent segments
4. $\triangle PAX \cong \triangle PBX$	4. SSS
5. $\angle PXA \cong \angle PXB$	5. Definition of congruent triangles
6. $\angle PXA$ and $\angle PXB$ form a linear pair	6. Definition of linear pair
7. $\angle PXA$ is a right angle	7. If congruent angles form a linear pair, then they are right angles
8. $\overrightarrow{PX} \perp \overleftrightarrow{AB}$	8. Definition of perpendicular

Can you construct a line perpendicular to *l* at *X* with a straightedge and a nonadjustable compass (a compass that has only one arc length)?

You can use SSS, SAS, or ASA to construct congruent triangles. SSS provides the easiest method since no angles need be constructed.

Construction 9

Copy a triangle.

> **Given:** △*ABC*
> **Construct:** A triangle congruent to
>   △*ABC*

1. Draw a line. Choose a point on the line and call it *A'*.
2. Using a compass, measure length *AB*. Place the point of the compass at *A'* and mark off a segment congruent to *AB* on the line. Call the point *B'*.
3. Using measure *AC* and using *A'* as center, construct an arc above *A'B'*.
4. Repeat step 3, using measure *BC* and with *B'* as center.
5. The arcs intersect at a point. Call it *C'*. Connect *C'* with *A'* and *B'* to form a triangle congruent to the original triangle.

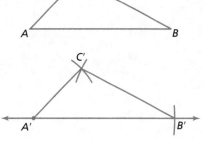

Figure 7.22

You can use this method to copy any polygon by subdividing the original into triangles.

Construction 10

Copy a polygon.

> **Construct:** A polygon congruent to a
>   given polygon

1. Subdivide the given polygon into triangles.
2. Copy one of the triangles (construction 9).
3. Copy adjacent triangles until the polygon is complete.

(2)

Figure 7.23

(1) 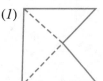   (3)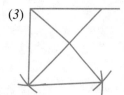

onstruction 11

Parallel through point

> ***Given:*** A line and a point outside that line
> ***Construct:*** A line containing the given point that is parallel to the given line

1. Draw a line that goes through point *A* and intersects the given line at a point *B*.
2. Construct an angle, ∠*BAE*, congruent to ∠*ABC*, so that *E* and *C* are in opposite half-planes (construction 4).
3. The two angles will be congruent alternate interior angles, thus $\overleftrightarrow{AE}$ is parallel to *k* by the Parallel Postulate.

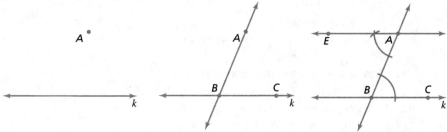

Figure 7.24

▶ A. Exercises

Construct polygons congruent to the given polygons by constructing congruent triangles.

1.

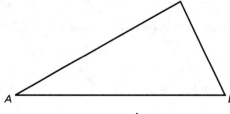

2.

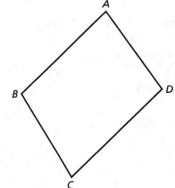

3.

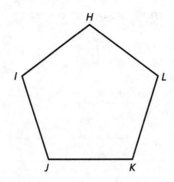

4.

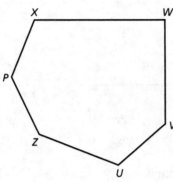

5.

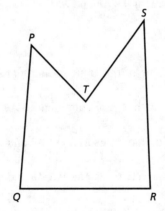

6. Construct a line parallel to a given line through a given point.

Use the figures below to construct the following.

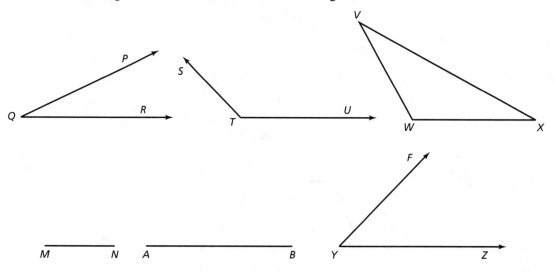

7. a triangle congruent to △*VWX*
8. a triangle that contains an angle congruent to ∠*STU* and a side congruent to $\overline{AB}$
9. a square with side length equal to *MN*
10. a rectangle with side lengths *AB* and *MN*

▶ B. Exercises

11. a right triangle with one of the angles congruent to ∠*PQR* and a leg congruent to *MN*
12. an isosceles triangle with a base congruent to *AB* and base angles congruent to ∠*PQR*
13. a triangle that has side measures *MN* and *AB* and an included angle congruent to ∠*PQR*
14. a triangle that has side measures *MN* and *AB* and an included angle that measures 45°
15. a parallelogram with side lengths *AB* and *MN* and with an angle congruent to ∠*FYZ*
16. a rhombus with side measure *MN* and with one angle congruent to ∠*FYZ*
17. a triangle that has a 75° angle and an angle congruent to ∠*PQR*

Tell how you would prove the constructions by answering the questions below.

18. What postulate justifies construction 11?

19. Construction 4: Bisecting $\angle A$ by marking off congruent segments: $\overline{AB}$ and $\overline{AC}$, $\overline{BD}$ and $\overline{CD}$. Explain why $\triangle ABD \cong \triangle ACD$ and $\angle BAD \cong \angle CAD$.

20. Construction 5: Copy $\angle ABC$; mark $\overline{A'B'} \cong \overline{AB}$, $\overline{B'C'} \cong \overline{BC}$, $\overline{A'C'} \cong \overline{AC}$. Explain why $\triangle ABC \cong \triangle A'B'C'$ and $\angle B \cong \angle B'$.

▶ C. Exercises

21. Construct a rectangle if one side has length MN and the diagonal has length PQ.

M   N   P   Q

▶ Dominion Thru Math

A customer wants a gazebo in the shape of a regular octagon. The carpenter knows only the desired length of a side. He has two choices: 1) construct a circle and use a ruler and protractor to find the vertices on it, or 2) draw a square and cut isosceles right triangles off each corner.

22. Using the carpenter's rule of thumb that the required radius is the side times 1.307, construct an octagon with a 6 cm side.

23. Construct an octagon with side $s = 2\frac{3}{8}$ inches. The square has sides $e = 2x + s$, where the right triangles that are cut off have legs $x = \frac{s}{\sqrt{2}}$.

■ Cumulative Review

Give a method of proving that two angles are congruent as directed.

24. using parallel lines
25. using two triangles
26. using one triangle
27. using related angles
28. using a special kind of angle
29. using parallelograms
30. Give two methods of proving that two angles are not congruent.

$\mathcal{G}$eometry *and* $\mathcal{S}$cripture

Dimensions

$\mathcal{A}$ postulate or theorem seen in nature is often called a law. You studied many laws for triangles in plane geometry. Match the verses below to the law that God created.

1. Law of Cause and Effect
2. Law of Eternal Life
3. Law of Judgment
4. Law of Gravity
5. Law of Cosmology (explains the origin of the universe)
6. Law of Christ's Ultimate Authority
7. Law of Conservation of Energy or First Law of Thermodynamics (no new energy enters the universe)
8. Law of Decay or Second Law of Thermodynamics (everything runs down)

A. Hebrews 9:27
B. Romans 8:20-22
C. Genesis 1:1, 31
D. Galatians 6:7-9
E. Genesis 2:1-3
F. Colossians 1:17
G. I John 5:12
H. Philippians 2:9-11

Some of these laws have exceptions, described in these Bible passages: Luke 8:43-44, John 6:10-13, John 11:39-44, Acts 1:9-11. Match each passage to a law in exercises 9-11.

9. Law of Gravity
10. Law of Conservation of Energy
11. Law of Decay
12. What do we call events in which God supersedes His own order and laws?

How many dimensions does God give for each item below?

13. Exodus 27:1   **15.** Ezekiel 45:6

14. Exodus 28:16   **16.** Revelation 21:16

17. Does God ever record more dimensions in His Word?

Law of Dimensions: God created a three-dimensional universe.

HIGHER PLANE: If we conclude that only three dimensions exist based on these passages, what type of argument do we have (see Chapter 5)?

Some suggest that God lives in the fourth (or fifth) dimension. Remember that a dimension expresses the physical limitations of space. For example, a one-dimensional object on a line is limited to that one dimension, having no width or height.

18. According to Psalm 139, where is God present?

19. Is God limited to a three-dimensional universe?

20. Is the problem avoided if God is in the fourth dimension?

21. First Corinthians 15:27 says that God is "excepted" from one of His laws. Which one?

Just as God is excepted from that law, He is not limited by His creation either.

You may have heard that time is the fourth dimension. Those who teach this concept usually support their case by appealing to relativity theory, where this view has proved useful. The usefulness of a mathematical variable, however, does not make it a physical dimension. Temperature and profits are also useful variables, but that does not make them physical dimensions.

22. In this study, you have seen nine laws of Creation. Which one of them is taught in Colossians 1:16?

> ### Line upon Line
>
> FOR BY HIM were all things created, that are in heaven, and that are in earth, visible and invisible, whether they be thrones, or dominions, or principalities, or powers: all things were created by him, and for him. ❧
>
> COLOSSIANS 1:16

Chapter 7 Review

Prove the following statements.

1. *Given:* Rectangle *ABCD* with diagonal $\overline{BD}$
   Prove: $\triangle ABD \cong \triangle CDB$

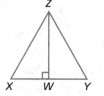

2. *Given:* $\triangle XYZ \cong \triangle LMN$; $\overrightarrow{ZW}$ bisects $\angle XZY$;
   $\angle NPL$ and $\angle XWZ$ are right angles
   Prove: $\overrightarrow{NP}$ bisects $\angle LNM$

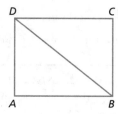
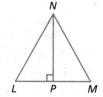

3. *Given:* $\overline{TU} \perp \overline{VU}$; $\overline{TS} \perp \overline{SV}$
   Prove: $\triangle VTS \cong \triangle VTU$

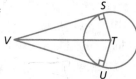

Trace $\triangle ABC$ onto your paper and then construct each point.

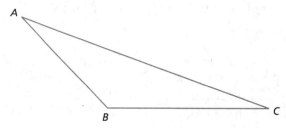

4. The orthocenter of $\triangle ABC$
5. The circumcenter of $\triangle ABC$
6. The incenter of $\triangle ABC$
7. The centroid of $\triangle ABC$

8. What is true about the distances from the circumcenter to each vertex?
9. What is true about the distances from the incenter to each side?
10. Construct a line perpendicular to a given line and justify the construction.
11. What are concurrent lines?
12. Prove the SS Congruence Theorem for rectangles.
   Given: *ABCD* and *PQRS* are rectangles; $\overline{AB} \cong \overline{PQ}$; $\overline{BC} \cong \overline{QR}$
   Prove: Rectangles *ABCD* and *PQRS* are congruent

13. Prove an SA Congruence Theorem for rhombi.
14. Prove the S Congruence Theorems for squares.

15. Tell which side of the following figure is the longest and which is the shortest; then explain why your answers are true.

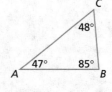

16. If the longest side of a triangle is 15 inches, then what is true about the sum of the lengths of the other two sides of the triangle?

Identify the theorem that proves each. Use the diagram shown.

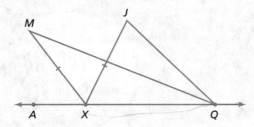

17. $m\angle AXM > m\angle XMQ$
18. $XJ + JQ > XQ$

19. $MQ > JQ$
20. $m\angle AXJ = m\angle XJQ + m\angle XQJ$

Consider the isosceles triangle shown with two sides of 7 cm. Consider each set of real numbers and decide if these numbers could be the length of the base of the triangle. Explain your answer including the type of triangle produced (if any).

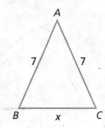

21. $x < 0$ or $x = 0$
22. $0 < x < 7$
23. $x = 7$
24. $7 < x < 7\sqrt{2}$
25. $x = 7\sqrt{2}$
26. $7\sqrt{2} < x < 14$
27. $x = 14$
28. $x > 14$
29. In which situation is $\angle B$ the smallest angle? Why?
30. Explain the mathematical significance of Colossians 1:16.

8 Area

God made water abundant on the earth for man's benefit. He made water before He made the land, and water still covers 75 percent of the surface of the earth. Our bodies are about 65 percent water, and we need to drink it to live. A person can live only a little over one week without water, but he can live weeks or months without anything else.

Most of the water God created is not easily available for drinking. A very small percentage (0.0001%) of all water is in the people and animals created by God. The oceans and seas of the world contain 96.5 percent of all water, and another 0.9 percent of all water is in lakes and springs that are also salty. The ice caps, glaciers, and permafrost contain another 1.8 percent, and soil moisture and the atmosphere each account for another 0.001 percent.

This means that only about 0.8 percent of all water is fresh water. This small percentage includes water in lakes, in rivers, and underground. Because fresh water is not easily procured, people store water in reservoirs and in water towers. Perhaps you will discover a way to remove the salt from sea water inexpensively.

A Christian can look forward to the day when he will be in heaven with the "pure river of water of life, clear as crystal, proceeding out of the throne of God and of the Lamb" (Rev. 22:1). God promised "I will give unto him that is athirst of the fountain of the water of life freely" (Rev. 21:6).

After this chapter you should be able to

1. define area.
2. state and use the postulates for area.
3. prove area formulas for various triangles and quadrilaterals.
4. prove and apply the Pythagorean theorem.
5. develop the formula for the area of a regular polygon.
6. develop the formula for the area of a circle.
7. find the surface areas of cylinders, cones, and spheres.
8. define the Platonic solids and find their surface areas.

8.1 Meaning of Area

Every believer should look forward to his future residence, the Holy City. In Revelation 21 God inspired John to give us a glimpse of the city whose residents' names are written in the Lamb's Book of Life (v. 27). Read Revelation 21 carefully. Notice that the measurement of the city is of great importance, for two entire verses are devoted to telling us about these measurements. From verses 16 and 17, we can find the volume of the city and the surface area of the city. Both surface area and volume are built on the basic concept of *area*.

What does *area* mean? On a line, which has one dimension, each segment corresponds to a number called the length of the segment. Likewise, in a plane, which has two dimensions, each region corresponds to a number called the *area*.

To find the length of a segment, you choose an arbitrary segment as a unit segment, and you measure segments with this arbitrary measure. When finding an area, you choose an arbitrary region to measure the area of a polygon. Any region can be chosen as the unit region. A triangular region, a rectangular region, or a hexagonal region can be used, but usually a square region is used. You can find the area by seeing how many square units will cover a given region.

The area of the rectangular region is the number of square units required to completely cover the region.

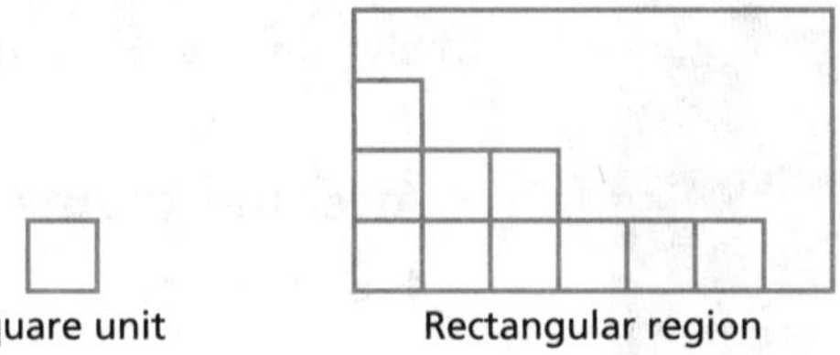

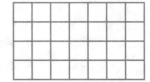

Square unit   Rectangular region

Figure 8.1

Notice that it is the rectangular region—not the rectangle itself—that you are finding the area of. Remember that a region is the union of the simple closed curve and its interior. What is the area of the rectangular region in figure 8.2?

Figure 8.2

The **area of a region** is the number of square units needed to cover it completely.

Postulate 8.1
Area Postulate. **Every region has an area given by a unique positive real number.**

Postulate 8.2
Congruent Regions Postulate. **Congruent regions have the same area.**

Postulate 8.3
Area of Square Postulate. **The area of a square is the square of the length of one side:** $A = s^2$.

The three postulates above give us a starting point for our study of area. The first one permits us to look for areas of regions. The second guarantees that the number will make sense. The Area of Square Postulate is the first postulate that tells us how to find an area without counting squares.

EXAMPLE 1   Find the area of square *ABCD*.

Answer $A = s^2$ 1. Apply the formula.
$A = (4)^2$
$A = 16$ square units

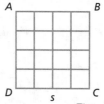

Figure 8.3

The web press at BJU Press prints 608,000 pages per hour for textbooks. This means that 3500 square feet of paper pass through the press every minute. Since it stores 292 square feet, it can continue printing while a new roll of paper is being installed.

Postulate 8.4

Area Addition Postulate. If the interiors of two regions do not intersect, then the area of the union is the sum of their areas.

This postulate tells us how to find the area of rectangle *ABPQ*.

The squares *ASTQ* and *SBPT* both have areas of 1600 square units according to Postulate 8.3. The Area Postulate tells us that

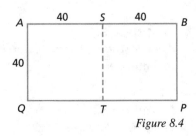

Figure 8.4

Area *ABPQ* = Area *ASTQ* + Area *SBPT*
$$= 1600 + 1600$$
$$= 3200 \text{ square units}$$

This computation was much faster than counting the 3200 squares, but you know that there is a simpler formula for areas of rectangles. $A = bh = 40 \cdot 80 = 3200$. Let's try to prove this formula.

Theorem 8.1

The *area of a rectangle* is the product of its base and height: $A = bh$.

Given: Rectangle *KLMN*

Auxiliary lines: Mark *P* on $\overrightarrow{KL}$ so *LP* = *h*

Mark *Q* on $\overrightarrow{NM}$ so *MQ* = *h*

Mark *R* on $\overrightarrow{KN}$ so *NR* = *b*

Mark *S* on $\overrightarrow{LM}$ so *MS* = *b*

Mark *T* where $\overrightarrow{RS}$ intersects $\overrightarrow{PQ}$

Prove: Area of *KLMN* = *b* · *h*

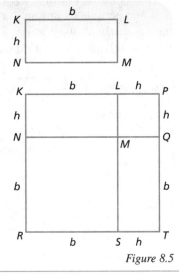

Figure 8.5

STATEMENTS	REASONS
1. Rectangle *KLMN*	1. Given
2. *LP* = *PQ* = *MQ* = *LM* = *ST* = *KN* = *h* and *NR* = *MS* = *QT* = *NM* = *RS* = *KL* = *b*	2. Auxiliary constructions
3. *LPQM*, *NMSR*, and *KPTR* are squares	3. Definition of square
4. *MQTS* is a rectangle	4. Definition of rectangle Continued ▶

5. *MQTS* and *KLMN* are congruent	**5.** Definition of congruent rectangles
6. Area *KPTR* = Area *KLMN* + Area *LPQM* + Area *NMSR* + Area *MQTS*	**6.** Area Addition Postulate
7. Area *KPTR* = $(b + h)^2 = b^2 + 2bh + h^2$ Area *LPQM* = h^2 Area *NMSR* = b^2	**7.** Area of Square Postulate
8. Area *KLMN* = Area *MQTS*	**8.** Congruent Regions Postulate
9. $b^2 + 2bh + h^2$ = Area *KLMN* + $h^2 + b^2$ + Area *KLMN*	**9.** Substitution (steps 7 and 8 into 6)
10. $2bh = 2 \cdot$ Area *KLMN*	**10.** Addition property of equality
11. bh = Area *KLMN*	**11.** Multiplication property of equality

Sometimes you will want to find the area of a polygonal region that is not convex or that has no special formula. To do this, you should separate the polygon into portions for which you can easily find the areas. After finding those areas, add them together to find the total area of the region. Notice how helpful the Area Addition Postulate is.

EXAMPLE 2 Find the area of the given polygonal region.

Answer

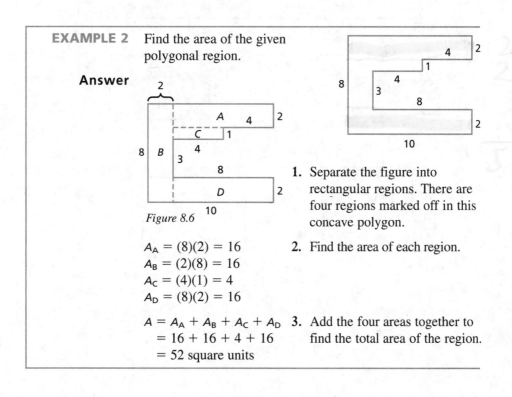

Figure 8.6

1. Separate the figure into rectangular regions. There are four regions marked off in this concave polygon.

2. Find the area of each region.

$A_A = (8)(2) = 16$
$A_B = (2)(8) = 16$
$A_C = (4)(1) = 4$
$A_D = (8)(2) = 16$

$A = A_A + A_B + A_C + A_D$
$\quad = 16 + 16 + 4 + 16$
$\quad = 52$ square units

3. Add the four areas together to find the total area of the region.

▶ A. Exercises

Complete the following tables.

	Square with side *s*	
	s	*A*
1.	15 yd.	
2.	6.7 in.	
3.	12 cm	
4.		324 sq. ft.
5.		32.49 sq. m

	Rectangles		
	b	*h*	*A*
6.	3 in.	12 in.	
7.	8 ft.	9 ft.	
8.	12 yd.	3.5 yd.	
9.	2.4 cm	10.9 cm	
10.	9 ft.		48 sq. ft.
11.		32 yd.	175 sq. yd.
12.	27 m		405 sq. m

Find the area of each polygonal region.

13.

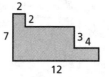

15.

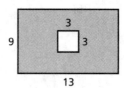

14.

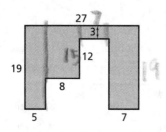

16.

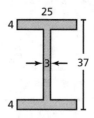

▶ B. Exercises

17. Find the area of your classroom floor in square feet and in square yards.
18. Find the area of your bedroom in square feet.
19. How many 8-inch square tiles would be needed to cover the floor of a room that is 12 by 15 feet?
20. Find the area, in square feet, of the foundation of the Holy City, according to Revelation 21:16.

Find the area of each rectangle.
21. $b = \sqrt{5}$ $\qquad$ $h = \sqrt{7}$
22. $b = x$ $\qquad$ $h = x + 5$
23. $b = x + 7$ $\qquad$ $h = x - 7$
24. $b = y + 5$ $\qquad$ $h = y + 7$

▶ C. Exercises

25. The inner square has its vertices at the midpoint of the sides of the outer square. Prove that the area of the outer square is double the area of the inner square.

▇ Cumulative Review

Find the perimeter of each region.
26. Rectangular region of exercise 19
27. Polygonal region of exercise 14
28. Circular region with diameter of $\sqrt{3}$ in.

29. Give bounds for the measure of angle x.

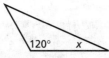

30. Give bounds for s.

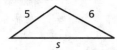

8.2 Other Polygons

Now that you understand the meaning of *area*, your next goal should be to find the area of familiar polygonal regions quickly.

This dolomite crystal from Monroe County, New York, has faces that are rhombi (rhombohedral).

This natrolite crystal from Bound Brook, New Jersey, is a right prism with rectangular bases (orthorhombic).

Theorem 8.2

The *area of a right triangle* is one-half the product of the lengths of the legs.

Given: Right △*ABC* with right angle at *C*

Draw: Auxiliary lines at *A* and *B* perpendicular to $\overleftrightarrow{AC}$ and $\overleftrightarrow{BC}$ respectively, thus forming quadrilateral *ACBD*

Prove: Area $= \frac{1}{2}bh$

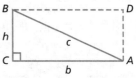

Figure 8.7

STATEMENTS	REASONS
1. △*ABC* with ∠*C* as the right angle	**1.** Given
2. $\overleftrightarrow{BC} \perp \overleftrightarrow{AC}$	**2.** Definition of right angle
3. $\overleftrightarrow{AC} \parallel \overleftrightarrow{BD}$; $\overleftrightarrow{BC} \parallel \overleftrightarrow{AD}$	**3.** Two lines perpendicular to the same line
4. $\overleftrightarrow{BD} \perp \overleftrightarrow{AD}$	**4.** Line perpendicular to one of two parallel lines is perpendicular to the other
5. ∠*D*, ∠*CAD*, ∠*CBD* are right angles	**5.** Definition of perpendicular
6. *ACBD* is a rectangle	**6.** Definition of rectangle
7. $\overline{BC} \cong \overline{AD}$, $\overline{AC} \cong \overline{BD}$	**7.** Opposite sides congruent
8. △*ABC* ≅ △*BAD*	**8.** LL (or SAS)
9. Area △*ABC* = Area △*BAD*	**9.** Congruent Regions Postulate
10. Area *ACBD* = Area △*ABC* + Area △*BAD*	**10.** Area Addition Postulate
11. Area *ACBD* = 2(Area △*ABC*)	**11.** Substitution (step 9 into 10)
12. Area *ACBD* = *bh*	**12.** Area of Rectangle Theorem
13. *bh* = 2(Area △*ABC*)	**13.** Substitution (step 12 into 11)
14. $\frac{1}{2}bh$ = Area △*ABC*	**14.** Multiplication property of equality

The formula for the area of a rectangle is $A = bh$. Likewise the formula for the area of a parallelogram region is $A = bh$, where *b* is the length of the base and *h* is the height of the parallelogram. The height, *h*, is always the perpendicular distance between the base and the opposite side.

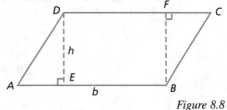

Figure 8.8

Notice that the area of the right triangles will be helpful in proving this area formula.

Theorem 8.3

The *area of a parallelogram* is the product of the base and the altitude: $A = bh$.

> *Given:* Parallelogram *ABCD*
> *Draw:* Altitudes from *B* and *D* to $\overline{CD}$ and $\overline{AB}$ respectively
> *Prove:* $A = bh$

STATEMENTS	REASONS
1. Parallelogram *ABCD*	**1.** Given
2. $\overline{BC} \cong \overline{AD}$	**2.** Opposite sides congruent
3. $\angle A \cong \angle C$	**3.** Opposite angles congruent
4. $\overleftrightarrow{DE} \perp \overleftrightarrow{AB}$; $\overleftrightarrow{BF} \perp \overleftrightarrow{CD}$	**4.** Definition of altitude
5. $\angle AED$ and $\angle CFB$ are right angles	**5.** Definition of perpendicular
6. $\triangle ADE \cong \triangle CBF$	**6.** HA (or SAA)
7. Area $\triangle ADE = \frac{1}{2}(AE)h$	**7.** Area of Right Triangle Theorem
8. Area $\triangle CBF = \frac{1}{2}(AE)h$	**8.** Congruent Regions Postulate
9. Area $BEDF = (BE)h$	**9.** Area of Rectangle Theorem
10. Area $ABCD$ = Area $\triangle ADE +$ Area $\triangle CBF$ + Area $BEDF$	**10.** Area Addition Postulate
11. Area $ABCD = \frac{1}{2}(AE)h + \frac{1}{2}(AE)h + (BE)h$	**11.** Substitution (steps 7, 8, and 9 into 10)
12. Area $ABCD = (AE + BE)h$	**12.** Distributive property
13. $AE + BE = AB = b$	**13.** Definition of betweenness
14. Area $ABCD = bh$	**14.** Substitution (step 13 into 12)

EXAMPLE 1 What is the area of parallelogram *ABCD*?

Answer $A = bh$
           $A = (25)(10)$
           $A = 250$ sq. in.

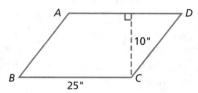

Figure 8.9

Thus 250 square-inch units could fit into this parallelogram region.
What about the area of a triangular region? Look at the following triangle.

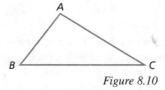

Figure 8.10

You have already learned the formula for the area of a right triangle, $A = \frac{1}{2}bh$.
This formula works for all triangles.

Theorem 8.4

The *area of a triangle* is one-half the base times the height: $A = \frac{1}{2}bh$.

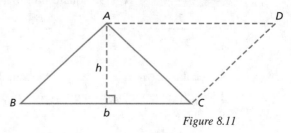

Figure 8.11

What kind of figure is polygon *ABCD*?

You know that the formula for the area of a parallelogram is $A = bh$. Notice that a diagonal such as $\overline{AC}$ divides the parallelogram into two congruent triangles, and thus the area is cut in half. So the area of a triangle is simply one-half the area of a parallelogram. The formal proof is exercise 18 and is similar to the proof of the area of a right triangle.

EXAMPLE 2   Find the area of $\triangle XYZ$.

Answer $A = \frac{1}{2}bh$

$A = \frac{1}{2}(12)(9)$

$A = 54$ square feet

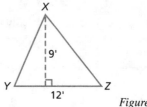

Figure 8.12

Notice that the height (altitude) must be perpendicular to the base and must intersect the opposite vertex. Similarly, the height of a trapezoid must be perpendicular to both of its bases.

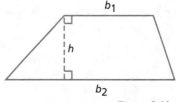

Figure 8.13

You will prove the formula for the area of a trapezoid in the exercises.

> ### Theorem 8.5
> The *area of a trapezoid* is one-half the product of the altitude and the sum of the lengths of the bases: $A = \frac{1}{2}h(b_1 + b_2)$.

EXAMPLE 3 Find the area of trapezoid *ABCD*.

Answer $A = \frac{1}{2}h(b_1 + b_2)$

$A = \frac{1}{2}(7)(18 + 10)$

$A = \frac{1}{2}(7)(28)$

$A = 98$ square millimeters

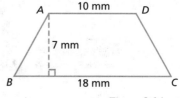

Figure 8.14

Notice that $\frac{1}{2}(b_1 + b_2)$ averages the lengths of the bases. For the trapezoid above, the average base length is $\frac{18 + 10}{2} = 14$. For the area, simply multiply this average by the height: $A = 14 \cdot 7 = 98$.

Since a rhombus is a parallelogram, you already know how to find its area.

EXAMPLE 4 Find the area of rhombus *FGHI*.

Answer $A = bh$

$A = 4(3)$

$A = 12$ square units

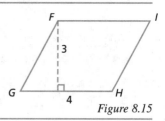

Figure 8.15

Here is another formula for the area of a rhombus.

In rhombus *ABCD* the lengths of the diagonals are d_1 and d_2. The area of the rhombus is $A = \frac{1}{2}d_1d_2$.

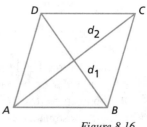

Figure 8.16

Theorem 8.6

The *area of a rhombus* is half the product of the lengths of the diagonals: $A = \frac{1}{2}d_1d_2$.

EXAMPLE 5   Find the area of rhombus *PQRS*.

    Answer $A = \frac{1}{2}d_1d_2$

              $A = \frac{1}{2}(12)(8)$

              $A = 48$ square units

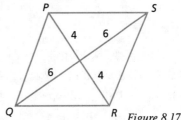

Figure 8.17

The trapezoid is sometimes called the "grandfather" of special quadrilaterals because rectangles, parallelograms, and squares are trapezoids. The formula for the area of a trapezoidal region should work for these other quadrilaterals. Do you think it does? Try it on one.

▶ A. Exercises

Make a summary table for the area formulas learned thus far.

	Figure	Formula
1.	Rectangle	
2.	Square	
3.	Triangle	
4.	Parallelogram	
5.	Trapezoid	
6.	Rhombus	

Find the area of each figure.

7.

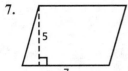

8.

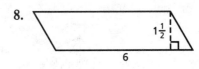

9.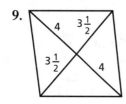

10.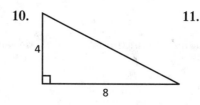

11.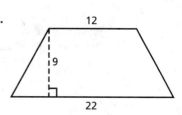

▶ B. Exercises

Find the area of the following:

12. A triangle with base 56 and height 24

13. A parallelogram with base 12 and height 6

14. A square with side 15

15. A rhombus with one diagonal equal to 27 and the other diagonal equal to 13

16. A trapezoid with bases equal to 32 and 14 and height measuring 25

17. Show how the formula for the area of a parallelogram can be obtained from the formula for the area of a trapezoid.

18. Prove Theorem 8.4: Area of triangle $= \frac{1}{2}bh$

19. Prove Theorem 8.5: Area of trapezoid $= \frac{1}{2}h(b_1 + b_2)$

20. Prove Theorem 8.6: Area of rhombus $= \frac{1}{2}d_1d_2$

21. Use the factoring method (from algebra) to solve the following quadratic equation: $x^2 + 2x - 15 = 0$.

22. The bases of a trapezoid are one and three feet longer than the height respectively. If the area is 63 square feet, find the height.

23. The base of a parallelogram is three more than the height. The area is $\frac{22}{9}$ square inches. Find the height.

▶ C. Exercises

24. Construct a rectangle with the same area as a given trapezoid.

25. Explain why a median of a triangle divides the triangular region into two regions of equal area.

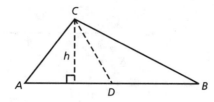

Sketch an example of each.
26. Octagon with two nonintersecting diagonals
27. Octahedron with vertices labeled
28. Closed curve that is not simple
29. Hexahedron that is a pyramid and has a concave base
30. Sphere with three radii, each perpendicular to the other two

8.3 Pythagorean Theorem

The Pythagorean theorem should be familiar to you. You have studied this relationship among the sides of a right triangle in previous math classes. Remember that this theorem works only with a right triangle.

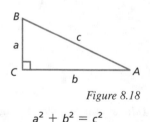

Figure 8.18

$$a^2 + b^2 = c^2$$

The second baseman at Camden Yards in Baltimore, Maryland, throws a ball to force a runner out at home. The distance thrown can be calculated from the baseline distances using the Pythagorean theorem.

The theorem is named after the ancient Greek mathematician Pythagoras, who first proved it.

Proof requires testing a statement against standards of truth. Math must be tested against Scripture and against God's laws of reasoning. In everyday life we must likewise prove everything to make sure that it is true (I Thess. 5:21). To do so, we must daily go to the Word of God (Acts 17:11).

Figure 8.19 illustrates the Pythagorean theorem.

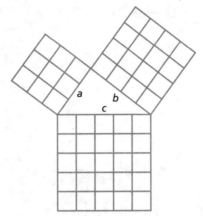

$$a^2 + b^2 = c^2$$

$$3^2 + 4^2 = 5^2$$

$$9 + 16 = 25$$

$$25 = 25$$

Figure 8.19

Theorem 8.7

Pythagorean Theorem. **In a right triangle, the sum of the squares of the lengths of the legs is equal to the square of the length of the hypotenuse:** $a^2 + b^2 = c^2$.

One proof of the Pythagorean theorem uses figure 8.20.

Find the areas of each part using formulas for the area of a rectangle ($A = bh$) and the area of a triangle ($A = \frac{1}{2}bh$).

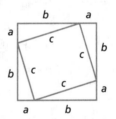

Figure 8.20

$$A_{\text{lg. square}} = (a + b)(a + b) = a^2 + 2ab + b^2$$

$$A_{\text{sm. square}} = c \cdot c = c^2$$

$$A_{\text{triangle}} = \frac{1}{2}ab$$

The sum of the areas of the four triangles and the smaller square equals the area of the larger square.

$$A_{\text{lg. square}} = 4A_{\text{triangle}} + A_{\text{sm. square}}$$

$$a^2 + 2ab + b^2 = 4(\tfrac{1}{2}ab) + c^2$$

$$a^2 + 2ab + b^2 = 2ab + c^2$$

$$a^2 + b^2 = c^2$$

The converse of this theorem is also true. The converse states, "If the sum of the squares of the lengths of two sides of a triangle equals the square of the length of the third side, then the triangle is a right triangle." You will prove this in the exercises, and you can always determine whether a triangle is a right triangle by using this theorem.

The Pythagorean theorem helps in the proof of a special formula for the area of an equilateral triangular region.

Theorem 8.8

The *area of an equilateral triangle* is $\frac{\sqrt{3}}{4}$ times the square of the length of one side: $A = s^2 \frac{\sqrt{3}}{4}$

EXAMPLE   Find the area of equilateral $\triangle LMN$.

Answer $A = s^2 \dfrac{\sqrt{3}}{4}$

$A = 7^2 \dfrac{\sqrt{3}}{4}$

$A = \dfrac{49\sqrt{3}}{4}$ square centimeters

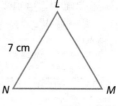

7 cm

Figure 8.21

$|+|$

▶ A. Exercises

Complete the following table. Consider $\triangle ABC$ to be a right triangle with c as the hypotenuse.

	a (units)	b (units)	c (units)
1.		3	5
2.		5	13
3.	9	2	
4.	3	10	
5.	$\sqrt{2}$	$\sqrt{3}$	
6.		32	40
7.	6	4	
8.	8		30
9.	x	5	
10.		x	$\sqrt{x^2 + 81}$

Tell which of the following triangles are right triangles.

11.

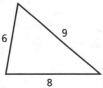

13.

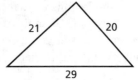

12.

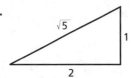

14.

▶ B. Exercises

Find each area. (*Hint:* Find the height first if necessary.)

15. A right triangle has a leg measuring 5 inches and a hypotenuse measuring 8 inches.

16.

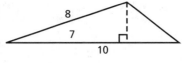

18.

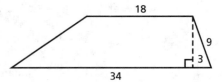

17.

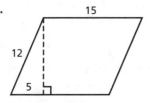

19.

20. Show that for an isosceles right triangle $c = a\sqrt{2}$.
21. The bases of an isosceles trapezoid are 6 inches and 12 inches. Find the area if the congruent sides are 5 inches.
22. The perimeter of a rhombus is 52 and one diagonal is 24. Find the area.

Prove the formulas. Use paragraphs or derivations instead of columns.

23. The altitude of an equilateral triangle is given by $h = \dfrac{\sqrt{3}}{2}s$ where s is the length of one side.

24. Theorem 8.8: $A = s^2 \dfrac{\sqrt{3}}{4}$ (*Hint:* Substitute result of exercise 23 into area of triangle formula.)

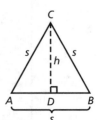

► C. Exercises

25. Prove the converse of the Pythagorean theorem: If the sum of the squares of two sides of a triangle equals the square of the third side, then the triangle is a right triangle.

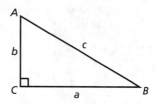

► Dominion Thru Math

When an electrician runs electrical conduit, he must often make the path shift to avoid an obstacle. To minimize friction when putting wire in the conduit, the shift is often made with two 30° (instead of 90° or 45°) angles.

26. What theorem or postulate assures the electrician that the new path of the conduit is parallel to the previous path?

27. If the shift in the path is x inches, explain why it takes $2x$ inches of conduit to make this shift.

28. How soon must the shift in the path begin in order to avoid the obstacle?

Cumulative Review

Find the area of each rectangle.

29.

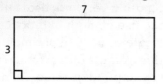

30.

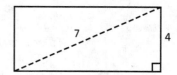

31. Give bounds for c if $b < 5$.

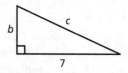

32. The consecutive sides of a rectangle have a ratio of 4:5. If the area is 5120 m², what are the dimensions of the rectangle?

33. The figure is made up of 8 congruent squares and has a total area of 968 cm². Find the perimeter.

HERON of ALEXANDRIA

Heron, also called Hero, was an ancient geometer who proved an amazing formula as you will soon see. Exactly when Heron lived is unknown, but guesses from 80 B.C. to A.D. 100 have all been proposed. For present purposes, you will most easily understand his circumstances by relating him to the time of Christ and the apostles (ca. 10 B.C. to A.D. 70). However, he probably lived in Alexandria, Egypt.

Heron considered math a very practical subject. All four of his books make use of math in practical ways. *Geometrics* presents calculations of perimeter, area, and volume—necessary for weights and measures in business. In *Pneumatics* he applied geometry to hydraulic engineering to design steam engines and water pumps. His book on optics, *Catoptrics*, applies geometry to show that the angle of incidence equals the angle of reflection and that light follows the shortest distance between two points. *Automata* explains the design of small machines and mechanical toys. Heron learned mathematics well from his teacher Ctesibus.

This practical application approach to geometry is called geodesy, while the earlier theoretical approach of Euclid, stressing reasoning, is called classical geometry. Heron, however, included proofs for some of his formulas

and applications. His famous book *Metrica* contains the proof of the area formula, now named in his honor.

Heron's formula: The area *A* of a triangle is

$$A = \sqrt{s(s - a)(s - b)(s - c)}$$

where *a*, *b*, *c* are the lengths of the sides and the

semiperimeter $s = \dfrac{(a + b + c)}{2}$.

Heron's geodesy has obvious benefits, but you should see that Euclid's geometry developed the theory that permitted Heron to accomplish his work. Geometry develops reasoning skills; geodesy applies those skills to specific problems. Classical geometry continued to flourish in Greek culture after Euclid until the time of Hipparchus of Nicaea (180-125 B.C.). No theoretical advances were made after Hipparchus until the time of Ptolemy (A.D. 100-168). During the interim, geodesy displaced classical geometry and works such as Heron's filled an important role.

The same tension between the theoretical and the practical occurs in other branches of math also. Classical arithmetic investigated the theoretical aspects of numbers, while logistics applied arithmetic to business problems. Are you surprised? In the Bible you can read epistles that begin with theory (doctrine) and end with practical applications (Ephesians, Galatians, Philippians, and Romans). Theory and practice go together. You should seek to develop a balance of theory and practice in all you do. Appreciate the classical geometers for their careful reasoning, but learn from Heron the value of applying what you know to life. Heron learned all he could and then he set about to use it. Will you?

His famous book Metrica contains the proof of the area formula, now named in his honor.

Heron's Formula

Heron's formula is used to find the area of a triangle when only the lengths of the sides are known. If you know the length of the altitude of a triangle, you can find the area by the formula $A = \frac{1}{2}bh$. But if you do not know the height, you can use Heron's formula to find the area. The formula uses a number called the semiperimeter.

Definition

The **semiperimeter** of a triangle is one-half of the perimeter of a triangle: $s = \frac{a+b+c}{2}$.

Heron's Formula

If $\triangle ABC$ has sides of lengths a, b, and c and semiperimeter s, then the area of the triangle is $A = \sqrt{s(s-a)(s-b)(s-c)}$.

EXAMPLE Find the area of the triangle shown.

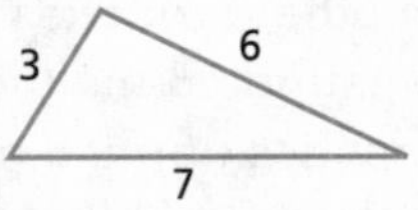

Answer $s = \dfrac{(3+6+7)}{2} = 8$

$A = \sqrt{s(s-a)(s-b)(s-c)}$

$ = \sqrt{8(8-3)(8-6)(8-7)}$

$ = \sqrt{8 \cdot 5 \cdot 2 \cdot 1}$

$ = \sqrt{80}$

$ = 4\sqrt{5}$

You can prove this formula using analytic geometry. The proof is based on the figure shown.

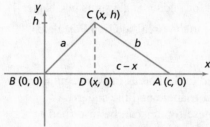

▶ Exercises

Find the area of each triangle that has the given side measures. Round your answers to the nearest tenth.

1. 3 units, 8 units, 9 units
2. 6 units, 18 units, 21 units
3. 27 units, 13 units, 18 units

Find the areas of the figures below. Round your numbers to the nearest tenth.

4.

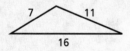

5.

8.4 Regular Polygons

Every polygon has an area. The area can be obtained by subdividing the polygon into triangles (as you did for finding angle sums in Chapter 4). Then add the areas of the triangles according to the Area Addition Postulate.

You have already learned to find areas of two regular polygons: the equilateral triangle and the square. You should remember that these figures can be inscribed or circumscribed in a circle. In fact, every regular polygon has both an inscribed and a circumscribed circle. Both circles are shown for the regular heptagon in figure 8.22. Notice that the circles have a common center. The incenter and circumcenter are the same point.

Figure 8.22

These amethyst quartz crystals from Idaho Springs, Colorado, are classified as hexagonal because of the cross sections.

These apophyllite crystals from Paterson, New Jersey, form square-based right prisms (tetragonal).

The **center of a regular polygon** is the common center of the inscribed and circumscribed circles of the regular polygon.

The **radius of a regular polygon** is a segment that joins the center of a regular polygon with one of its vertices.

The **apothem of a regular polygon** is the perpendicular segment that joins the center with a side of the polygon.

These three special parts of a regular polygon are illustrated in figure 8.23. The center is C, the radius is r, and the apothem is a. The angle formed by two radii drawn to consecutive vertices has a special name.

Figure 8.23

> A **central angle of a regular polygon** is the angle formed at the center of the polygon by two radii drawn to consecutive vertices.

$\angle RPQ$ is a central angle of the octagon. Notice that there are eight central angles in all, one for each side.

Figure 8.24

Theorem 8.9

The central angles of a regular n-gon are congruent and measure $\frac{360°}{n}$.

EXAMPLE 1   What are the measures of the central angles and the base angles of the regular octagon?

Answer $m\angle RPQ = \dfrac{360°}{8} = 45°$

1. By Theorem 8.9, the eight central angles divide the circle into eight congruent parts.

$180° - 45° = 135°$

2. Find the remaining degrees in the triangle.

$m\angle PRQ = m\angle PQR$
$= \dfrac{135°}{2} = 67.5°$

3. $\triangle RPQ$ is isosceles because the radii are congruent. Since the base angles must be congruent, each angle measures half of the remaining degrees in the triangle.

The next theorem enables you to find areas of regular polygons.

Theorem 8.10

The *area of a regular polygon* is one-half the product of its apothem and its perimeter: $A = \frac{1}{2}ap$.

Theorem 8.11

The apothem of an equilateral triangle is one-third the length of the altitude: $a = \frac{1}{3}h$.

Theorem 8.12

The apothem of an equilateral triangle is $\sqrt{3}$ times one-sixth the length of the side: $a = \dfrac{\sqrt{3}}{6}s.$

EXAMPLE 2 Find the area of the following regular octagon.

Answer $p = ns$ **1.** Find the perimeter.
$\quad\quad = 8 \cdot 10$
$\quad\quad = 80$

$A = \frac{1}{2}ap$ **2.** Find the area.

$A = \frac{1}{2}(12)(80)$

$A = 480$

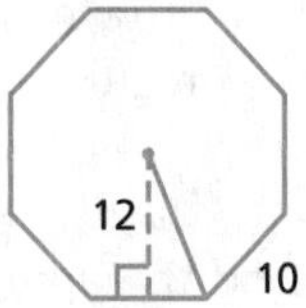

Figure 8.25

EXAMPLE 3 Find the area of the regular hexagon.

Figure 8.26

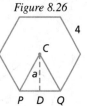

Answer $p = 6 \cdot 4 = 24$ units   **1.** The hexagon has six sides, each having a length of 4 units.

$\dfrac{360°}{6} = 60°$ **2.** Find the central angle.

$\dfrac{180 - 60}{2} = 60°$ **3.** Find the base angles.

$CP = CQ = PQ = 4$ **4.** $\triangle CPQ$ is equiangular and therefore equilateral.

$PD = QD$ **5.** You can see that the apothem always bisects the side of a regular polygon (use HA).

$2^2 + a^2 = 4^2$ **6.** To find the apothem, use the
$a^2 = 16 - 4 = 12$ Pythagorean theorem.
$a = \sqrt{12}$ or $2\sqrt{3}$

$A = \frac{1}{2}ap$ **7.** To find the area, use the formula.

$A = \frac{1}{2}(2\sqrt{3})(24)$

$A = 24\sqrt{3}$
(or $A \approx 41.6$ square units)

As you calculate areas of regular polygons, remember that the apothem is always the perpendicular bisector of a side and that the radius always bisects an angle.

▶ **A. Exercises**

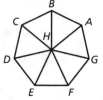

1. Name the central angles in this heptagon.
2. How many degrees are in a central angle of a regular heptagon?

Find the area of each regular polygon using Theorem 8.10.

3.

14

4

5.

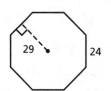

29   24

7.

4

4.

9.6   14

6.

27

26

8.

12

▶ **B. Exercises**

Complete the following table.

	Number of sides of a regular polygon	Length of apothem (units)	Length of side (units)	Length of radius (units)	Area (sq. units)
9.	4		12		
10.	6			15	
11.	5	8		9.9	
12.	9	24.7	18		
13.	3		6		

Prove the formulas in exercises 14-16. However, use derivations with formulas rather than two-column proofs.

14. Prove Theorem 8.10 for the case of a regular quadrilateral by showing that the formula for the area of a regular quadrilateral reduces to the formula for the area of a square.

15. Theorem 8.11. Use the two formulas for the area of a triangle to show that $a = \frac{h}{3}$ for an equilateral triangle.

16. Theorem 8.12. For an equilateral triangle $a = \frac{s\sqrt{3}}{6}$. (*Hint:* Exercise 23 in section 8.3 may help.)

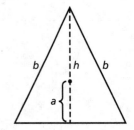

17. Each central angle of a regular polygon forms a triangle with a side of the polygon. Why must the base angles of these triangles be congruent?

Use the portion of the regular *n*-gon shown at the right to help you complete the following proofs.

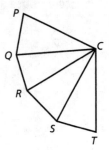

18. The triangles formed by the central angles of a regular *n*-gon are congruent. (Prove $\triangle CPQ \cong \triangle CRQ$.)

19. Use exercise 18 and example 3 to find a formula for the area of a regular hexagon.

▶ C. Exercises

20. Prove Theorem 8.10 that for any regular *n*-gon its area can be found by the formula $A = \frac{1}{2}ap$. (*Hint:* Form *n* congruent triangles in the *n*-gon; then find the area of each triangle.)

■ Cumulative Review

Give the name for each.
21. A cylinder with polygonal bases
22. A cone with a polygonal base
23. A pair of lines that are neither parallel nor intersecting
24. The form of argument shown

    $p \rightarrow q$

    p

    Therefore q.
25. Adjacent angles in which the noncommon sides form opposite rays

8.5 Circles

Umbrellas display radial symmetry, congruent parts around a central axis.

In the last section you learned that the formula for the area of a regular polygon is $A = \frac{1}{2}ap$, where a is the apothem and p represents the perimeter. Look at a regular polygon and a circle that is circumscribed about the polygon.

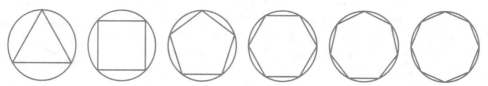

Figure 8.27

As you look from the left to the right, notice that the length of the apothem gets closer to the radius of the circle and that the perimeter gets closer to the circumference of the circle. Furthermore, the area of the polygon approaches the area of the circumscribed circle. This relationship enables us to predict the formula for the area of a circle from the formula for the area of the inscribed regular polygon.

In summary,

$a \rightarrow r$ (apothem approaches radius)

$p \rightarrow c$ (perimeter approaches circumference)

$A_{n\text{-gon}} \rightarrow A_{circle}$ (area of n-gon approaches area of circle)

So the formula $A_{n\text{-gon}} = \frac{1}{2}ap$ becomes $A_{circle} = \frac{1}{2}rc$. To simplify, substitute the formula for circumference, $c = 2\pi r$.

$$A = \frac{1}{2}rc$$

$$A = \frac{1}{2}r(2\pi r)$$

$$A = \pi r^2$$

Theorem 8.13

The *area of a circle* is pi times the square of the radius: $A = \pi r^2$.

▶ **A. Exercises**

Complete the table.

	c (units)	r (units)	d (units)	A (sq. units)
1.		9		
2.		7		
3.			4	
4.	π			
5.			1.6	
6.				0.04π
7.	30π			
8.		22		
9.				25π
10.				0.56π

▶ **B. Exercises**

Find the area of the shaded portion of each figure.

11.

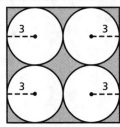

14.

12.

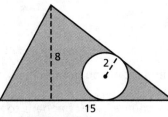

15.

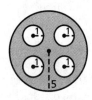

13.

16.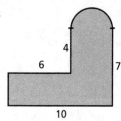

17.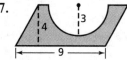

18. Find the shaded area if the pentagon is regular.

19. An *annulus* is the ring formed by two concentric circles with different radii. Derive a formula for the area of an annulus formed by two circles with radii *x* and *y*, where *x* > *y*.

▶ C. Exercises

20. Find the shaded area if *ABCD* is a square. (*Hint:* The arcs are semicircles.)

■ Cumulative Review

Define each.

21. Between
22. Parallelogram
23. Congruent triangles
24. Perpendicular lines
25. Median of a triangle

1. To double the area of a given rectangle, can you double the lengths of consecutive sides? Explain. What should you do?
2. If you triple the length of a side of a square, what happens to the area?
3. If you multiply the radius of a circle by a factor of five, what happens to the area?
4. What percentage should Bill select on the photocopier to achieve an enlargement of his picture that doubles its area?

8.6 Surface Areas of Prisms and Cylinders

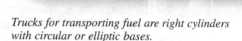

In Chapter 2 you learned that cylinders and cones with polygonal bases are called prisms and pyramids respectively. In fact, prisms and pyramids are classified by their bases. A triangular prism has triangles as bases, while a pentagonal pyramid has a pentagon for a base. Now you will learn to find the surface areas of these three-dimensional solids.

Trucks for transporting fuel are right cylinders with circular or elliptic bases.

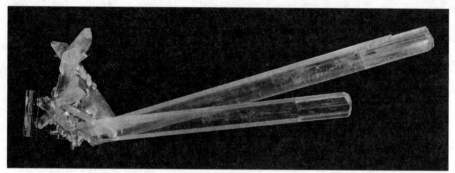

Gypsum from the Cave of Swords, Mexico, forms right prisms with parallelogram bases (monoclinic).

This chalcanthite crystal from Utah forms an oblique prism with parallelogram bases (triclinic).

You already know that the word *area* describes the number of squares needed to cover a region. Similarly, *surface area* is the number of square units needed to cover the outer shell of a solid in space.

Just what does the word *surface* mean? Mathematically, *surface* means "the boundary of a three-dimensional figure." Another meaning of *surface* is "the outward appearance." Are you more than a surface Christian? People who show only an outward appearance of Christianity but do not have a changed heart are not really saved. Surface Christianity is just a pretense. As Romans 10:10 clearly states, true Christianity is an inner change of the heart. "For with the heart man believeth unto righteousness," and man becomes a new creature (II Cor. 5:17). Make sure that you are not just a surface Christian.

To find the surface area of a prism, you must find the number of square units that would cover the two bases and all the lateral faces. Each lateral face is a parallelogram, and the height of a prism is always the perpendicular distance between the two bases.

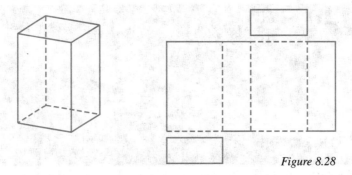

Figure 8.28

For a right prism, the height H of the prism is the same as the length of a lateral edge. This makes it easy to derive a formula for surface area. The sum of the areas of all the lateral faces is called the lateral surface area (L). In figure 8.28, the bases were cut out by cutting along the perimeter p of the base, and then an edge of the prism was slit and laid open. The union of all the lateral faces forms one large rectangle. The base of this rectangle is the perimeter p of the base of the prism. The height of the rectangle is the height H of the prism. The base times the height gives the lateral area.

$$L = pH$$

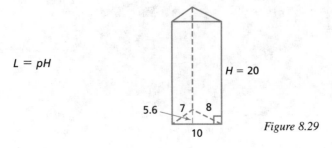

$H = 20$

5.6   7   8

10

Figure 8.29

The perimeter of the base of the triangular prism shown is $p = 10 + 8 + 7 = 25$. The lateral surface area is as follows.

$L = pH$

$L = 25(20)$

$L = 500$ square units

To find the total surface area, add the area of the two bases to the lateral surface area. This results in the following formula, where B represents the area of each base.

$S = L + 2B$

Now how do you find the area of the bases? The bases are triangles, so use the triangle formulas that you have learned.

$B = \frac{1}{2}bh$

Remember that h (lowercase) here represents the height of the triangle, not the height of the prism.

$B = \frac{1}{2}(10)(5.6)$

$B = 28$ square units

The surface area of this triangular prism is calculated below.

$S = L + 2B$

$S = 500 + 2(28)$

$S = 500 + 56$

$S = 556$ square units

This means that you could cover the triangular prism with 556 square units.

Theorem 8.14

The *surface area of a prism* is the sum of the lateral surface area and the area of the bases: $S = L + 2B$.

The lateral surface area of a right prism is the product of its height and the perimeter of its base: $L = pH$.

A cube is a special case of a rectangular prism. Since a cube has six congruent faces, you should be able to apply this theorem to show that the surface area of a cube is $6s^2$ (see exercise 17). Another special case is that of a regular prism. A *regular prism* is a right prism with a regular polygon as its base.

The development of the formula for the surface area of a right circular cylinder is similar to that of a right prism.

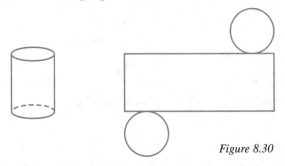

Figure 8.30

When the lateral surface of a cylinder is unrolled and laid out in a plane, you again obtain a rectangle. The base and height of the rectangle still correspond to the perimeter and height of the cylinder. However, the correct term for the perimeter of the base is *circumference* since the base is a circle:

$$L = cH$$

where c is the circumference of the cylinder. Remember that the height (H) is always the perpendicular distance between the two bases.

To find the total surface area of a cylinder, simply add the area of the two bases to the lateral surface area.

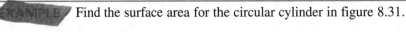

EXAMPLE Find the surface area for the circular cylinder in figure 8.31.

Answer
$S = L + 2B$
$S = cH + 2B$
$S = 2\pi rH + 2\pi r^2$
$S = 2\pi(6)(9) + 2(36)\pi$
$S = 108\pi + 72\pi$
$S = 180\pi$ square units ≈ 565 square units

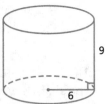

9

6

Figure 8.31

Theorem 8.15

The *surface area of a cylinder* is the sum of the lateral surface area and the area of the bases: $S = L + 2B$.

The lateral surface area of a right cylinder is the product of its circumference and height: $L = cH$.

▶ A. Exercises

1. Find the lateral surface area of the right prism if the base is a square.

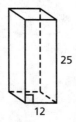

25

12

2. Find the surface area of the right prism shown.

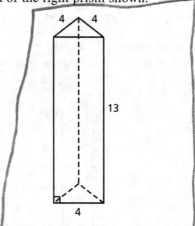

4   4

13

4

Find the lateral surface areas and total surface areas of the following figures. The bases in exercises 3, 5, and 8 are regular.

3.

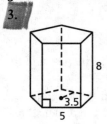

8

3.5

5

5.

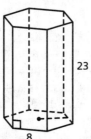

23

8

7.

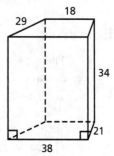

18

29

34

21

38

4.

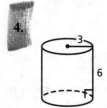

3

6

6.

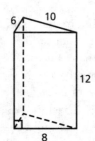

6   10

12

8

8.

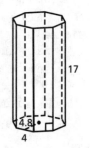

17

4.8

4

9. The local water tower is a right circular cylinder standing 21 meters high and having a radius of 7 meters. Its exterior needs to be painted. How many square meters of paint will you need for one coat?

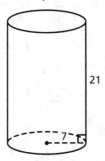

10. The diagram shows a vertical mine shaft from which rocks have been falling. How much sheet metal is needed to cover the sides of the shaft as a safety measure?

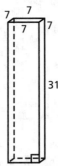

▶ B. Exercises

11. The diameter of the base of a right circular cylinder is 12 inches, and the height is 29 inches. What is the surface area of the cylinder?

12. The surface area of a right prism is 224 square feet, and the length of a side of the square base is one-third the height. What are the dimensions of the prism?

13. The surface area of a cube is 1350 square inches. Find the dimensions of this cube.

14. The surface area of a right circular cylinder is 1248π square feet. The radius of the base of the cylinder is 7 feet less than the height. Find the radius and the height of the cylinder.

15. Find the lateral surface area of a right circular cylinder whose diameter is $10\sqrt{3}$ feet and whose height is 27 feet.

16. The surface area of a cylinder is 500 square feet; the lateral surface area is 320 square feet. Give the area of a base.

17. Prove that the surface area of a cube is $6s^2$.
18. Prove that the surface area of a right circular cylinder is
    $$S = 2\pi r(H + r).$$

▶ C. Exercises

19. Prove that the surface area of a right prism with bases that are regular
    n-gons is $ns(H + a)$.
20. Find the surface area of the napkin ring.

▶ Dominion Thru Math

Chemical pipes need to be lined on the inside
with a chemical-resistant material like glass,
polypropylene, or Teflon. Since pipes are
elongated cylinders, knowing the interior lateral
surface area is important for finding the
amount of lining needed.

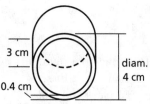

21. How much polypropylene lining is needed
    for 110 m of piping with an interior diameter of 150 mm?
22. How much Teflon lining is needed for 290 m of piping with an interior
    diameter of 100 mm?
23. If polypropylene lining material costs $45/sq. m and Teflon costs
    $95/sq. m, how much will the lining for the piping system cost?

■ Cumulative Review

Define each term.
24. circle
25. tangent
26. supplementary angles
27. congruent angles
28. circumcenter

8.7 Surface Areas of Pyramids and Cones

In Chapter 2 you studied cones and pyramids. Since a pyramid is a special type of cone, we will begin with the pyramid. Before finding its surface area, you must first learn its essential parts.

This triangular pyramid has four faces: three lateral faces and a base. The common point above the base where the lateral faces intersect is the *vertex*. The *altitude* (*H*) of a pyramid is a segment that extends from the vertex, perpendicular to

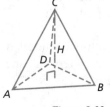

Figure 8.32

The conveyor dumps processed materials into the top of the hopper for storage. Tapering at the bottom into a pyramid, the hopper functions like a funnel to dispense the product as necessary.

the plane of the base. The lateral edges are the segments formed by the intersection of the lateral faces. In figure 8.32, $\triangle ABD$ is the base, point *C* is the vertex, $\overline{AC}$ is a lateral edge, and $\triangle CBD$ is one of the lateral faces. The surface area consists of the lateral surface area and the area of the base: $S = L + B$.

A *regular pyramid* is a right pyramid that has a regular polygon as its base. The formula that we will develop for the surface area of a pyramid can be used only for regular pyramids. What kind of figure is each lateral face of a regular pyramid?

For a regular pyramid, you can calculate the base area and the lateral surface area easily. To find the lateral surface area, find the area of one of the congruent isosceles triangles and multiply that area by the number of sides of the base. The slant height is labeled *l* in figure 8.33. The *slant height* of a regular pyramid is the height of a triangular face of the pyramid.

Figure 8.33

EXAMPLE 1 Find the surface area of the regular pyramid.

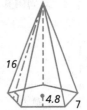

Figure 8.34

Answer	
$A = \frac{1}{2}bh$	**1.** Use the slant height to find the area of a triangular face.
$A = \frac{1}{2}(7)(16)$	
$A = 56$	
$L = nA$	**2.** The lateral surface area consists of five such triangles.
$= 5 \cdot 56$	
$= 280$ sq. units	
$p = ns = 5 \cdot 7 = 35$	**3.** Find the perimeter of the base.
$B = \frac{1}{2}ap$	**4.** Find the base area.
$= \frac{1}{2}(4.8)(35) = 84$	
$S = L + B = 280$	**5.** Find the surface area.
$+ 84 = 364$ sq. units	

In general, the formula for the lateral surface area of a regular pyramid is

$$L = \frac{1}{2}bln,$$

where b is the length of a side of the base, n is the number of sides of the base, and l is the slant height. Since bn is the same as the perimeter, p, the formula could be written as follows:

$$L = \frac{1}{2}pl$$

Theorem 8.16

The *surface area of a pyramid* is the sum of the lateral surface area and the area of the base: $S = L + B$.

By using this theorem and some algebra, we can derive a practical formula for the surface area of a regular pyramid. Notice the use of factoring with the distributive property.

$$S = \frac{1}{2}pl + \frac{1}{2}ap$$

$S = \frac{1}{2}p(l + a)$, where p is the perimeter of the base, l is the slant height, and a is the length of the apothem.

In some regular pyramids, you can figure out some of the lengths without being told.

EXAMPLE 2 Find the surface area of this regular pyramid.

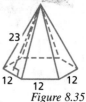

23

12   12   12

Figure 8.35

Answer

$p = ns$

    $= 6(12)$

    $= 72$

1. Find the perimeter of the hexagonal base.

$L = \frac{1}{2}pl$

    $= \frac{1}{2}(72)(23)$

    $= 828$ sq. units

2. Find the lateral surface area. The slant height is 23 units.

$a^2 + b^2 = c^2$

$a^2 + 6^2 = 12^2$

$a^2 = 144 - 36$

$a^2 = 108$

$a = 6\sqrt{3}$ or 10.4

3. Apply the Pythagorean theorem to find the apothem a, which is a leg of a right triangle. The other leg is 6 (half the base), and the hypotenuse (radius) is 12.

$B = \frac{1}{2}ap$

    $= \frac{1}{2}(10.4)(72)$

    $= 374.4$ sq. units

4. Find the area of the base.

$S = L + B$

$S = 828 + 374.4$

$S = 1202.4$ sq. units

5. Find the surface area.

Triangular and hexagonal pyramids are classes of pyramids, while pyramids themselves are a type of cone. Classification systems help you organize information. You will classify other figures later in this book, but systems of classification are also important outside mathematics. Christians, for example, should display the fruit of the Spirit (Gal. 5:22-26). They are to be a peculiar class of people so that they can draw other people to Christ (I Pet. 2:9).

Do you recall how the formula to find the area of a circle was developed from the formula for the area of a regular polygon? The same procedure will be used here to develop the formula for the surface area of a cone. As the number of sides of the base of the regular pyramid increases, the pyramid approaches the shape of a cone. The formulas for surface area are also very similar. Figure 8.36 shows the development and then the essential elements of the cone.

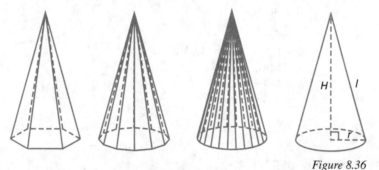

Figure 8.36

In the last figure, H is the height of the cone, l is the slant height, and r is the radius of the circular base. Now compare the surface-area formulas. Remember that perimeter approaches circumference, and apothem approaches radius.

	Pyramid	Cone
Lateral area	$L = \frac{1}{2}pl$	$L = \frac{1}{2}cl$
Surface area	$S = L + B$	$S = L + B$
	$S = \frac{1}{2}pl + \frac{1}{2}ap$	$S = \frac{1}{2}cl + \frac{1}{2}rc$
	$S = \frac{1}{2}p(l + a)$	$S = \frac{1}{2}c(l + r)$

Theorem 8.17

The *surface area of a cone* is the sum of the lateral surface area and the area of the base: $S = L + B$; the lateral surface area of a circular cone is half the product of the circumference and slant height: $L = \frac{1}{2}cl$.

Using the theorem above and some algebra, you can derive a practical formula for the area of a cone. For a circular cone, $c = 2\pi r$ and $B = \pi r^2$.

$$S = L + B$$
$$S = \frac{1}{2}cl + B$$
$$S = \frac{1}{2}(2\pi r)l + \pi r^2$$
$$S = \pi rl + \pi r^2$$

EXAMPLE Find the surface area of the cone.

Answer

$L = \pi r l$ 1. Find the lateral area.
$L = \pi(6)(18)$
$L = 108\pi$

$B = \pi r^2$ 2. Find the area of the base.
$B = \pi \cdot 6^2$
$B = 36\pi$

$S = L + B$ 3. Add L and B.
$S = 108\pi + 36\pi$
$S = 144\pi$

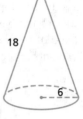

Figure 8.37

▶ **A. Exercises**

Complete the following tables.

	Apothem length (units)	Slant height (units)	Number of sides in the base	Length of one side of the base (units)	Lateral area (sq. units)	Surface area (sq. units)
Regular Right Pyramids						
1.	9.3	23	7	9		
2.	4.2	9	7	4		
3.		15	4	24		
4.	1.4	12	9	1		
5.	$4\sqrt{3}$	8	6			
6.	3.4	10	5		125	
7.	3.7	9		2		152.4
8.		15	3	7		
9.		8		11	132	
10.	4.8		8	4		288

Right Circular Cones				
	Radius (units)	Slant height (units)	Lateral area (sq. units)	Surface area (sq. units)
11.	9	12		
12.	5	6		
13.		6	18π	
14.	10			220π
15.	4		112π	

▶ B. Exercises

16. Tepees are made with buffalo hides in the shape of a cone and have dirt floors. One buffalo hide provides 60 square feet of leather. How many hides are necessary to make a tepee that is 12 feet high and 12 feet in diameter?

17. Give the dimensions of the square piece of paper needed to make the dunce's cap shown below.

 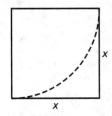

18. How much paint is needed to paint the steeple of Heritage Bible Church if one gallon covers 250 square feet per coat? The steeple needs three coats.

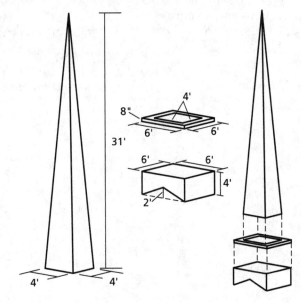

19. Derive and prove a factored formula for the surface area of a right pyramid with a square base, where l = slant height and s = length of a side of the square.

▶ C. Exercises

20. Slice a cone parallel to its base to obtain a *frustum* of the cone. Find the total surface area of the frustum. (*Hint:* The part removed is proportional to the whole.)

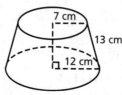

21. Rotate a square about one of its diagonals to form a conical figure. If one side of the square is 6 m, what is the surface area of the solid?

■ Cumulative Review

Match. Assume that each pair of rays has at least one point in common.
Use each answer once.

22. segment   **A.** union of two collinear rays

23. angle   **B.** intersection of two collinear rays

24. line   **C.** union of two noncollinear rays

25. point   **D.** intersection of two noncollinear rays

8.8 Surface Areas of Polyhedra and Spheres

You are now ready to find surface areas for two more polyhedra and spheres.

*Galena crystals from Galena, Kansas, form in cubes (*isometric*).*

Definition

A **sphere** is the set of all points in space equidistant from a given point.

The *center* of the sphere is the given point that is used as a reference to all other points of the sphere. The *radius* of a sphere is a segment that connects the center of the sphere to any point of the sphere. A plane that passes through the center of the sphere intersects it in a *great circle*. The great circle and its sphere have the same radius.

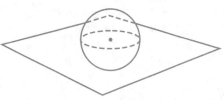

Figure 8.38

Figure 8.39 shows an example of a great circle. Two great circles separate the sphere into four sections, or *lunes*. (You can think of an orange section as a lune.)

Figure 8.39

In the diagram, all four lunes are the same size: each is one quarter of the sphere. It can be proved that the area of a great circle is numerically the same as the surface area of one of these four lunes. This proof is beyond the scope of high school geometry. Since the formula for the area of a great circle with radius r is πr^2, the surface area of each of the four lunes is also πr^2. Therefore, the surface area of the sphere is $4\pi r^2$.

Theorem 8.18
The *surface area of a sphere* is 4π times the square of the radius: $S = 4\pi r^2$.

EXAMPLE 1 Find the surface area of the following sphere.

Answer $S = 4\pi r^2$
$S = 4\pi 8^2$
$S = 256\pi$ square units ≈ 804 sq. units

Figure 8.40

Just as a regular polygon can be inscribed in a circle, so a regular polyhedron can be inscribed in a sphere.

Definition

A **regular polyhedron** is a polyhedron with faces bounded by congruent regular polygons and with the same number of faces intersecting at each vertex.

A regular hexahedron is a cube, made up of six congruent regular quadrilaterals (squares). Four other regular polyhedra are shown below.

Regular hexahedron (cube)
Figure 8.41

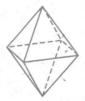
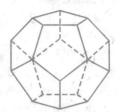
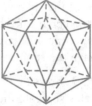

Regular tetrahedron   Regular octahedron   Regular dodecahedron   Regular icosahedron   *Figure 8.42*

The five regular polyhedra shown are the only possible regular polyhedra. Each of these polyhedra determines a convex solid called a *Platonic solid*. Since Plato discovered that these were the only regular polyhedra, they were given the name *Platonic solids* in his honor.

Do you remember the formula for the area of a regular polygon? Since all the faces of a regular polyhedron are congruent, it is easy to find the surface area.

Theorem 8.19

The *surface area of a regular polyhedron* is the product of the number of faces and the area of one face: $S = nA$.

EXAMPLE 2   Blake sells calendars in the shape of a regular dodecahedron. To make sure the calendar for a month would fit on a side, he designed the area of one face to cover 3 square inches. Find the total surface area.

Answer   The dodecahedron has 12 sides, so $S = nA = 12 \cdot 3 = 36$ square inches.

▶ A. Exercises

Find the total surface area of each sphere.

1.

2.

3.

4. The surface area of a sphere is 676π square feet. What is the length of its diameter?

5. The surface area of a sphere is 320π square yards. What is the length of its radius?

	Regular polyhedron	Number of faces	Number of edges	Number of vertices
6.	tetrahedron			
7.	dodecahedron			
8.	hexahedron			
9.	octahedron			
10.	icosahedron			

Give the surface area of each regular polyhedron.

11. An octahedron with one face of area 10 square inches
12. A hexahedron with an edge 3 feet long
13. An icosahedron with an edge six centimeters long
14. A tetrahedron with an edge three meters long
15. A dodecahedron with an edge fourteen feet long and an apothem for one face eleven feet long

▶ B. Exercises

16. What is the surface area of a soccer ball that has a circumference of 28 inches?
17. What is the diameter of a softball that has a surface area of 324π square centimeters?
18. If the maintenance men are going to paint a spherical water tower whose radius is 30 feet and each gallon of paint covers 150 square feet of metal, how many gallons of paint must they have to give the tower one coat of paint?
19. Prove that for a sphere $S = \pi d^2$.
20. Prove that for a regular polyhedron $S = \frac{1}{2}nap$, where n is the number of faces.

▶ C. Exercises

21. For each of the five regular polyhedra, give the simplest formula that you can for its surface area. Use e for the length of one edge.
22. If a cube has edge e, a cylinder has diameter and height e, and a sphere has diameter e, which has the least surface area?
23. Two spherical balls are sitting on a table touching each other. If their radii are 8 inches and 5 inches, how far apart are their points of contact with the table?

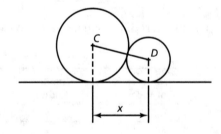

24. If a sphere has a radius of 26 inches, how far from the center should a plane cut through the sphere to give a circle whose area is 576π in.^{2}?

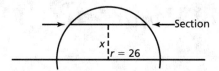

▶ Dominion Thru Math

A 100-foot diameter sphere holds water in a 186-foot water tower. The tower is a cylinder with congruent cones sticking into both ends. The cones have a slant height of 50 feet. The cylinder diameter is 25 feet. (A sketch may help.)

25. Find the radius and altitude of the cones.
26. Find the height of the cylinder.
27. Find the amount of steel needed to build the sphere.
28. Find the amount of steel needed to build the water tower.

■ Cumulative Review

Identify each set by name.
29. The intersection of all faces of a polyhedron
30. The intersection of two sides of a triangle
31. The intersection of two faces of a tetrahedron
32. The intersection of a right pyramid with a plane containing the altitude of the pyramid
33. The intersection of the lateral surface of a cylinder with a plane parallel to the bases

Geometry and Scripture

Area

You know that twelve inches is equal to one foot. However, twelve square inches is not the same as one square foot. This concept is not hard to understand, but it can be confusing when you are converting units.

1. How many square inches are in one square foot?

The Jews measured areas differently depending upon what was being measured. A field was measured by stating either the number of yokes of oxen needed for plowing it or the amount of seed needed for sowing it.

Which method is used in each verse?

2. I Samuel 14:14
3. Isaiah 5:10
4. If a man sanctified a field using five and a half homers of barley, what would the land's value be (Lev. 27:16)?

You can understand why most things could not be measured the way fields were measured. Does a yoke of oxen plow a city or a curtain? Would you sow a city or a curtain with seed? In such cases dimensions were given.

5. Review the dimensions given for New Jerusalem. Give the dimensions of the area that the city will cover. How many square furlongs will that be?

Read Exodus 36, which describes the curtains for the tabernacle.

6. Give the dimensions of each linen curtain (vv. 8-13).

7. How many square cubits of material were used for each curtain? All together?

8. Give the dimensions of each goat-hair curtain (vv. 14-18).

9. How many square cubits were there in each curtain? All together?

HIGHER PLANE: Convert the dimensions in questions 5, 6, and 8 to modern units (use the table on page 113).

First Samuel 14:14 is a key verse since it defines the term *acre*. It is one of only two Bible verses that use this area measurement. In the context, the purpose of including the measurement is to show the courage and power of Jonathan and his armourbearer in defeating twenty enemies in a small open area. All twenty men had freedom of movement to attack the valiant pair from all sides.

Line upon Line

AND THAT FIRST SLAUGHTER, which Jonathan and his armourbearer made, was about twenty men, within as it were an half acre of land, which a yoke of oxen might plow. ✒

I SAMUEL 14:14

In the following exercises, b = base length, h = height, A = area, H = altitude, L = lateral area, l = slant height, r = radius, c = circumference, d = diameter length, a = apothem, and s = side length.

Draw pictures and find the following areas (measurements are in units).

1. Rectangle with $b = 17$ and $h = 9$
2. Square with $s = 24$
3. Square with $s = \sqrt{7}$
4. Rectangle with $b = x^2$ and $h = x - 9$
5. Parallelogram with $b = 12$ and $h = 6$
6. Parallelogram with $b = 26$ and $h = 8$
7. Triangle with $b = 6$ and $h = 5$
8. Triangle with $b = 19$ and $h = 24$
9. Equilateral triangle with $s = 18$
10. Trapezoid with $b_1 = 10$, $b_2 = 14$, $h = 8$
11. Trapezoid with $b_1 = 24$, $b_2 = 10$, $h = 6$
12. Circle with $r = 15$
13. Circle with $c = 18\pi$
14. Circle with $d = 12$
15.

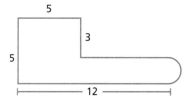

16. Regular pentagon with $s = 9$ and $a = 6.2$
17. Regular hexagon with $s = 44$

Find the surface area of each figure.

18. Sphere with $r = 14$
19. Regular icosahedron with a face covering 7 square units
20. Regular dodecahedron with $s = 2\sqrt{3}$ and $a = 2.4$

Find the lateral surface area and the total surface area for each surface below.

21. A right prism having a square base with $s = 8$ and $H = 22$
22. A right prism having a rhombus as a base with $H = 29$ and diagonals of lengths 8 and 13
23. A right circular cylinder with $d = 18$ and $H = 36$
24. Equilateral triangular pyramid with $b = 34$ and $l = 43$
25. A right circular cone with $r = 6$ and $l = 13$
26. Give the meaning of area.
27. In any regular polygon, why must the base angles of the triangles formed by consecutive radii be congruent?
28. Prove that for a sphere $S = cd$.
29. Determine the altitude of a regular tetrahedron if the edge has length s.

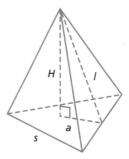

30. Explain the mathematical significance of I Samuel 14:14.

▶ Dominion Thru Math

31. A stairway connects two floors 10 feet apart with 10-inch treads (steps) and 7.5-inch risers (vertical pieces that hold up the front of each step). Find the horizontal distance needed for the stairs and the length of the stringers (supporting pieces along each side that the steps connect to). The stringers are like the hypotenuse of a big triangle.
32. How long is a finished stringer for the stairway in exercise 31? The stair requires three stringers—one for each side and one for the center with right triangles removed from one side of a 2 × 12 board (using a carpenter's square or a premade right-triangle pattern). Find the hypotenuse of a right triangle with sides of 7.5 inches and 10 inches. The stringers require 15 risers and 15 treads.

9 Circles

Farms on the Great Plains, such as this one in Texas, use center-pivot irrigation. The arm becomes the radius of a circle as it pivots on one end. Sometimes the circles are externally tangent circles, while at other times they do not touch anywhere. Land that is not irrigated produces roughly 20 percent as much as irrigated land because the Great Plains are dry and naturally covered with sagebrush.

Besides circular irrigation, circles are used in wheels, windows, umbrellas, and water towers. God used the geometry of circles in creation also. The sand dollar, sea urchin, and starfish all display circles. Such creatures provide only a glimpse into the wonders of God's creation. "For my thoughts are not your thoughts, neither are your ways my ways, saith the Lord" (Isa. 55:8).

After this chapter you should be able to

1. define and correctly use terms related to circles.

2. use the Law of Contradiction to do an indirect proof.

3. determine arc measures using central or inscribed angles.

4. determine angle measures between intersecting secant and/or tangent lines.

5. find areas and perimeters of segments and sectors.

6. prove relationships involving circles.

7. construct figures using circles.

9.1 Circles and Chords

The wheel is one of many applications of circles. If a point revolves around a circle, the point never reaches an end. Circles are often used to represent eternity because the point can continue around forever. You, too, are an eternal being and will spend eternity in heaven or hell. Accept Jesus Christ as your Savior and look forward to a glorious eternity in heaven. First John 5:11-13 explains how you can know for sure that your eternal life will be spent in heaven with God.

The rod joining the back wheels of a locomotive shows a portion of a secant line. The train shown is at the Railroad Museum in Sacramento, California.

Before studying the circle in more detail, review the definitions on page 56 in Chapter 2. Make sure that you understand each definition.

In ⊙C the center is C, and the circle has radius $\overline{CX}$ and a chord $\overline{LM}$. $\overline{QR}$ is a diameter, and $\overarc{LM}$ is an arc of the circle. From the definition of circle, you can clearly see that all radii of a circle are congruent. What are congruent circles?

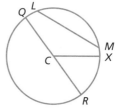

Figure 9.1

Definition

Congruent circles are circles whose radii are congruent.

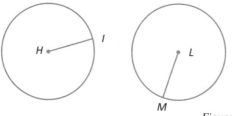

Figure 9.2

If you know that $\overline{HI} \cong \overline{LM}$, then you know that $\odot H \cong \odot L$. In this section we will examine one postulate and three theorems about chords and circles.

Postulate 9.1

Chord Postulate. **If a line intersects the interior of a circle, then it contains a chord of the circle.**

Theorem 9.1

In a circle, if a radius is perpendicular to a chord of a circle, then it bisects the chord.

    Given: $\odot O$ with radius $\overline{OC}$ and chord $\overline{AB}$; $\overline{OC} \perp \overline{AB}$
    Prove: $\overline{OC}$ bisects $\overline{AB}$

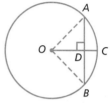

Figure 9.3

STATEMENTS	REASONS
1. $\odot O$ with radius $\overline{OC}$ and chord $\overline{AB}$; $\overline{OC} \perp \overline{AB}$	1. Given
2. Draw radii $\overline{OA}$ and $\overline{OB}$	2. Auxiliary lines
3. $\angle ODA$ and $\angle ODB$ are right angles	3. Definition of perpendicular lines
4. $\triangle ODA$ and $\triangle ODB$ are right triangles	4. Definition of right triangles
5. $\overline{OA} \cong \overline{OB}$	5. Radii of a circle are congruent
6. $\overline{OD} \cong \overline{OD}$	6. Reflexive property of congruent segments
7. $\triangle ODA \cong \triangle ODB$	7. HL
8. $\overline{AD} \cong \overline{BD}$	8. Definition of congruent triangles
9. $AD = BD$	9. Definition of congruent segments
10. D is the midpoint of $\overline{AB}$	10. Definition of midpoint
11. $\overline{OC}$ bisects $\overline{AB}$	11. Definition of segment bisector
12. If $\overline{OC} \perp \overline{AB}$, then $\overline{OC}$ bisects $\overline{AB}$	12. Law of Deduction

Here is another theorem about chords.

Theorem 9.2

In a circle or in congruent circles, if two chords are the same distance from the center(s), the chords are congruent.

Given: ⊙A ≅ ⊙B; $\overleftrightarrow{AZ} \perp \overleftrightarrow{QR}$; $\overleftrightarrow{BU} \perp \overleftrightarrow{PN}$; AX = BY
Prove: $\overline{QR} \cong \overline{PN}$

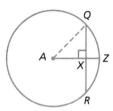 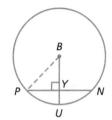

Figure 9.4

STATEMENTS	REASONS
1. ⊙A ≅ ⊙B; $\overleftrightarrow{AZ} \perp \overleftrightarrow{QR}$; $\overleftrightarrow{BU} \perp \overleftrightarrow{PN}$; AX = BY	1. Given
2. $\overline{AX} \cong \overline{BY}$	2. Definition of congruent segments
3. Draw $\overline{AQ}$ and $\overline{BP}$	3. Line Postulate
4. $\overline{AQ} \cong \overline{BP}$	4. Definition of congruent circles
5. △AQX ≅ △BPY	5. HL
6. $\overline{QX} \cong \overline{PY}$	6. Definition of congruent triangles
7. QX = PY	7. Definition of congruent segments
8. $\overleftrightarrow{AZ}$ bisects $\overline{QR}$; $\overleftrightarrow{BU}$ bisects $\overline{PN}$	8. A radius perpendicular to a chord bisects the chord
9. X is the midpoint of $\overline{QR}$; Y is the midpoint of $\overline{PN}$	9. Definition of segment bisector
10. QX = $\frac{1}{2}$QR; PY = $\frac{1}{2}$PN	10. Midpoint Theorem
11. $\frac{1}{2}$QR = $\frac{1}{2}$PN	11. Substitution (step 10 into 7)
12. QR = PN	12. Multiplication property of equality
13. $\overline{QR} \cong \overline{PN}$	13. Definition of congruent segments
14. If $\overleftrightarrow{AZ} \perp \overleftrightarrow{QR}$, $\overleftrightarrow{BU} \perp \overleftrightarrow{PN}$, and AX = BY, then $\overline{QR} \cong \overline{PN}$	14. Law of Deduction

The converse of this theorem is also true. It is stated here, and the proof will be done as an exercise.

Theorem 9.3

In a circle or in congruent circles, if two chords are congruent, then they are the same distance from the center(s).

A. Exercises

Use the figure for exercises 1-4.
1. Name the circle, the center of the circle, and a diameter.
2. Name two chords.
3. Name three radii.
4. Name all the arcs.

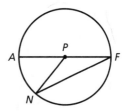

Find the indicated measures. Use the figure for exercises 5-16.

5. *Given:* $\overline{AG} \cong \overline{AC}$; $BE = 12$ units
   Find: IF
6. *Given:* $BE = 18$ units
   Find: BC
7. *Given:* $IF = 16$ units; $BE = 16$ units; $AC = 6$ units
   Find: AG
8. *Given:* $AD = 4$ units
   Find: AH
9. *Given:* $AC = 3$ units; $BE = 8$ units
   Find: AB
10. *Given:* $GH = 6$ units; $IG = 12$ units
    Find: HF
11. *Given:* $\overline{BE} \cong \overline{IF}$; $BC = 4$ units; $AB = 7$ units
    Find: AG
12. *Given:* $\overline{AG} \cong \overline{AC}$; $IG = 10$ units
    Find: BE
13. *Given:* $AI = 8$ units; $AG = 6$ units
    Find: IF
14. *Given:* $\overline{AC} \cong \overline{AG}$; $CE = 3x + 5$ units; $GI = 8$ units
    Find: x
15. *Given:* $\overline{BE} \cong \overline{IF}$; $AC = 2x + 6$ units; $AG = 4x - 10$ units
    Find: AC
16. *Given:* $\overline{AC} \cong \overline{AG}$; $IF = x^2 + 7x - 12$ units; $BE = x^2 + 3x + 8$ units
    Find: BC

▶ B. Exercises

Use the diagram for the proofs in exercises 17-19.

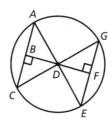

17. *Given:* ⊙D with $\overleftrightarrow{DB} \perp \overleftrightarrow{AC}$; $\overleftrightarrow{DF} \perp \overleftrightarrow{GE}$; $\overline{BD} \cong \overline{DF}$
    Prove: ∠CAD ≅ ∠DGE
18. *Given:* ⊙D with $\overleftrightarrow{DB} \perp \overleftrightarrow{AC}$
    Prove: △ADB ≅ △CDB
19. *Given:* ⊙D with $\overleftrightarrow{DB} \perp \overleftrightarrow{AC}$; $\overleftrightarrow{DF} \perp \overleftrightarrow{GE}$; $\overline{AC} \cong \overline{GE}$
    Prove: △DBC ≅ △DFG

Prove the following statements.

20. *Given:* ⊙X with $\overleftrightarrow{XA} \perp \overleftrightarrow{LM}$; $\overleftrightarrow{XB} \perp \overleftrightarrow{MN}$; $\overline{XA} \cong \overline{XB}$
    Prove: △LNM is an isosceles triangle

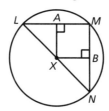

21. *Given:* ⊙M with $\overleftrightarrow{MQ} \perp \overleftrightarrow{PR}$; $\overleftrightarrow{MT} \perp \overleftrightarrow{PS}$; ∠RPM ≅ ∠SPM
    Prove: $\overline{PR} \cong \overline{PS}$

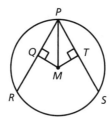

22. *Prove:* The perpendicular bisector of a chord contains the center of the circle.

23. *Given:* $\odot S$; $\overleftrightarrow{AB} \parallel \overleftrightarrow{CD}$; $\overline{CX} \cong \overline{XD}$; $\overleftrightarrow{PQ} \perp \overleftrightarrow{CD}$
  Prove: $\overline{SP}$ bisects $\overline{AB}$

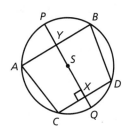

24. *Given:* $\overleftrightarrow{AC} \parallel \overleftrightarrow{BD}$, $\overleftrightarrow{CE} \parallel \overleftrightarrow{DF}$, $\overline{AC} \cong \overline{BD}$
  Prove: $\odot E \cong \odot F$

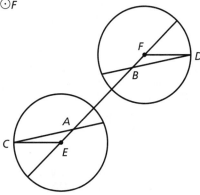

▶ C. Exercises

25. *Prove:* If chords from each of two circles are congruent and are the same distance from the center of their respective circles, then the two circles are congruent.

◼ Cumulative Review

Complete each definition.

26. Congruent segments: $\overline{AB} \cong \overline{CD}$ if . . .

27. Congruent angles: $\angle ABC \cong \angle DEF$ if . . .

28. Congruent circles: $\odot A \cong \odot B$ if . . .

29. Congruent triangles: $\triangle ABC \cong \triangle DEF$ if . . .

30. Congruent polygons: . . .

9 Analytic Geometry

Graphing Circles

A circle is a conic section formed by the intersection of a right circular cone with a plane that is perpendicular to the axis. A circle is also a locus of points that are a given distance from a given point in a plane. From this latter definition we can develop an equation for a circle.

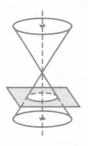 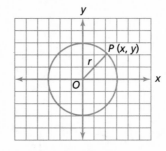

Circle O in the figure above has its center at the origin and radius r. The radius intersects $\odot O$ at P, which has coordinates (x, y). Since a circle is the set of points equidistant from the center O, this set can be described by the distance formula. Substitute the radius for d and the coordinates for the two points.

$$d = \sqrt{(x_1 - x_2)^2 + (y_2 - y_2)^2}$$
$$r = \sqrt{(x - 0)^2 + (y - 0)^2}$$
$$r = \sqrt{x^2 + y^2}$$
$$r^2 = \sqrt{(x^2 + y^2)^2}$$
$$r^2 = x^2 + y^2$$

This is the standard form of the equation of a circle with its center at the origin. Notice that the equation contains the square of the radius rather than the length of the radius itself. Now graph a circle from its equation.

EXAMPLE   Graph $x^2 + y^2 = 16$.

Answer   Use the formula $x^2 + y^2 = r^2$ to find r.

Continued ▶

$r^2 = 16$
$r = 4$

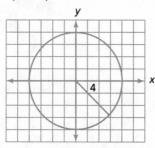

1. Apply the transitive property.

2. Count four units from the origin along each axis and connect the points with a circle.

You should also be able to look at the graph of a circle in standard position and write its equation.

▶ Exercises

Graph each:
1. $x^2 + y^2 = 25$
2. $x^2 + y^2 = \frac{1}{4}$

Write the equation of each graph:

3.

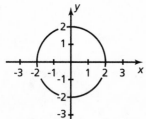

4.

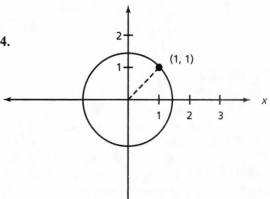

(1, 1)

Prove the theorem:
5. The midpoint of the hypotenuse of a right triangle is equidistant from each vertex.

9.2 Tangents

The word *tangent* is derived from the Latin word *tangere*, which means "to touch." This derivation is logical when you think about the relationship between a tangent line and a circle. A tangent line intersects the circle in one and only one point, therefore giving the effect of simply touching the circle. $\overleftrightarrow{FG}$ is tangent to circle O at point F.

The belts on this 2.5 liter Duratech engine used in Ford vehicles form tangent segments to the circles.

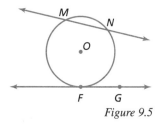

Figure 9.5

$\overline{MN}$ is a chord of $\odot O$, and the line $\overleftrightarrow{MN}$ that contains chord $\overline{MN}$ is called a *secant*.

Definitions

A **secant** is a line that is in the same plane as the circle and intersects the circle in exactly two points.

A **tangent** is a line that is in the same plane as the circle and intersects the circle in exactly one point.

The **point of tangency** is the point at which the tangent intersects the circle.

You will study theorems related to circles and tangents in this section. Many of the main ideas of the theorems were developed in several of the exercises in the previous section.

Theorem 9.4

If a line is tangent to a circle, then it is perpendicular to the radius drawn to the point of tangency.

Given: $\overleftrightarrow{AB}$ is a tangent to $\odot O$ at point A
Prove: $\overline{OA} \perp \overleftrightarrow{AB}$

Figure 9.6

This theorem can be proved using indirect proof. This is the first time that you have seen an indirect proof; pay close attention to the method used here so that you can use it in the future.

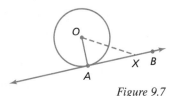

Figure 9.7

Suppose that $\overleftrightarrow{OA}$ is not perpendicular to $\overleftrightarrow{AB}$. Then there must be some other $\overleftrightarrow{OX}$ that is perpendicular to $\overleftrightarrow{AB}$ (where $X \in \overleftrightarrow{AB}$). Since we know that the shortest distance from a point to a line is the perpendicular distance, then $OX < OA$. This implies that X is in the interior of $\odot O$, thus making $\overleftrightarrow{AB}$ a secant, which intersects a circle in two points. But this is a contradiction of the given information that $\overleftrightarrow{AB}$ is a tangent. Thus the assumption that $\overleftrightarrow{OA}$ is not perpendicular to $\overleftrightarrow{AB}$ must be false. Hence $\overleftrightarrow{OA} \perp \overleftrightarrow{AB}$.

This type of reasoning is different from the normal deductive proof that you have been studying. The main steps in an indirect proof are as follows:

1. Assume the opposite of what you are trying to prove.
2. Reason deductively from the assumption.
3. Reason to a conclusion that contradicts the assumption, the given, or some theorem.
4. Conclude that the assumption is false and therefore the statement you are trying to prove is true.

These types of proofs are also called proofs by contradiction, because the Law of Contradiction is used to draw the conclusion from the first three steps. This law will be proved in the review exercises.

Theorem 9.5

Law of Contradiction. **If an assumption leads to a contradiction, then the assumption is false and its negation is true.**

Now let us return to Theorem 9.4 and rewrite the proof of it in a two column form, using the Law of Contradiction.

STATEMENTS	REASONS
1. $\overleftrightarrow{AB}$ is tangent to $\odot O$ at A	1. Given
2. $\overleftrightarrow{AB}$ intersects $\odot O$ in exactly one point	2. Definition of tangent
3. Assume $\overleftrightarrow{OA}$ is not perpendicular to $\overleftrightarrow{AB}$	3. Assumption
4. Draw the line perpendicular to $\overleftrightarrow{AB}$ that passes through O; let X be the point of intersection	4. Auxiliary line
5. $OX < OA$	5. Longest Side Inequality ($\angle X$ is right)
6. X is interior to $\odot O$	6. Definition of interior of a circle
7. $\overleftrightarrow{AB}$ contains a chord of $\odot O$	7. Chord Postulate
8. $\overleftrightarrow{AB}$ intersects $\odot O$ in two points	8. Definition of chord
9. $\overleftrightarrow{OA}$ is perpendicular to $\overleftrightarrow{AB}$	9. Law of Contradiction (compare steps 2 and 8)

This type of proof makes proofs of some theorems easier. With practice you will learn when to use the indirect type of proof. The converse of Theorem 9.4 is also true. You will be asked to prove this theorem in the exercises.

Theorem 9.6

If a line is perpendicular to a radius at a point on the circle, then the line is tangent to the circle.

In Chapter 3 you studied tangent line segments. If you are given a circle and a point in the exterior of the circle, you can find two line segments that are tangent to the circle. For example, given $\odot P$ and point X, you can find $\overline{XA}$ and $\overline{XB}$ tangent to $\odot P$.

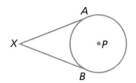

Figure 9.8

The next theorem proves that such segments are congruent.

Theorem 9.7

Tangent segments extending from a given exterior point to a circle are congruent.

Given: ⊙A with exterior point X; $\overline{XM}$ and $\overline{XN}$ are tangent to ⊙A
Prove: $\overline{XM} \cong \overline{XN}$

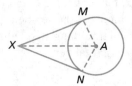

Figure 9.9

STATEMENTS	REASONS
1. ⊙A with exterior point X; $\overline{XM}$ and $\overline{XN}$ are tangent to ⊙A	1. Given
2. Draw $\overleftrightarrow{AM}$, $\overleftrightarrow{AN}$, and $\overleftrightarrow{AX}$	2. Auxiliary lines
3. $\overline{AM} \cong \overline{AN}$	3. Radii of a circle are congruent
4. $\overleftrightarrow{AM} \perp \overleftrightarrow{XM}$; $\overleftrightarrow{AN} \perp \overleftrightarrow{XN}$	4. A radius to the point of tangency is perpendicular to the tangent segment
5. ∠XMA and ∠XNA are right angles	5. Definition of perpendicular
6. △XAM and △XAN are right triangles	6. Definition of right triangles
7. $\overline{AX} \cong \overline{AX}$	7. Reflexive property of congruent segments
8. △XAM ≅ △XAN	8. HL
9. $\overline{XM} \cong \overline{XN}$	9. Definition of congruent triangles

Definitions

A **common tangent** is a line that is tangent to each of two coplanar circles.

Tangent circles are coplanar circles that are tangent to the same line at the same point.

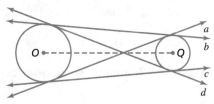

Figure 9.10

Figure 9.10 shows four common tangents of ⊙O and ⊙Q. Common tangents are classified as internal if they intersect the segment joining the centers. In the figure, *a* and *d* are common internal tangents. Other common tangents, such as *b* and *c*, are common external tangents.

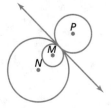

Figure 9.11

Circles can be tangent in two ways. *Internally tangent circles* are tangent circles on the same side of the common tangent. *Externally tangent circles* are tangent circles on opposite sides of the common tangent. ⊙P and ⊙M are externally tangent circles, while ⊙N and ⊙M are internally tangent circles.

▶ A. Exercises

Given: ⊙L with line *m* tangent to ⊙L at C; $\overline{BF}$ and $\overline{FE}$ are tangent segments; find the indicated information.

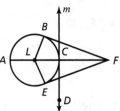

1. m∠LCD
2. What kind of triangle is △LBF?
3. If LC = 8 units and CD = 14 units, find LD (not drawn).
4. If AC = 16 units, find LB.
5. m∠LBF
6. If FE = 23 units, find BF.
7. If BL = 6 units and EF = 9 units, find LF.
8. If BL = 12 units, find AC.
9. If LC = 7 units and LF = 18 units, find EF.
10. Consider $\overline{BE}$ (not drawn). What kind of triangle is △BEF?

Consider the following diagrams for exercises 11-15.

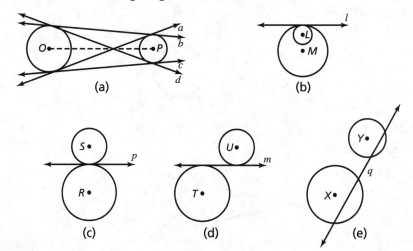

(a) (b) (c) (d) (e)

11. Which diagram shows a secant? Name the secant.
12. Name two common internal tangents in diagram (a).
13. Name the common external tangents in diagrams (a) and (b).
14. Name all internally tangent circles.
15. Name all externally tangent circles.

▶ **B. Exercises**

Draw the following figures.
16. Two internally tangent circles, one having a radius that is half the length of the other; show the common tangent line and a line that is a secant of only one of them
17. Two circles having only one common internal tangent and two common external tangents
18. Two circles having two common external tangents and no common internal tangents
19. Two circles for which no common tangent is possible
20. Two circles having two common internal tangents and two common external tangents

Prove the following statements.
21. If a line is perpendicular to a tangent line at the point of tangency, then the line passes through the center of the circle.

22. *Given:* ⊙X and ⊙Y; common tangents $\overleftrightarrow{AD}$ and $\overleftrightarrow{EC}$
    Prove: $\overline{EC} \cong \overline{AD}$

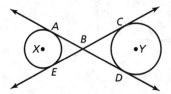

23. *Given:* ⊙M and ⊙N; common tangents $\overleftrightarrow{XW}$ and $\overleftrightarrow{XZ}$
    Prove: $\overline{YZ} \cong \overline{VW}$

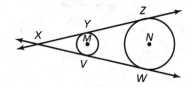

▶ **C. Exercises**

24. *Prove:* Theorem 9.6
25. *Given:* $\overline{JF} \cong \overline{GI}$; ⊙F and ⊙G are tangent to line *a* at point *L*
    Prove: △FKG is an isosceles triangle

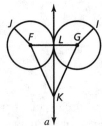

■ **Cumulative Review**

Justify the steps in the proof of the Law of Contradiction.

	Step	Reason
26.	p	
	q	previously known (given or proved)
27.	p → q	
	~q	previously known (given or proved)
28.	~p	

29. Make a truth table to prove the Law of Contradiction: $[p \rightarrow (q \wedge \sim q)] \rightarrow \sim p$.
30. Give an alternate symbolic form of the Law of Contradiction.
    (*Hint:* Consider exercises 26-28.)

9.3 Arcs

An arc was defined earlier to be a curve that is a subset of a circle. How do we measure arcs? Arcs are measured in degrees, just as angles are, but you will need to know more about the relationship between angles and arcs in order to measure them.

Two special types of angles are associated with a circle. These angles are called *central angles* and *inscribed angles*. The figure at right shows an example of each. ∠*LKM* is a central angle; notice that it intersects the circle in two points and has its vertex at the center of ⊙*K*.

∠*UVW* is an inscribed angle. It has a point of the circle as a vertex, and it intersects the circle in two other points.

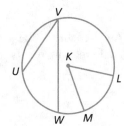

Figure 9.12

Natural features such as Delicate Arch, Utah, also exhibit an arc for support.

A **central angle** is an angle that is in the same plane as the circle and whose vertex is the center of the circle.

An **inscribed angle** is an angle with its vertex on a circle and with sides containing chords of the circle.

Note: Each of these angles determines a pair of arcs of the circle. For example, ∠*LKM* determines $\overarc{LM}$ and $\overarc{LUM}$. Similarly, ∠*UVW* is inscribed in $\overarc{UVW}$ and intercepts $\overarc{UW}$. Here is the basic relationship between the measures of central angles and of the arcs that they intercept.

Arc measure is the same measure as the degree measure of the central angle that intercepts the arc.

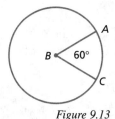

Figure 9.13

In ⊙*B*, if the measure of the central angle, ∠*ABC*, is 60, then $m\overarc{AC} = 60$ also. If an arc is cut off by a diameter, then the arc has a special name—semicircle. The measure of a semicircle is 180°, since it is intercepted by a straight angle, whose measure is 180°. Because a circle is made of two semicircles, the total degree measure of a circle is 360°. There are also some other arcs that have special names.

A **minor arc** is an arc measuring less than 180 degrees. Minor arcs are denoted with two letters, such as $\overarc{AB}$, where *A* and *B* are the endpoints of the arc.

A **major arc** is an arc measuring more than 180 degrees. Major arcs are denoted with three letters, such as $\overarc{ABC}$, where *A* and *C* are the endpoints and *B* is another point on the arc.

A **semicircle** is an arc measuring 180 degrees.

The Arc Addition Postulate is similar to the Angle Addition Postulate.

Postulate 9.2
Arc Addition Postulate. If B is a point on $\overarc{AC}$, then $m\overarc{AB} + m\overarc{BC} = m\overarc{AC}$.

Figure 9.14

This postulate is used often in proving theorems that involve arcs. You will prove as an exercise the theorem below, which shows that there is a relationship between the measures of minor and major arcs. The degree measure of a major arc can be given in terms of its associated minor arc. Look at figure 9.15 and remember that $m\odot P = 360$.

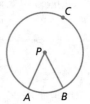

Figure 9.15

Theorem 9.8
Major Arc Theorem. $m\overarc{ACB} = 360 - m\overarc{AB}$.

EXAMPLE   If $m\overarc{AB} = 50$, find $m\overarc{ACB}$.

Answer   By the theorem, $m\overarc{ACB} = 360 - m\overarc{AB} = 360 - 50 = 310$.

Definition

Congruent arcs are arcs on congruent circles that have the same measure.

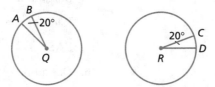

Figure 9.16

Since $m\overarc{AB} = 20$ and $m\overarc{CD} = 20$, you can say that $\overarc{AB} \cong \overarc{CD}$.

The next theorem introduces arcs that are subtended (cut off) by chords. Remember that proving a biconditional statement requires two parts.

Theorem 9.9

Chords of congruent circles are congruent if and only if they subtend congruent arcs.

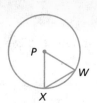 

Figure 9.17

STATEMENTS	REASONS
Part 1	
1. $\odot P$ with chord $\overline{WX}$; $\odot Q$ with chord $\overline{UV}$; $\odot P \cong \odot Q$; $\overline{UV} \cong \overline{WX}$	1. Given
2. $\overline{QU} \cong \overline{PX}$; $\overline{QV} \cong \overline{PW}$	2. Radii of a circle are congruent
3. $\triangle UQV \cong \triangle XPW$	3. SSS
4. $\angle UQV \cong \angle XPW$	4. Definition of congruent triangles
5. $m\angle UQV = m\angle XPW$	5. Definition of congruent angles
6. $m\angle UQV = m\widehat{UV}$; $m\angle XPW = m\widehat{WX}$	6. Definition of arc measure
7. $m\widehat{UV} = m\widehat{WX}$	7. Substitution (step 6 into 5)
8. $\widehat{UV} \cong \widehat{WX}$	8. Definition of congruent arcs
9. If $\overline{UV} \cong \overline{WX}$, then $\widehat{UV} \cong \widehat{WX}$	9. Law of Deduction
Part 2	
10. $\widehat{UV} \cong \widehat{WX}$; $\odot P \cong \odot Q$	10. Given
11. $m\widehat{UV} = m\widehat{WX}$	11. Definition of congruent arcs
12. $m\angle UQV = m\widehat{UV}$; $m\angle WPX = m\widehat{WX}$	12. Definition of arc measure
13. $m\angle UQV = m\angle WPX$	13. Substitution (step 12 into 11)
14. $\overline{QU} \cong \overline{PX}$; $\overline{QV} \cong \overline{PW}$	14. Radii of a circle are congruent
15. $\triangle UQV \cong \triangle XPW$	15. SAS
16. $\overline{UV} \cong \overline{WX}$	16. Definition of congruent triangles
17. If $\widehat{UV} \cong \widehat{WX}$, then $\overline{UV} \cong \overline{WX}$	17. Law of Deduction
Part 3	
18. In congruent circles, chords are congruent if and only if the arcs are congruent	18. Definition of biconditional (see steps 9 and 17)

Three other biconditional theorems follow (exercises 18-21).

Theorem 9.10
In congruent circles, chords are congruent if and only if the corresponding central angles are congruent.

Theorem 9.11
In congruent circles, minor arcs are congruent if and only if their corresponding central angles are congruent.

Theorem 9.12
In congruent circles, two minor arcs are congruent if and only if the corresponding major arcs are congruent.

▶ A. Exercises

Use the diagram for exercises 1-10. In $\odot O$, $\overline{AC}$ is a diameter.

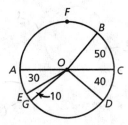

1. Name at least ten minor arcs.
2. Name at least five major arcs.
3. Name all semicircles.

Find each of the following.

4. $m\overarc{ED}$
5. $m\overarc{AB}$
6. $m\overarc{BD}$
7. $m\angle BOD$
8. $m\overarc{AD}$
9. $m\overarc{BC} + m\overarc{BA}$
10. Name all congruent arcs.

Use the figure for exercises 11-13. $\odot P \cong \odot Q$.

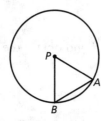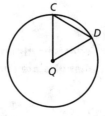

11. If $\overline{AB} \cong \overline{CD}$ and $m\angle BPA = 80$, find $m\angle CQD$.
12. If $m\angle APB = m\angle CQD$ and $AB = 12$, find CD.
13. If $m\angle APB = 75$ and $m\angle CQD = 75$, what is true about $\overline{AB}$ and $\overline{CD}$? Why?

▶ B. Exercises

Prove the following theorems.
14. Theorem 9.8
15. *Given:* $\odot U$ with $\widehat{XY} \cong \widehat{YZ} \cong \widehat{ZX}$
    Prove: $\triangle XYZ$ is an equilateral triangle

16. *Given:* Points M, N, O, and P on $\odot L$; $\widehat{MO} \cong \widehat{NP}$
    Prove: $\widehat{MP} \cong \widehat{NO}$

17. *Given:* $\odot O$; E is the midpoint of $\overline{BD}$ and $\overline{AC}$; $\overline{BE} \cong \overline{AE}$
    Prove: $\overline{AC} \cong \overline{BD}$

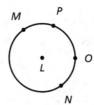

18. Theorem 9.10a. If two central angles of congruent circles are congruent, then the chords they subtend are congruent.

19. Theorem 9.10b. If two chords in congruent circles are congruent, then the central angles that intercept them are congruent.
20. Theorem 9.11

▶ C. Exercises

21. *Prove:* Theorem 9.12

▶ Dominion Thru Math

Printing presses, copiers, and automated packaging machines use a series of rollers to feed paper. The smallest roller often drives the others as it turns. A certain copier has three rollers whose radii are 2 inches for *A*, 3 inches for *B*, and 4 inches for *C*. In the coordinate plane, their centers are *A*(12, 2), *B*(3, 3), and *C* so that the roller rests on the other two. Roller *A* is the drive, and circle *C* is tangent to both *A* and *B*.

22. Draw the circles on graph paper. How can you find the center (and shaft) of roller *C*?
23. Roller *A* turns counterclockwise at 120 revolutions per minute. Find the rate and direction of rotation of the other two rollers.

■ Cumulative Review

24. State the Triangle Inequality.
25. State the Exterior Angle Inequality.
26. State the Hinge Theorem.
27. State the greater than property.
28. Prove that the surface area of a cone is always greater than its lateral surface area.

Find the perimeter and area of the shaded region in the marked figure.

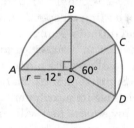

GEOMETRY AND TRANSPORTATION

We can hardly imagine the days when people had to walk or ride a horse. The invention of the circular wheel has changed everything. Bicycles, motorcycles, cars, trucks, and trains all rely on these circles. In fact, even airplanes rely on circles for takeoffs and landings. The revolution in transportation has not stopped with the invention of the wheel. Bicycles have been adapted for racing and for mountain trails. Cars have come a long way since the Model T. Commercial jets, fighter planes, and the Concorde would all look strange to Wilbur and Orville Wright.

Those who design vehicles must use the geometry of circles to make sure that the brakes enable the wheels to stop fast enough and to make appropriate gear ratios. Even designing the little wheels or pulleys for trolleys and cable cars requires knowledge of circles. The people who design these vehicles use simple geometric principles, but the calculations can get complicated.

Airplanes cannot fly unless they have enough lift. The force of the air flowing over the wings must create a vacuum or low pressure area above the wings, giving enough upward force to support the weight of the airplane. The Anglo-French supersonic transport, Concorde, has 3856 square feet of wing surface. The maximum takeoff weight is 408,000 pounds, depending on speed, atmospheric pressure, temperature, and altitude. Many supersonic aircraft are equipped with devices on their wings to increase low speed lift, but retracting these allows the plane to fly at supersonic speeds.

Suppose that as the designer of the aircraft you want to increase its maximum takeoff weight by 16,000 pounds. How many square feet will you have to add to the wings to increase the lift by that much?

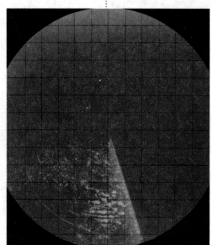

Radar screens track weather and airplanes on a circular grid.

$$\frac{16{,}000 \text{ lb.}}{106 \text{ lb./ft.}^2} = 151 \text{ square feet}$$

Thus you will need to add 151 square feet to the total wing area, or 75.5 square feet to each wing. Because part of the increased capacity will be the structure of the aircraft, the increased carrying capacity will be only part of the 16,000 pounds.

Besides vehicle design, transportation also depends on routing. On the ground, construction crews build roads, tunnels, and bridges. These, too, require geometry. Tunnel builders plan dynamite blasts to clear more tunnel in a certain radius without destroying previous work. Air traffic controllers and submarine captains both use circular range markers. They must be able to read the circular grid (polar coordinates) to locate other planes or vessels in the area.

Some designers are experimenting with solar-powered cars; a solar-powered airplane has already crossed the English Channel. How much of a problem would it be to power a simple golf cart with solar cells?

A cart that travels 25 miles per hour requires about 900 watts of power. Simple solar (or photovoltaic) cells produce 0.084 watts.

$$\frac{900 \text{ watts}}{0.084 \text{ watts per cell}} = 10{,}714 \text{ cells}$$

Since solar cells cost about $10.00 apiece, 10,714 cells would cost $107,140. Since each cell is 5 by 2.5 centimeters, each cell covers $5 \cdot 2.5 = 12.5$ sq. cm. Altogether, they would cover 133,925 cm$^2 \approx 13.4$ m^2 (10,714 cells by 12.5 cm^2/cell).

Thus the cells would cover an area of about 3.7 meters on a side. Do you think your golf cart would look a little top-heavy? Perhaps your solar panel could double as a sail.

Those who design vehicles must use the geometry of circles to make sure that the brakes enable the wheels to stop fast enough and to make appropriate gear ratios.

The Concorde is the fastest commercial aircraft, flying at an average speed of Mach 2.2.

9.4 Inscribed Angles

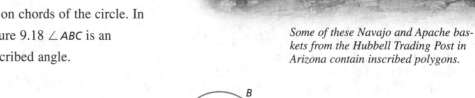

Some of these Navajo and Apache baskets from the Hubbell Trading Post in Arizona contain inscribed polygons.

Remember that an inscribed angle is an angle whose vertex is on a circle and whose sides lie on chords of the circle. In figure 9.18 ∠ABC is an inscribed angle.

Can you guess the relationship between the measure of an inscribed angle and the measure of its intercepted arc?

Figure 9.18

Theorem 9.13

The measure of an inscribed angle is equal to one-half the measure of its intercepted arc.

Given: ⊙K with inscribed ∠ABC that intercepts $\overset{\frown}{AC}$

Prove: $m\angle ABC = \frac{1}{2}m\overset{\frown}{AC}$

To prove this theorem, you must consider three possible cases.

Case 1: The center *K* lies on ∠ABC.

Figure 9.19

Case 2: The center *K* lies in the interior of ∠ABC.

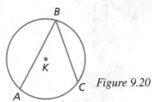

Figure 9.20

Case 3: The center *K* lies in the exterior of ∠*ABC*.

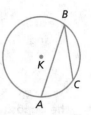

Figure 9.21

The first case will be proved here, and the other two cases will be done as exercises. The proofs of the other two parts use this case as a reason.

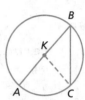

STATEMENTS	REASONS
1. ⊙*K* with inscribed ∠*ABC* that intercepts $\widehat{AC}$	1. Given
2. *K* lies on ∠*ABC*	2. Given for Case 1
3. Draw $\overleftrightarrow{KC}$	3. Auxiliary line
4. $\overline{KB} \cong \overline{KC}$	4. Radii of a circle are congruent
5. △*KBC* is an isosceles triangle	5. Definition of isosceles triangle
6. ∠*KBC* ≅ ∠*BCK*	6. Isosceles Triangle Theorem
7. *m*∠*KBC* = *m*∠*BCK*	7. Definition of congruent angles
8. *m*∠*KBC* + *m*∠*BCK* = *m*∠*CKA*	8. Exterior Angle Theorem
9. *m*∠*CKA* = $m\widehat{AC}$	9. Definition of arc measure
10. *m*∠*KBC* + *m*∠*KBC* = $m\widehat{AC}$	10. Substitution (steps 7 and 9 into 8)
11. 2*m*∠*KBC* = $m\widehat{AC}$	11. Distributive property
12. *m*∠*KBC* = $\frac{1}{2}m\widehat{AC}$	12. Multiplication property of equality

This theorem has some corollaries. A *corollary* is a theorem that follows immediately from some other theorem. The following corollaries are exercises.

Theorem 9.14
If two inscribed angles intercept congruent arcs, then the angles are congruent.

Theorem 9.15
An angle inscribed in a semicircle is a right angle.

Theorem 9.16
The opposite angles of an inscribed quadrilateral are supplementary.

▶ **A. Exercises**

Using the diagram, find the indicated measures in exercises 1-10.

1. If $m\widehat{DC} = 60°$, find $m\angle 4$.
2. If $m\angle 1 = 40°$, find $m\widehat{AB}$.
3. If $m\angle 3 = 25°$, find $m\widehat{DC}$.
4. If $m\widehat{AB} = 130°$, find $m\angle 2$.
5. If $m\angle 3 = 28°$, find $m\angle 4$.
6. If $m\angle 3 = 47°$, find $m\widehat{DC}$.
7. If $m\widehat{DC} = 55°$, find $m\widehat{DBC}$.
8. If $m\widehat{AD} = 142°$, find $m\widehat{ACD}$.
9. If $m\widehat{ADB} = 290°$, find $m\angle 1$.
10. If $m\angle CBD = 40°$, find $m\widehat{CAD}$.

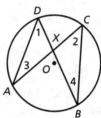

▶ **B. Exercises**

Use the diagram above for exercise 11.

11. If $m\widehat{DC} = 68°$ and $m\widehat{AB} = 134°$, find $m\angle DXA$.

Use the following figure for exercises 12-16.

12. If $m\widehat{LN} = 138°$, find $m\widehat{LMN}$.
13. If $m\widehat{MLO} = 240°$, find $m\angle MLO$.
14. If $m\angle LOM = 82°$, find $m\widehat{MNL}$.
15. If $m\widehat{MLO} = 212°$, find $m\angle MNO$ and $m\angle MLO$.
16. If $m\widehat{MN} = 98°$ and $m\widehat{LO} = 86°$, find $m\angle MYL$.

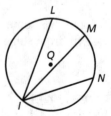

Prove the following theorems and statements.

17. Case 2 of Theorem 9.13
18. Case 3 of Theorem 9.13
19. Theorem 9.14
20. *Given:* ⊙Q; $\overrightarrow{IM}$ bisects $\angle LIN$
    Prove: $m\widehat{LM} = \frac{1}{2}m\widehat{LN}$

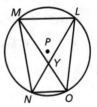

21. *Given:* ⊙R; $m\angle TSU = \frac{1}{2}m\widehat{UV}$

Prove: $\widehat{UV} \cong \widehat{TU}$

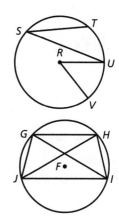

22. *Given:* ⊙F; $\overleftrightarrow{GH} \parallel \overleftrightarrow{JI}$

Prove: $\widehat{GJ} \cong \widehat{HI}$

23. Theorem 9.15

24. Theorem 9.16

▶ C. Exercises

25. *Prove:* If a trapezoid is inscribed in a circle, then it is isosceles.

▶ Dominion Thru Math

26. Explain how to use a carpenter's framing square and Theorem 9.15 (an angle inscribed in a semicircle is a right angle) to find the center of the end of a circular wooden column being installed on the porch of a house.

■ Cumulative Review

Justify each statement with a reason.

Given: △ABC is obtuse, and $\overleftrightarrow{CD} \perp \overline{AB}$ at D

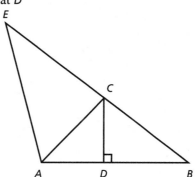

27. ∠CDA and ∠CDB are right angles

28. BC > BD

29. $m\angle ACE = m\angle B + m\angle CAD$

30. AE + AB > BE

31. $m\angle BAE + m\angle ABE + m\angle E = 180°$

9.5 Lines and Circles

In figure 9.22 you can see several intersecting lines. A tangent and two secants intersect at *F*. In this section you will see special relationships involving the angles formed by specific intersecting lines.

The two cables carrying the ski gondolas form tangent segments to the circular gears at the top and bottom of the lift.

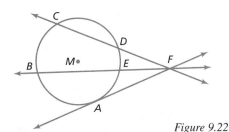

Figure 9.22

Theorem 9.17a

The measure of an angle formed by two secants that intersect in the exterior of a circle is one-half the difference of the measures of the intercepted arcs.

Given: ⊙*O* with secants $\overleftrightarrow{CA}$ and $\overleftrightarrow{EA}$

Prove: $m\angle 1 = \frac{1}{2}(m\overarc{CE} - m\overarc{BD})$

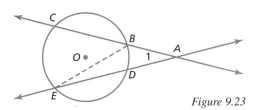

Figure 9.23

STATEMENTS	REASONS
1. $\odot O$ with secants $\overleftrightarrow{CA}$ and $\overleftrightarrow{EA}$	1. Given
2. Draw $\overleftrightarrow{BE}$	2. Auxiliary line
3. $m\angle CBE = m\angle BEA + m\angle 1$	3. Exterior Angle Theorem
4. $m\angle 1 = m\angle CBE - m\angle BEA$	4. Addition property of equality
5. $m\angle CBE = \frac{1}{2}m\widehat{CE}$; $m\angle BEA = \frac{1}{2}m\widehat{BD}$	5. An inscribed angle measures half the measure of its intercepted arc
6. $m\angle 1 = \frac{1}{2}m\widehat{CE} - \frac{1}{2}m\widehat{BD}$	6. Substitution (step 5 into 4)
7. $m\angle 1 = \frac{1}{2}(m\widehat{CE} - m\widehat{BD})$	7. Distributive property

Two secants may intersect in the interior of a circle. If this happens, then the angle measure between the secants is not half the difference of the arc measures but half the sum of the arc measures.

> **Theorem 9.18**
> The measure of an angle formed by two secants that intersect in the interior of a circle is one-half the sum of the measures of the intercepted arcs.

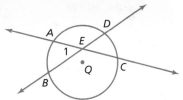

Figure 9.24

By Theorem 9.18 you know that the $m\angle 1 = \frac{1}{2}(m\widehat{AB} + m\widehat{DC})$. This theorem will be proved as an exercise.

The only other possibility for two secants is that they intersect on the circle itself. In this case, an inscribed angle is formed, which you have already studied in the previous section.

What happens to the measure of the angle if one of the lines is a tangent line?

> **Theorem 9.19**
> The measure of an angle formed by a tangent and secant that intersect at the point of tangency is one-half the measure of the intercepted arc:
> $m\angle HIJ = \frac{1}{2}m\widehat{HI}$.

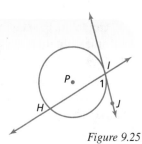

Figure 9.25

We will prove this theorem in an exercise.

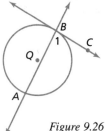
Figure 9.26

EXAMPLE Find the measure of ∠ABC if $\overleftrightarrow{BC}$ is tangent to ⊙Q at B and $\overleftrightarrow{AB}$ is a secant. $m\widehat{AB} = 170$.

Answer $m\angle ABC = \frac{1}{2}m\widehat{AB}$

$m\angle ABC = \frac{1}{2}(170)$

$m\angle ABC = 85$

In the exercises you will see how to combine Theorems 9.17b and 9.17c with Theorem 9.17a.

Theorem 9.17b

The measure of an angle formed by a secant and a tangent that intersect in the exterior of a circle is one-half the difference of the measures of the intercepted arcs.

Theorem 9.17c

The measure of an angle formed by the intersection of two tangents is one-half the difference of the measures of the intercepted arcs.

The theorems that you have seen in this section will help you to compute angle measures.

▶ A. Exercises

1. Theorems 9.17a, 9.17b, and 9.17c are the three cases for a single theorem. Complete the theorem.

Theorem 9.17

If two lines intersect a circle and intersect each other in the exterior of the circle, then . . .

2. Complete the statement: If two lines that intersect a circle intersect each other at a point on the circle, then the angle formed measures . . .
3. Complete the statement: If two lines that intersect a circle intersect each other in the interior of the circle, then the angle formed measures . . .
4. Sketch the three cases of exercise 1. Label them.
5. Sketch the two cases for exercise 2. Label them (include theorem numbers).
6. Sketch the only case for exercise 3. Label it.

7.

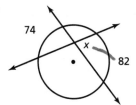

74

x

82

12.

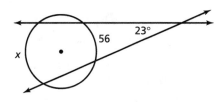

23°

56

x

8.

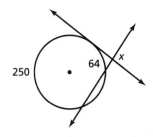

250

64

x

13.

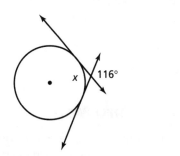

x

116°

9.

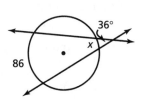

36°

x

86

14.

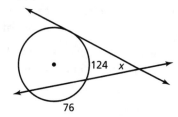

124 *x*

76

10.

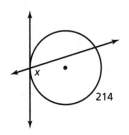

x

214

15.

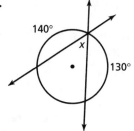

140°

x

130°

11.

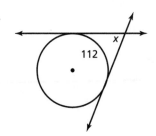

112

x

▶ B. Exercises

16. Explain why there are no other cases in exercise 6.

17. Explain why there is no third case in exercise 5.

Prove the following theorems.

18. Theorem 9.18

19. Theorem 9.17b

20. Theorem 9.17c

▶ C. Exercises

21. Prove Theorem 9.19.

▶ Dominion Thru Math

One room of a museum includes archeological artifacts in a circular cylinder case. Three hidden security cameras evenly placed around the room constantly monitor the case. The case has a radius of 9 feet, and each camera lens has a view angle of 40°.

22. How many degrees of the case can each camera view, and how many degrees of overlap are there? (*Hint*: Use exterior tangents.)

23. How many feet of the circumference of the case does each camera view that the other cameras cannot view?

▣ Cumulative Review

Use the quadrilateral shown for the following questions.

24. Name the type of quadrilateral.

25. How do the diagonals relate?

If the diagonals of the quadrilateral shown are 10 and 6 inches respectively, give the

26. perimeter.

27. area.

28. Name the type of polyhedron formed by using the quadrilateral shown as the base of a pyramid.

9.6 Sectors and Segments

To find the distance around a circle, you find the circumference of the circle, which is given by the following formula.

The spokes of this antique fertilizer from the Badlands of South Dakota determine congruent sectors. The front wheel has 12 sectors; how many are on the back wheels?

$$c = 2\pi r$$

But how can you find the length of an arc? You already know that the measure of the arc equals the degree measure of the corresponding central angle. But this is the degree measure of the arc and not the length of the arc. The next theorem gives a formula for finding the length of the arc. The Greek letter θ (theta) is often used as a variable for measures of arcs and angles.

Theorem 9.20

If the degree measure of an arc is θ and the circumference of the circle is c, then the length of the arc is l, given by $\dfrac{l}{c} = \dfrac{\theta}{360}$, or $l = \dfrac{c\theta}{360}$.

You can convert this formula to a more useful form by substituting $c = 2\pi r$ into the formula.

$$l = \frac{2\pi r\theta}{360}$$

$$l = \frac{\pi r\theta}{180}$$

We will now look at some special subsets of a circle and its interior.

Definitions

A **sector of a circle** is the region bounded by two radii and the intercepted arc.

A **segment of a circle** is the region bounded by a chord and its intercepted arc.

The first diagram below shows a sector, and the second diagram shows a segment.

Figure 9.27

(*a*)   (*b*)

The area of a sector is to the area of the circle as the arc measure is to the degree measure of the circle.

Theorem 9.21

The area of a sector is given by the proportion
$\frac{A}{A_c} = \frac{\theta}{360}$**, or** $A = A_c \frac{\theta}{360}$**, where** *A* **is the area of the sector,** A_c **is the area of the circle, and** θ **is the arc measure in degrees.**

By substituting the formula for the area of a circle, we obtain
$$A = \frac{\pi r^2 \theta}{360}.$$

To find the area of a segment, you need to use some common sense and some basic arithmetic operations.

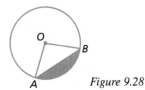

Figure 9.28

Find the area of the sector intercepting $\overset{\frown}{AB}$ and subtract the area of the triangle, $\triangle ABO$. An example will help you understand.

EXAMPLE 1   Find the area of the segment formed by $\overline{CD}$.

Answer $A = \frac{\pi r^2 \theta}{360}$ **1.** Find the area of the sector intercepting $\overset{\frown}{CD}$.

$= \frac{\pi 8^2 (60)}{360}$

$= \frac{64\pi}{6}$

$= \frac{32\pi}{3}$ sq. units

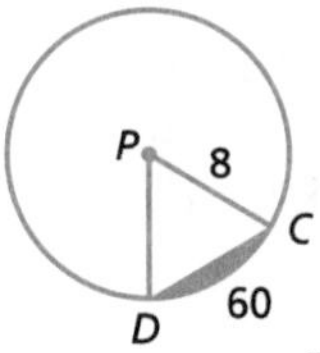

Figure 9.29

Continued ▶

$$A = \frac{s^2\sqrt{3}}{4}$$

$$= 8^2 \frac{\sqrt{3}}{4}$$

$$= \frac{64}{4}\sqrt{3}$$

$$= 16\sqrt{3} \text{ sq. units}$$

2. Find the area of equilateral $\triangle PDC$.

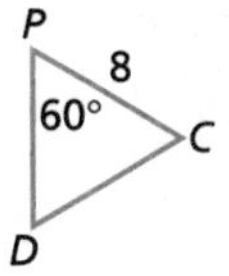

$A_{segment}$

$$= A_{sector} - A_{triangle}$$

$$= \frac{32\pi}{3} - 16\sqrt{3}$$

$$\approx 5.8 \text{ sq. units}$$

3. Subtract to find the area of the segment.

To find the perimeter of a sector, find the length of the arc and add the lengths of the two radii.

EXAMPLE 2   Find the perimeter of the sector intercepting $\overset{\frown}{XY}$.

Answer $l = \frac{\pi r \theta}{180}$

$$= \frac{\pi(12)(110)}{180}$$

$$= \frac{22\pi}{3}$$

$$\approx 23 \text{ units}$$

1. Find the length of $\overset{\frown}{XY}$.

$p = l + r + r$
$\approx 23 + 12 + 12$
$\approx 47 \text{ units}$

2. Add the two radii lengths to the arc length.

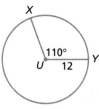

Figure 9.30

▶ A. Exercises

Find the indicated measure in each diagram. The radius is *r*, the angle measure is θ, and the arc length is *l*.

1.

4.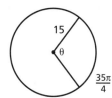

2.

5.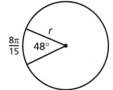

3.

Copy and complete this table.

	Radius (units)	Central angle measure (degrees)	Arc length (units)
6.	3	50	
7.		16	$\frac{8\pi}{9}$
8.	7	90	
9.	7		$\frac{35\pi}{3}$
10.		220	$\frac{55\pi}{9}$

Copy and complete the table.

	Radius (units)	Arc measure (degrees)	Area of the sector	Perimeter of the sector
11.	2	55		
12.	12	80		
13.	4		$\frac{16\pi}{3}$	
14.		40	4π	$\frac{4\pi}{3} + 12$
15.	8		$\frac{128\pi}{3}$	$\frac{32\pi}{3} + 16$
16.		180	$\frac{\pi}{2}$	
17.	15	26		
18.	3		$\frac{9\pi}{8}$	

19. A farmer in the Great Plains irrigates a circular field with a radius of 500 ft. If he plants a 90° sector of that field with corn, how much area will he need to seed?

20. A 10-inch pizza is cut into 8 slices. How many square inches of topping does Janelle consume if she has just one slice?

Find the area of the shaded regions.

21.

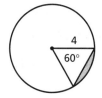

23.

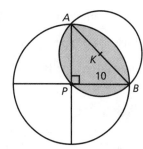

22.

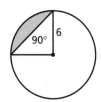

24. Develop a formula for the area of the sector of a circle with radius *r* and arc length *l*.

▶ C. Exercises

25. Develop a formula for the area of a segment of a circle with radius *r*, chord length 2*a*, and central angle *d*.

▶ Dominion Thru Math

In the Great Plains, where land is watered by center-pivot irrigation, the part that does not get watered has a reduced yield per acre of just 20% of the yield for watered acreage. Irrigated corn yields 250 bushels/acre, and one acre is 43,560 square feet.

26. If one center-pivot irrigation arm irrigates a circle with a radius of 800 ft., find the number of center-pivot arms needed for a 1600 ft. × 3200 ft. field.

27. Find the acreage not watered by center-pivot arms.

28. Find the loss in yield due to incomplete coverage when the field is watered with center pivot arms instead of a full coverage system.

Motion-detecting floodlights generally have a rating for the distance they illuminate and the angle of illumination. The farther they illuminate, the narrower the angle of illumination. Type A bulbs reach 30 feet through an arc of 180° (semicircle). Type B bulbs reach 40 feet through an arc of 100° (sector of a circle). Bill wants to illuminate completely the outside corner of his house. Give each angle in degrees.

29. What is the angle at the outside corner that must be illuminated?

30. If Bill uses two type A floodlights, what is the overlap of light patterns?

31. If he uses three type B floodlights equally spaced, what is the overlap?

■ Cumulative Review

Match to each set the best description. Use each answer once.

32. $\overleftrightarrow{FG} \cap \overleftrightarrow{RS}$

33. $\overleftrightarrow{BC} \cap \odot A$ where *B* is on $\odot A$

34. $\overleftrightarrow{DH} \cap \overline{KL}$ where *K* and *L* are in opposite half-planes

35. $\overrightarrow{AB} \cap \overrightarrow{BA}$

36. *PQRS* $\cap \overline{MN}$ where *M* and *N* are in the interior of the convex polygon

A. at least one point

B. at most one point

C. exactly one point

D. less than one point

E. more than one point

9.7 Circle Constructions

Now you will see some constructions involving inscribed and circumscribed circles. You should remember the theorems from this chapter that justify them.

 onstruction 12

Circumscribed circle

Construct: A circle circumscribed about a given triangle

1. Draw any triangle, △*ABC*.
2. Construct the perpendicular bisector of two sides of the triangle and call their point of intersection *D*.
3. Measure the distance from *D* to one of the vertices and construct a circle with this radius and center *D*.

The circular design of the Wheelie at Six Flags over Georgia tips passengers upside down.

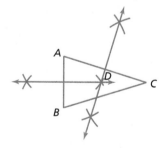
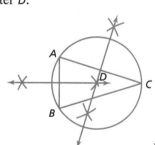

Figure 9.31

In the next example only the construction drawing is given. Look at the drawing and give the required steps for the construction.

Construction 13

Inscribed circle

Construct: A circle inscribed in a given triangle

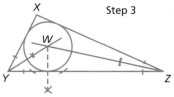

Figure 9.32

Construction 14

Regular pentagon

Construct: A regular pentagon

Figure 9.33

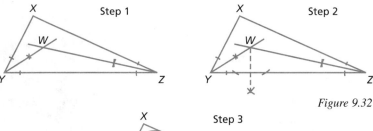

1. Draw a circle *M*. Draw a diameter of ⊙*M*, and construct its perpendicular bisector.
2. Bisect one of the radii. Point *X* is the midpoint.
3. Measure $\overline{XA}$ (midpoint of radius to the endpoint of another radius).
   Mark off the same length on the original diameter from the midpoint of the radius. Call this point *Y*.
4. Measure the length of $\overline{AY}$. Using this length, mark off consecutive arcs on the circle.
5. Connect consecutive marks with segments to form a regular pentagon.

The next construction enables you to double the number of sides of a regular polygon. For instance you could use a regular decagon to construct a regular 20-sided polygon. Remember that an *n*-gon is a polygon with *n* sides.

 onstruction 15

Double the number of sides of a regular polygon

Given: A regular *n*-gon
Construct: A regular 2*n*-gon

1. Construct the perpendicular bisector of two sides.
2. The bisectors intersect at the center of the regular polygon.
3. Use the distance from the center to a vertex to construct the circumscribed circle.
4. Open your compass to measure the distance from the point where one perpendicular bisector intersects the circle to an adjacent vertex of the original polygon.
5. Use this distance on your compass to mark off equal arcs all the way around the circle (you should obtain all the original vertices and the arc midpoints between them).
6. Connect consecutive marks to form a regular 2*n*-gon.

▶ A. Exercises

Construct the following.
1. The circle circumscribed about △*PQR*

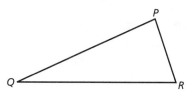

2. The circle inscribed in △*ABC*

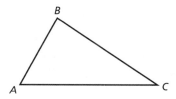

3. A regular pentagon
4. A regular hexagon from an equilateral triangle
5. A regular octagon (*Hint:* Construct a square first.)

What theorem justifies
6. construction 12?
7. construction 13?

Use △ABC to construct the following.

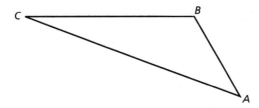

8. The altitude of △ABC through vertex A
9. The perpendicular bisector of $\overline{BC}$
10. The median of △ABC through vertex B
11. The angle bisector of ∠B

Use the two triangles for exercises 12-15.

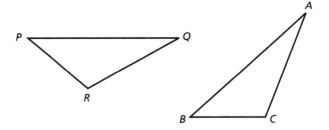

12. Construct the circumcenter of △PQR.
13. Construct the incenter of △PQR.
14. Construct the centroid of △ABC.
15. Construct the orthocenter of △ABC.

▶ **B. Exercises**

16. Construct a circle that passes through points P, Q, and R.

Q•   •R

P •

17. Inscribe a right triangle in ⊙A.

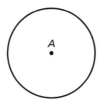

18. Construct a regular decagon.
19. What regular polygons can you construct so far (up to 25 sides)?

▶ C. Exercises

20. Construct a five-pointed star.

▪ Cumulative Review

Construct the following and then justify your constructions.

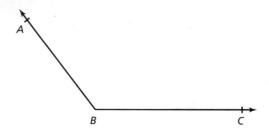

21. An angle congruent to ∠ABC
22. A line perpendicular to $\overleftrightarrow{AB}$ that passes through point C
23. A perpendicular bisector of $\overline{AB}$
24. A 45° angle
25. A 60° angle

Geometry *and* Scripture

Pi

Do you recall constructing a circle with a compass? The Bible mentions the use of a compass for this purpose.

1. Read Isaiah 44:13. What occupation is mentioned as making use of compass constructions?

2. The word *circle* appears only once in the Bible. Find it.

3. In the verse above, what is described as circular?

4. Read Psalm 19. What do verses 4-6 describe?

5. What is described as (roughly) circular in these verses?

The circle in I Kings 7:23-26 is the most important geometric figure for mathematical studies in the Bible.

6. Give the diameter of the circle.

7. Give the circumference of the circle.

HIGHER PLANE: Can you find another passage in the Bible that describes the same circle?

Notice that the Jews measured circumference by stretching a line around the circle. Compare this with the discussion of circumference (p. 98). Notice also that since $\pi = \frac{c}{d}$, we can calculate that $\pi = \frac{30}{10} = 3$ from these verses.

Be careful, though! It does seem close to 3.14159, but some ancient civilizations already had better estimates of π. Egypt used

$4 \cdot \left(\frac{8}{9}\right)^2$ or 3.16 and Babylon used $3\frac{1}{8}$ or 3.125. Because some civilizations had more accurate estimates, some modern critics have called this an error in the Bible.

God knows everything. He knows more than the Egyptians, the Babylonians, and the modern critics. Let's look more closely at what God said.

8. Reread the diameter measurement. It measures from where to where?

9. Reread the circumference measurement. What did it measure around?

10. Does the sea itself have the same diameter as the brim according to verse 26? What is the width of the brim?

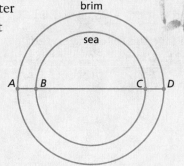

11. Use the conversions from Chapter 3 (p. 113) to give these measurements in inches: *AD, AB, AC, BC,* and the sea's circumference.

12. Using problem 11, what do you get for the value of π? Is it better or worse than the ancient values listed above?

This value of π is the only irrational number approximated by Bible measurements. The theme verse lays the basis for the real number system that includes both rational and irrational numbers.

Line upon Line

And He made a molten sea, ten cubits from the one brim to the other: it was round all about, and his height was five cubits: and a line of thirty cubits did compass it round about. ❧

I Kings 7:23

Chapter 9 Review

Use the figure shown for exercises 1-10.

1. Name the center of the circle.
2. Name a radius.
3. Name a chord.
4. Name a secant.
5. Name a tangent.
6. Find $m\angle OCD$.
7. If $PQ = 12$ units, find XQ.
8. Name a central angle.
9. Name an inscribed angle.
10. If $OQ = 26$ units, find OC.

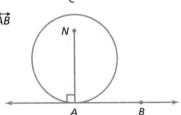

9.1

Prove the following statements.

11. *Given:* $\odot P$, $\angle PDC$ is a right angle
    Prove: $\overrightarrow{BP}$ bisects $\angle ABC$

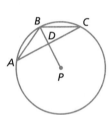

12. *Given:* $\odot M$; $\overline{AB} \cong \overline{CD}$
    Prove: $\angle ABC \cong \angle BCD$

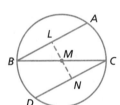

13. *Given:* $\overleftrightarrow{AB}$ and $\overleftrightarrow{AC}$ are tangent to $\odot X$;
    $\overrightarrow{AX}$ bisects $\angle BAC$
    Prove: $m\angle BDA = 90$

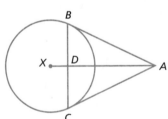

14. *Given:* $\overleftrightarrow{AB}$ is tangent to $\odot C$ at A. $\overleftrightarrow{AN} \perp \overleftrightarrow{AB}$
    Prove: $\overleftrightarrow{AN}$ contains center C

Use the diagram for exercises 15-16. Find the following measures if
$m\overset{\frown}{MN} = 50$ and $m\overset{\frown}{PQ} = 30$.

15. $m\angle POQ$

16. $m\angle NOQ$

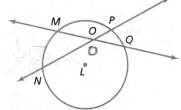

Use the figure for exercises 17-21. $\overline{HI}$ and $\overline{HJ}$ are tangent to $\odot A$.

17. If $HI = 18$ units, find HJ.

18. If $m\overset{\frown}{BC} = 50$, find $m\angle BAC$ and $m\angle BDC$.

19. Find $m\overset{\frown}{DJB}$. What special name does $\overset{\frown}{DJB}$ have?

20. Use the Arc Addition Postulate to express $m\overset{\frown}{JC} + m\overset{\frown}{CB}$.

21. If $m\angle ACD = 48$, find $m\overset{\frown}{CB}$.

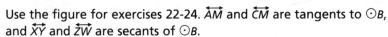

Use the figure for exercises 22-24. $\overleftrightarrow{AM}$ and $\overleftrightarrow{CM}$ are tangents to $\odot B$,
and $\overleftrightarrow{XY}$ and $\overleftrightarrow{ZW}$ are secants of $\odot B$.

22. If $m\overset{\frown}{XZ} = 80$ and $m\overset{\frown}{WY} = 55$, find $m\angle WMY$.

23. If $m\angle AMZ = 30$ and $m\overset{\frown}{AW} = 70$, find $m\overset{\frown}{ZA}$.

24. If $m\overset{\frown}{CY} = 100$ and $m\overset{\frown}{CX} = 120$, find $m\angle CMX$.

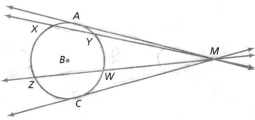

Use the figure for exercises 25-27.

25. Find the length of $\overset{\frown}{AC}$.

26. Find the area of the sector intercepting $\overset{\frown}{AC}$.

27. Find the perimeter of the sector intercepting $\overset{\frown}{AC}$.

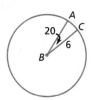

28. Construct a square circumscribed about $\odot C$.

29. Given two intersecting lines, first construct a circle tangent to both lines, then construct a pentagon inscribed in the circle.

30. Explain the mathematical significance of I Kings 7:23.

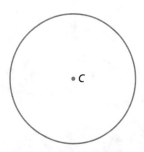

10 Space

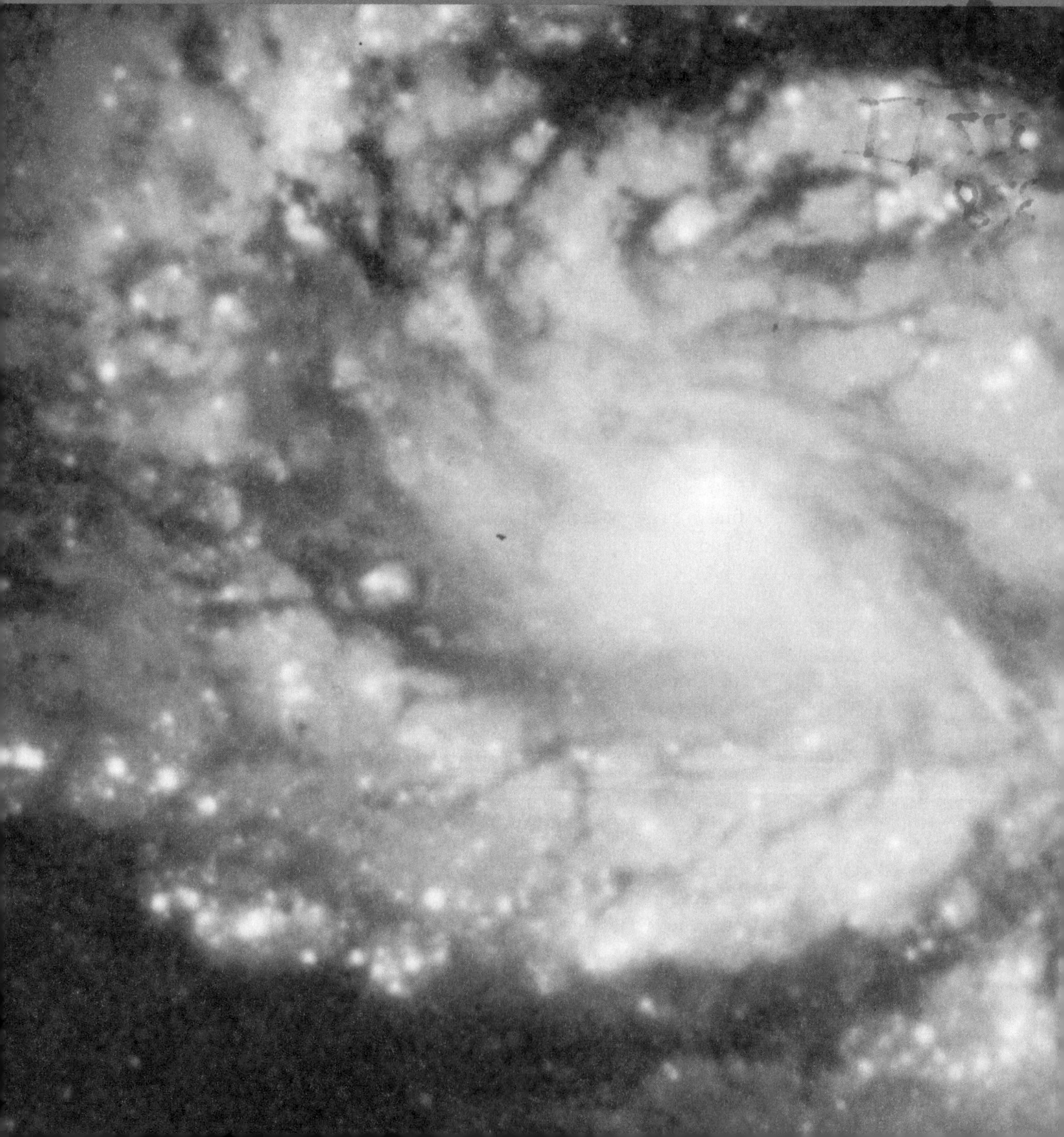

God created space orderly and precise. The stars and planets that astronomers track form geometric shapes.

A combination of spin, mass, and gravity causes the planets to be *spheroids* (almost spheres). The earth, in particular, is an oblate spheroid, being flattened at the poles. The orbits of the planets are elliptical, and each planet attains maximum speed when it is nearest the sun. The earth and the moon cast shadows, and the *umbra* (darkened portion of space) forms a cone. The larger conical shadow where the sunlight is only partly obscured is called the *penumbra*. Galaxies such as M100 shown form spirals.

Astronomers use geometry in their study of space and have developed a system for locating the heavenly bodies. The path of the sun forms a celestial equator for the heavens. The *declination* (positive or negative) measures the position of a heavenly body above or below the celestial equator, much like north or south latitudes. The *right ascension* describes the eastward distance along the celestial equator much like E and W longitude. In fact, some astronomers disagree on whether the universe is Euclidean or non-Euclidean. Such issues affect distance measurements and show how calculations depend on underlying geometric assumptions.

After this chapter you should be able to

1. sketch three-dimensional figures using various perspective techniques.

2. define and measure dihedral angles.

3. prove theorems involving perpendicular lines and planes in space.

4. prove theorems involving parallel lines and planes in space.

5. classify polyhedra and apply Euler's formula.

6. prove theorems involving spheres and apply the results to latitude and longitude.

7. contrast spherical geometry to Euclidean geometry.

10.1 Perspective in Space

Many times in geometry you will need to draw three-dimensional figures. To draw a good three-dimensional figure, you must understand the principle of perspective and must practice the skill. You

These identical Canadian fighter jets appear smaller in the background—an example of perspective.

have already seen many three-dimensional figures in this book, but can you draw them effectively? This section will give you some guidelines to follow for drawing them. Can you name the kinds of figures that are represented below?

 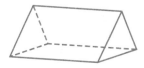 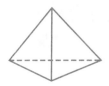

Figure 10.1

Notice that the dotted lines indicate lines that are unseen by the viewer. Many times you will need to visualize the intersection of two three-dimensional objects, so it will help if you can draw these dotted lines. Practice sketching a sphere by first drawing a circle as the outline of the sphere. Next draw an ellipse as in figure 10.1 with the nearest half solid and the farthest half dotted. This ellipse is like the equator of the earth viewed in perspective. For pyramids and cones, simply sketch the base, place a dot for the vertex, and connect the vertex to the base. For prisms and cylinders, draw both bases and connect them. Parallel lines are drawn parallel in these conceptual drawings.

In real life, however, parallel lines do not always look parallel. When you stand on railroad tracks and look into the distance, the two parallel rails appear to meet at the horizon. A drawing that captures this real-life appearance is a *perspective drawing.* Such drawings represent objects as they appear to the eye. Perspective drawing takes into account that an object will appear smaller in the distance than up close.

Figure 10.2

There are two main components of perspective drawings. One is the vanishing point(s) and the other is the horizon. The *horizon* is at the eye level of the observer, and the *vanishing point,* where parallel lines meet, is located on the horizon. There are three types of perspective drawings distinguished by the number of vanishing points: *one-point perspective, two-point perspective,* and *three-point perspective.* Each of these is shown in the following examples.

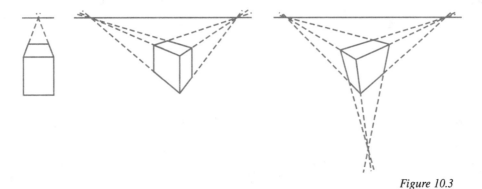

Figure 10.3

Notice that two vanishing points determine the horizon but can be in any position along the horizon. Notice also that in each dimension, parallel lines either meet at the vanishing point or appear parallel.

Both conceptual drawings and perspective drawings are tools for depicting spatial relationships. You will need to picture figures in space throughout this chapter, especially for proofs. You already know how to prove the following theorem for points *A, B, P,* and *Q* if they are coplanar. However, seeing that the proof works also for noncoplanar points requires perspective.

Theorem 10.1
If the endpoints of a segment are equidistant from two other points, then every point between the endpoints is also equidistant from the two other points.

Given: $\overline{AB}$ with points *P, Q*; *AP = AQ, BP = BQ,* and *A-X-B*
Prove: *XP = XQ*

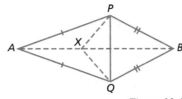

Figure 10.4

STATEMENTS	REASONS
1. $\overline{AB}$ with points P, Q; $AP = AQ$, $BP = BQ$, and A-X-B	1. Given
2. $\overline{AP} \cong \overline{AQ}$, $\overline{BP} \cong \overline{BQ}$	2. Definition of congruent segments
3. $\overline{AB} \cong \overline{AB}$, $\overline{AX} \cong \overline{AX}$	3. Reflexive property of congruent segments
4. $\triangle ABP \cong \triangle ABQ$	4. SSS
5. $\angle PAB \cong \angle QAB$	5. Definition of congruent triangles
6. Draw $\overline{XP}$ and $\overline{XQ}$	6. Line Postulate
7. $\triangle PAX \cong \triangle QAX$	7. SAS
8. $\overline{XP} \cong \overline{XQ}$	8. Definition of congruent triangles
9. $XP = XQ$	9. Definition of congruent segments

Work on your perspective sketches so that you will be ready for future three-dimensional proofs.

▶ A. Exercises

Draw each of the following figures on graph paper.

1. A sphere
2. A cone
3. A cylinder
4. A pentagonal prism
5. Two parallel planes
6. Two intersecting planes
7. An octahedron

Draw a picture of the intersection of the surface and the plane described.

8. A horizontal plane   **9.** A vertical plane   **10.** A horizontal plane

  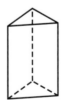

11. A plane can intersect a circular cylinder to form a circle or a rectangle. Draw a picture to show these different cases.

▶ B. Exercises

Make a sketch as directed in exercises 12-15.

12. Three cubes, all with different one-point perspectives
13. Two cubes, each with different two-point perspectives
14. A rectangular prism drawn from a three-point perspective
15. Segments representing a row of trees extending into the distance from a one-point perspective
16. Find a picture in a book or magazine that uses one-point perspective. Find the vanishing point.
17. Find a picture in a book or magazine that uses two-point or three-point perspective. Find all the vanishing points.

▶ C. Exercises

Each frame below represents an unfurnished room in perspective. Maintain the perspective (find vanishing points first!) and draw a couch along one wall and a window on the adjacent wall (include curtains if you wish). Try to draw the same room in all three perspectives.

18.

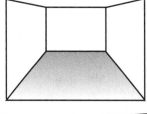

20.

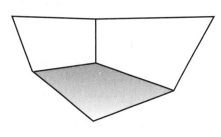

19.

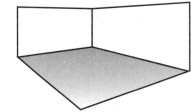

■ Cumulative Review

True/False

21. Two distinct lines can intersect at two distinct points.
22. Two parallel lines are contained in exactly one plane.
23. If $ST = \frac{1}{2}SV$, then T is the midpoint of $\overline{SV}$.
24. A line and a point not on that line are contained in exactly one half-plane.
25. Two intersecting lines are always coplanar.

10.2 Separation in Space

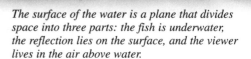

In Chapter 2 we saw that according to the Line Separation Postulate, a point separates a line into three disjoint sets. The three sets are the point and the two half-lines. We also saw that according to the Plane Separation Postulate, a line separates a plane into three disjoint sets:

The surface of the water is a plane that divides space into three parts: the fish is underwater, the reflection lies on the surface, and the viewer lives in the air above water.

the line and two half-planes. Space can also be separated into three disjoint sets. What figure do you think would separate space into three disjoint sets? What do you think those sets would be called?

Postulate 10.1

Space Separation Postulate. **Every plane separates space into three disjoint sets: the plane and two half-spaces.**

In the figure, space is called S, and plane p divides S into two half-spaces: s_1 and s_2.

The plane is called the *face* of the half-space. So the face of half-space s_1 is plane p. Each half-space is convex because when you connect any two points of a half-space, the entire segment is contained in the half-space. If $A \in s_1$ and $B \in s_2$, what is true about $\overline{AB}$?

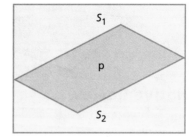

Figure 10.5

In a plane two intersecting lines form four angles as shown.

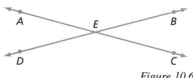

Figure 10.6

What happens when two planes intersect in space? What is the intersection of two planes?

Two planes that intersect in space also form four angles. These angles are called *dihedral* angles. The notation is similar to other angles except that the vertex is replaced by two points on the edge. The four dihedral angles above are ∠*P-MN-R*, ∠*R-MN-Q*, ∠*Q-MN-O*, and ∠*O-MN-P*. ∠*O-MN-R* and ∠*P-MN-Q* are straight dihedral angles.

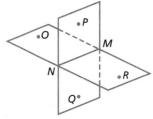

Figure 10.7

Definitions

A **dihedral angle** is the union of any line and two half-planes that have the line as a common edge.

The **edge of a dihedral angle** is the common edge of two half-planes that form a dihedral angle.

A **face of a dihedral angle** is the union of the edge of a dihedral angle and one of the half-planes that form the dihedral angle.

Dihedral angles abound in the world around us. For example, an open booklet forms a dihedral angle with the covers as faces and the spine as the edge. The walls in your classroom also form dihedral angles at the corners.

The *interior of a dihedral angle* can be described as an infinite wedge formed by the intersection of two half-spaces. Can you visualize the interior of a dihedral angle?

How do we measure dihedral ∠*L-PQ-M*?

∠*LNM* is a *plane angle* of the dihedral angle if $\overrightarrow{NM} \perp \overleftrightarrow{PQ}$ and $\overrightarrow{NL} \perp \overleftrightarrow{PQ}$. The *measure* of the dihedral angle is the measure of the plane angle. Thus if the plane angle is a right angle, the dihedral angle is a right dihedral angle. Dihedral angles are *congruent* if they have equal measures.

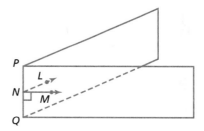

Figure 10.8

The closed surface of a polyhedron determines several dihedral angles but also determines a solid. Remember that a solid is the union of a closed surface and its interior.

▶ A. Exercises

Use the proper notation to describe all the dihedral angles in exercises 1-3.

1.
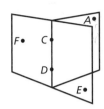

2.
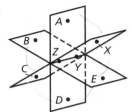

3.
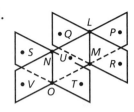

Use the figure for exercises 4-8.

4. Name the faces of the dihedral angle.
5. What is the edge of the dihedral angle?
6. Name a plane angle.
7. If $m\angle MXL = 90$, what is the dihedral angle called?
8. Find $a \cap b$.

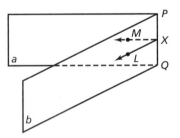

▶ B. Exercises

Use the figure shown for exercises 9-16.

9. Describe half-plane *XLM* ∪ half-plane *YLM*.
10. Describe $s_1 \cup p$.

Perform each set operation.

11. $s_1 \cap s_2$
12. $s_1 \cup s_2 \cup p$
13. p'
14. $(s_1 \cup p)'$
15. $(s_1 \cup s_2 \cup p)'$
16. Draw the plane determined by *S*, *T*, and *Y*.
17. Is a dihedral angle a closed surface? Explain.
18. How many parts do two intersecting planes separate space into?

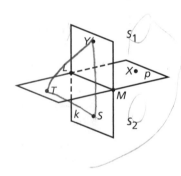

▶ C. Exercises

19. How can three planes separate space into four parts? seven parts? (include diagrams)
20. Show two ways in which three planes can separate space into six parts.

Tell whether these statements are true or false.

21. If $\angle LMN \cong \angle PQR$, then $m\angle LMN = m\angle PQR$.

22. If $\angle LMN \cong \angle PQR$, then $\angle LMN \cong \angle PRQ$.

23. Vertical angles are congruent.

24. Supplementary angles form a linear pair.

25. A transversal perpendicular to one line is perpendicular to the other.

10.3 Perpendiculars in Space

Proofs of theorems in space are very much like proofs in the plane. The theorems you have learned can be used in these proofs as long as each is applied in a single plane. Recall that perpendicular lines form right angles.

The roof of this covered bridge in Bucks County, Pennsylvania, illustrates a dihedral angle.

Definitions

Perpendicular planes are two planes that form right dihedral angles.

A **line perpendicular to a plane** is a line that intersects the plane and is perpendicular to every line in the plane that passes through the point of intersection.

A **perpendicular bisecting plane** of a segment is a plane that bisects a segment and is perpendicular to the line containing the segment.

Study the following proofs to see how to do proofs in space.

Theorem 10.2

A line perpendicular to two intersecting lines in a plane is perpendicular to the plane containing them.

Given: Line *l* is perpendicular to lines *m* and *n* at *A*; plane *p* contains *m* and *n*

Prove: Line *l* is perpendicular to plane *p*

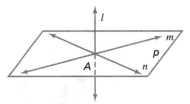

Figure 10.9

According to the definition of a line perpendicular to a plane, we must prove that *l* is perpendicular to every line in plane *p* that passes through *A*. Let $\overleftrightarrow{AK}$ be a representative line in plane *p*. Since *K* is interior to one of the four angles formed by lines *m* and *n*, find auxiliary points on *m* and *n* respectively so that B-K-C.

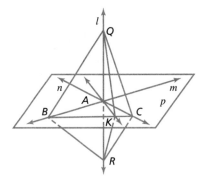

Figure 10.10

STATEMENTS	REASONS
1. $l \perp m$ at A, $l \perp n$ at A	**1.** Given
2. Lines m and n determine a plane p	**2.** Two intersecting lines lie in the same plane
3. $Q \in l$; let R be on the opposite ray to $\overrightarrow{AQ}$ so that $AR = AQ$	**3.** Completeness Postulate
4. $\overleftrightarrow{AK}$ is another line in plane p; $B \in m$; $C \in n$; B-K-C	**4.** Auxiliary lines
5. m is the perpendicular bisector of $\overline{QR}$ in the plane of points Q, B, and R; n is the perpendicular bisector of $\overline{QR}$ in the plane of points Q, C, and R	**5.** Definition of perpendicular bisector
6. $BQ = BR$; $CQ = CR$	**6.** Points on a perpendicular bisector are equidistant from the endpoints
7. $QK = RK$	**7.** B and C are equidistant from Q and R, so point K between B and C is also (Theorem 10.1)
8. K is on the perpendicular bisector of $\overline{QR}$ (in the plane of points K, Q, and R)	**8.** Points equidistant from the endpoint lie on the perpendicular bisector
9. $\overleftrightarrow{AK} \perp l$ at point A	**9.** Definition of perpendicular bisector
10. $l \perp p$	**10.** Definition of line perpendicular to plane

Theorem 10.3

If a plane contains a line perpendicular to another plane, then the planes are perpendicular.

Given: $\overleftrightarrow{AB} \perp m$ at B, plane n contains $\overleftrightarrow{AB}$
Prove: $m \perp n$

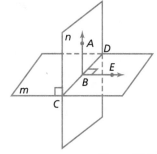

Figure 10.11

STATEMENTS	REASONS
1. $\overleftrightarrow{AB} \perp m$ at B	1. Given
2. $m \cap n = \overleftrightarrow{CD}$	2. Plane Intersection Postulate
3. $\overleftrightarrow{BE} \perp \overleftrightarrow{CD}$ at B	3. Auxiliary perpendicular
4. $\overleftrightarrow{AB} \perp \overleftrightarrow{BE}$ and $\overleftrightarrow{AB} \perp \overleftrightarrow{CD}$	4. Definition of line perpendicular to plane
5. $\angle ABE$ is right	5. Definition of perpendicular bisector
6. $\angle ABE$ is a plane angle	6. Definition of plane angle
7. $\angle A\text{-}CD\text{-}E$ is a right dihedral angle	7. Definition of dihedral angle measure
8. $m \perp n$	8. Definition of perpendicular planes

The following theorems are discussed in the exercises. (Proofs of the first three are exercises 16-18 and the fourth is in the chapter review.)

Theorem 10.4
If intersecting planes are each perpendicular to a third plane, then the line of intersection of the first two is perpendicular to the third plane.

Theorem 10.5
If $\overleftrightarrow{AB}$ is perpendicular to plane p at B, and $\overline{BC} \cong \overline{BD}$ in plane p, then $\overline{AC} \cong \overline{AD}$.

Theorem 10.6
Every point in the perpendicular bisecting plane of segment $\overline{AB}$ is equidistant from A and B.

Theorem 10.7
The perpendicular is the shortest segment from a point to a plane.

▶ A. Exercises

Make a sketch of the following.
1. A dihedral linear pair
2. Theorem 10.4
3. Theorem 10.5
4. Theorem 10.6
5. Theorem 10.7

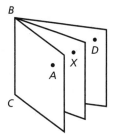

Use the diagram for exercises 6-7.
6. Name 2 dihedral angles.
7. State an Angle Addition Theorem for Dihedral Angles.

Explain or define each term below.
8. Supplementary dihedral angles
9. Complementary dihedral angles

▶ B. Exercises

Draw conclusions about dihedral angles (based on your knowledge of plane angles).
10. All right dihedral angles are . . .
11. Vertical dihedral angles are . . .
12. Supplementary dihedral angles that are congruent are . . .
13. Supplementary dihedral angles that are adjacent . . .
14. The dihedral angles in a dihedral linear pair are . . .
15. If one dihedral angle of a dihedral linear pair is a right angle, then . . .

▶ C. Exercises

Prove each theorem below.
16. Theorem 10.4
17. Theorem 10.5
18. Theorem 10.6

■ Cumulative Review

19. Name two postulates for proving triangles congruent.
20. Name two theorems for proving triangles congruent.
21. Name a fifth method for proving triangles congruent that works only for right triangles.
22. What is SAS called when applied to right triangles?
23. What is ASA called when applied to right triangles?

10.4 Parallels in Space

The study of parallels builds on the study of perpendiculars. You have studied perpendicular lines, perpendicular planes, and lines perpendicular to planes. Likewise, you must know when lines are parallel, planes are parallel, and lines are parallel to planes. Recall that parallel lines are two coplanar lines that do not intersect. Skew lines are not coplanar, which means that they are neither concurrent nor parallel.

The solar panels of the International Space Station illustrate parallel planes. The station is one of 2,465 functional man-made satellites orbiting Earth.

Parallel planes are two planes that do not intersect.

A **line parallel to a plane** is a line that does not intersect the plane.

Theorem 10.8

Two lines perpendicular to the same plane are parallel.

Given: Plane m containing points B and D,
             $\overleftrightarrow{AB} \perp m$ and $\overleftrightarrow{CD} \perp m$

Prove: $\overleftrightarrow{AB} \parallel \overleftrightarrow{CD}$

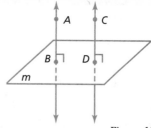

Figure 10.12

To prove that lines are parallel, we must prove two things: they must be coplanar and they must not intersect. You can see that $\overleftrightarrow{AB} \perp \overleftrightarrow{BD}$ and $\overleftrightarrow{CD} \perp \overleftrightarrow{BD}$, but lines perpendicular to the same line are not parallel unless they are also coplanar. This theorem is important because it enables you to prove lines parallel without first having to prove them coplanar.

Some exercises require indirect proofs. Study the proof of Theorem 10.9 to review this method, which assumes the theorem is false and finds a contradiction.

Theorem 10.9

If two lines are parallel, then any plane containing exactly one of the two lines is parallel to the other line.

Given: $\overleftrightarrow{AB} \parallel \overleftrightarrow{CD}$, plane n contains $\overleftrightarrow{AB}$
           but not $\overleftrightarrow{CD}$

Prove: Plane n is parallel to $\overleftrightarrow{CD}$

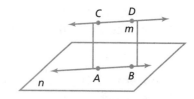

Figure 10.13

STATEMENTS	REASONS
1. $\overleftrightarrow{AB} \parallel \overleftrightarrow{CD}$, n contains $\overleftrightarrow{AB}$, n does not contain $\overleftrightarrow{CD}$	1. Given
2. n intersects $\overleftrightarrow{CD}$ at point P	2. Assumption
3. Let m be the plane containing $\overleftrightarrow{AB}$ and $\overleftrightarrow{CD}$	3. Definition of parallel lines (they are coplanar)
4. Planes m (containing $\overleftrightarrow{CD}$) and n intersect in exactly one line, $\overleftrightarrow{AB}$	4. Plane Intersection Postulate
5. $\overleftrightarrow{AB}$ intersects $\overleftrightarrow{CD}$ at point P	5. Two intersecting lines intersect in exactly one point (compare steps 2 and 4)
6. Plane n is parallel to $\overleftrightarrow{CD}$	6. Law of Contradiction (see steps 1 and 5)
7. If two lines are parallel and a plane contains exactly one of them, then the plane is parallel to the other line	7. Law of Deduction

Theorem 10.10
A plane perpendicular to one of two parallel lines is perpendicular to the other line also.

Theorem 10.11
Two lines parallel to the same line are parallel.

Theorem 10.12
A plane intersects two parallel planes in parallel lines.

Theorem 10.13
Two planes perpendicular to the same line are parallel.

Theorem 10.14
A line perpendicular to one of two parallel planes is perpendicular to the other also.

Theorem 10.15
Two parallel planes are everywhere equidistant.

▶ A. Exercises

Make a sketch for each theorem below.
1. Theorem 10.10
2. Theorem 10.11
3. Theorem 10.12
4. Theorem 10.13
5. Theorem 10.14
6. Theorem 10.15

▶ B. Exercises

Disprove each of these false statements by sketching a counterexample.
7. Two planes parallel to the same line are parallel.
8. Two lines parallel to the same plane are parallel.
9. If two planes are parallel, then any line in the first plane is parallel to any line in the second.
10. If a line is parallel to a plane, then the line is parallel to every line in the plane.
11. Lines perpendicular to parallel lines are parallel.

Sketch each.
12. A plane containing exactly one of two parallel lines
13. A line perpendicular to both of two skew lines
14. A plane containing one of two skew lines and parallel to the other
15. A plane parallel to a line so that a point of the plane is farther from the line than the distance between the line and the plane

▶ C. Exercises

Prove the theorems.
16. Theorem 10.10 (*Hint:* Use proof by contradiction and the Historic Parallel Postulate.)
17. Theorem 10.11
18. Theorem 10.12

▮ Cumulative Review

Answer true or false. Refer to the prism shown.
19. Point G is interior to the prism.
20. $\triangle DEF$ is a base of the prism.
21. $\overline{CD}$ is an edge of the prism.
22. $\triangle DEF \cong \triangle ABC$
23. If Q is between G and H, then Q is interior to the prism.

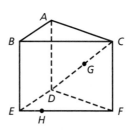

Slopes of Parallel Lines

Do you remember what slope means? In Chapter 4 we defined slope as the ratio of vertical to horizontal change. Slope measures the angle that a line makes with the horizontal axis. The angle between the line and x-axis is the angle of inclination. The relationship between slope and angle of inclination can be used to prove that lines with the same slope are parallel.

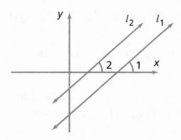

STATEMENTS	REASONS
1. $m_1 = m_2$	1. Given
2. $m\angle 1 = m\angle 2$	2. Same slopes define same angle of inclination
3. $\angle 1 \cong \angle 2$	3. Definition of congruent angles
4. $l_1 \parallel l_2$	4. Corresponding Angle Theorem
5. If the slopes of two lines are equal, then the lines are parallel	5. Law of Deduction

You can also prove the converse that parallel lines have equal slopes. You can find equations of lines using this relationship together with point-slope form.

EXAMPLE Find the equation of the line through (2, 5) and parallel to $3x + y = 7$.

Answer $3x + y = 7$ 1. Find the slope of the given line.
$$y = -3x + 7$$ The slope of the parallel line is
$$m = -3$$ the same.

$$y - y_1 = m(x - x_1)$$ 2. Use point-slope form to obtain
$$y - 5 = -3(x - 2)$$ the desired equation.
$$y - 5 = -3x + 6$$
$$y = -3x + 11$$

▶ Exercises

1. Find the slope of a line parallel to $2x + 5y = 3$.

Find the equation of the line
2. parallel to $x - 3y = 9$ and having the y-intercept of (0, 4).
3. parallel to $2x - y = 4$ and passing through $(-1, 3)$.
4. Prove that parallel lines have the same slope.
5. Combine the two theorems about parallel lines and slopes by writing a biconditional.

10.5 Polyhedra

Do you remember the relationship between a polygon and its angles? Quadrilateral *ABCD* determines ∠*ABC*, but ∠*ABC* is not a subset of the quadrilateral. In the same way polyhedra determine dihedral angles.

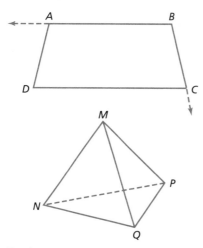

Figure 10.14

The top half of the lunar module Orion *from Apollo 16 forms a concave polyhedron.*

For instance, ∠*M-PQ-N* is a dihedral angle determined by the tetrahedron. ∠*M-PQ-N* is called an angle of the tetrahedron. Can you name other angles of this tetrahedron?

Another similarity between polygons and polyhedra is classification. In Chapter 2, you learned to classify by the number of sides or faces, respectively. Just as you further classified quadrilaterals, you can also further classify *hexahedra*.

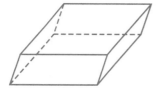 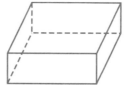 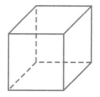

Figure 10.15

Remember that a prism is *rectangular* if its bases are rectangles and that a cube is a right rectangular prism with all sides congruent.

A **parallelepiped** is a hexahedron in which all faces are parallelograms.

A **diagonal of a hexahedron** is any segment joining vertices that do not lie on the same face.

Opposite faces of a hexahedron are two faces with no common vertices.

Opposite edges of a hexahedron are two edges of opposite faces that are not joined by an edge of the hexahedron.

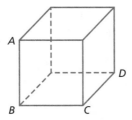

Figure 10.16

Be sure you do not confuse these concepts: $\overline{AB}$ is an edge of the cube; $\overline{AC}$ is a diagonal of the square face of the cube; $\overline{AD}$ is a diagonal of the cube. Similarly, do not confuse opposite edges of the cube with opposite edges of one face of the cube.

Here are some other theorems on parallelepipeds.

Theorem 10.16
Opposite edges of a parallelepiped are parallel and congruent.

Theorem 10.17
Diagonals of a parallelepiped bisect each other.

Theorem 10.18
Diagonals of a right rectangular prism are congruent.

One last theorem, called Euler's formula, applies not only to parallelepipeds but to all convex polyhedra. The following diagrams will help you discover this important formula for polyhedra. Which polyhedra are not regular?

(a)   (b)   (c)   (d)

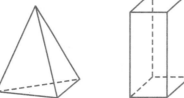

Figure 10.17

Look again at diagram (a) in figure 10.17. How many vertices can you count? How many edges? How many faces?

    Let *V* = number of vertices
      E = number of edges
      F = number of faces
    Compute $V - E + F$

The table below summarizes this computation for all four diagrams in figure 10.17. Find the value for the last column in each case.

Diagram	V	E	F	$V - E + F$
(a)	4	6	4	2
(b)	8	12	6	
(c)	6	12	8	
(d)	12	30	20	

The Swiss mathematician Leonard Euler (OY–ler) was the first to discover the numerical relationship that you just discovered about a polyhedron. For any convex polyhedron the following equation is true.

Euler's Formula
$V - E + F = 2$ where *V*, *E*, and *F* represent the number of vertices, edges, and faces of a convex polyhedron respectively.

This formula works for some other polyhedra also, but it always works for convex polyhedra. Although the relationship can be proved, the proof involves concepts you have not yet studied.

You can *truncate* a solid by slicing off part of it. Begin with a regular polyhedron and slice off all the corners so that the faces of the truncated polyhedron are regular polygons. Can you describe two possible truncated cubes (give the number and types of faces)?

▶ A. Exercises

Tell whether the following statements are true or false.
1. The faces of a regular octahedron are equilateral triangles.
2. Euler's formula can be written $V + F = E + 2$.
3. A hexagonal prism is an octahedron.
4. A regular dodecahedron has hexagonal faces.
5. A regular tetrahedron is a pyramid.

For each decahedron below, determine the number of faces, edges, and vertices. Check Euler's formula for each.

6.
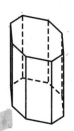

8.

10.

7.
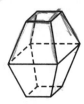

9.

11.

▶ B. Exercises

Each exercise below refers to a prism having the given number of faces, vertices, edges, or sides of the base. Determine the missing numbers to complete the table below. Draw the prism when necessary; find some general relationships between these parts of the prism to complete exercise 18.

	F	V	S	E
	Faces	Vertices	Sides of the base	Edges
Example	14	24	12	36
12.		6		9
13.	7	10		
14.		8	4	
15.			7	
16.		28		
17.	8			
18.			n	

19. Based on this section, draw two conclusions about the diagonals of a cube.

Use the figure to prove the following theorems.

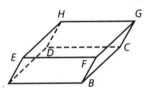

20. Opposite edges of a parallelepiped are parallel (part of Theorem 10.16).
21. Opposite edges of a parallelepiped are congruent (part of Theorem 10.16).

▶ C. Exercises

Prove each theorem. You may write a paragraph rather than a two-column proof.

22. Theorem 10.17

23. Theorem 10.18

▪ Cumulative Review

Do not solve exercises 24-27 below, but write (in complete sentences) what you would do to solve them.

24. Find the area.

25. Prove that $\angle A \cong \angle B$.

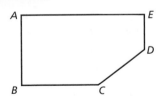

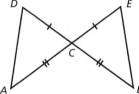

26. Find the distance between two numbers a and b on a number line.
27. True/False: Water contains helium or hydrogen.
28. When are the remote interior angles of a triangle complementary?

10.6 Spheres

Many of the concepts that you have learned about circles also apply to spheres. So if you understand the basic definitions and theorems of circles, you will understand how these same ideas apply to spheres. From this section you should see how a strong foundation will help you be better prepared for the future. The same

In 1978, astronomers at the U.S. Naval Observatory near Flagstaff, Arizona, found Charon, the only moon of Pluto, with a 61-inch reflecting telescope. The hemispherical dome rotates on its cylindrical base to track the heavenly bodies.

principle applies to your spiritual life. If you build a strong Bible-based foundation while you are young, you will be better prepared for the adult life you have before you. In I Timothy 6:17-19 Paul admonishes Christians that they should trust "in the living God" (v. 17) and lay a "good foundation against the time to come" (v. 19).

Let us see how your knowledge of circles helps you in the study of spheres. Earlier in this book you saw the definition of *sphere*. A sphere is the set of points in space that are a given distance from a given point. A sphere is a surface or a shell. The given point is the center of the sphere, and the segment from the center to a point on the surface is the radius. In sphere *S*, *O* is the center, and $\overline{OA}$ is a radius.

Figure 10.18

A chord of a sphere is a segment with endpoints on the sphere. A diameter of a sphere is a chord containing the center, and a secant of a sphere is a line that contains a chord. A tangent is a line containing exactly one point of the sphere.

What results when a plane and a sphere intersect?

Voyager 2 viewed Saturn from a distance of only 13 million miles. The three icy moons below the planet are Tethys, Dione, and Rhea (left to right). Do you remember the name for the shape made by the rings?

Definitions

A **secant plane** to a sphere is a plane that intersects a sphere in more than one point.

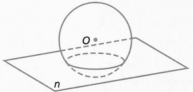

Figure 10.19

A **tangent plane** to a sphere is a plane that intersects a sphere in exactly one point. The point is called the **point of tangency**.

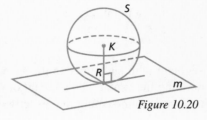

Figure 10.20

In the diagram above, sphere *S* with center *K* is tangent to plane *m* at point *R*.

Theorem 10.19

The intersection of a sphere and a secant plane is a circle.

Given: Sphere S with center C and secant plane n

Prove: $n \cap S$ is a circle

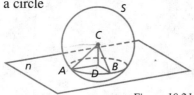

Figure 10.21

STATEMENTS	REASONS
1. Sphere S with center C and secant plane n	1. Given
2. S intersects n in at least two points A and B	2. Definition of secant plane
3. $AC = BC$; $\overline{AC} \cong \overline{BC}$	3. Definitions of sphere, congruent segments
4. If $C \in n$, then A, B, and C are coplanar and A and B are equidistant from C, which is the center of a circle of radius AC	4. Definition of circle
5. If $C \notin n$, then there is a perpendicular $\overleftrightarrow{CD}$ to n at D	5. Exactly one line passes through a point perpendicular to a plane
6. $\overleftrightarrow{AD}$ and $\overleftrightarrow{BD}$ exist	6. Line Postulate
7. $\overleftrightarrow{CD} \perp \overleftrightarrow{AD}$, $\overleftrightarrow{CD} \perp \overleftrightarrow{BD}$	7. Definition of line perpendicular to a plane
8. $\angle ADC$ and $\angle BDC$ are right angles	8. Definition of perpendicular
9. $\triangle ACD$ and $\triangle BCD$ are right triangles	9. Definition of right triangle
10. $\overline{CD} \cong \overline{CD}$	10. Reflexive property of congruent segments
11. $\triangle ACD \cong \triangle BCD$	11. HL
12. $\overline{AD} \cong \overline{BD}$	12. Definition of congruent triangles
13. $AD = BD$	13. Definition of congruent segments
14. A and B (and every point of intersection) are on a circle centered at D with radius AD	14. Definition of circle

The intersection of the sphere with center O and plane m is a circle. If a plane passes through the center of the sphere, it intersects the sphere in what is called a great circle.

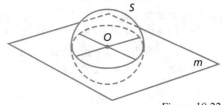

Figure 10.22

Definition

A **great circle** of a sphere is the intersection of the sphere and a secant plane that contains the center of the sphere.

The following theorems give important facts concerning spheres.

Theorem 10.20
Two points on a sphere that are not on the same diameter lie on exactly one great circle of the sphere.

Theorem 10.21
Two great circles of a sphere intersect at two points that are endpoints of a diameter of the sphere.

Theorem 10.22
All great circles of a sphere are congruent.

The next theorems are similar to theorems about circles and can be proved in a similar manner.

Theorem 10.23
A secant plane of a sphere is perpendicular to the line containing the center of the circle of intersection and the center of the sphere.

Theorem 10.24
A plane is tangent to a sphere if and only if it is perpendicular to the radius at the point of tangency.

When traveling on the surface of the earth, it is useful to find the shortest distance between two points on the surface of a sphere—for example, the distance between points A and B.

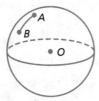

Figure 10.23

Notice that the distance could not be a straight segment because the sphere has curvature. To find the distance between A and B, you must find the length of the arc on the great circle that contains A and B. The symbol for the distance is $d\widehat{AB}$.

If $m\widehat{AB}$ is 3°, the arc is $\frac{3}{360}$ of the circumference of the great circle. Use $c = 2\pi r$ with $r = 8$.

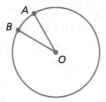

Figure 10.24

$$d\widehat{AB} = \frac{m\widehat{AB}}{360}(2\pi r)$$

$$d\widehat{AB} = \frac{3}{360}(2\pi \cdot 8)$$

$$= \frac{1}{120}(16\pi)$$

$$= \frac{2\pi}{15} \text{ or about } 0.4189 \text{ units}$$

So the distance from A to B is a little less than half a unit.

This method can be used to find the distance between two places on the earth. Although the earth is not a perfect sphere, it is so close to the shape of a sphere that the variations are usually insignificant.

► A. Exercises

1. Draw three diagrams showing the three possible cases of the intersection of a sphere and a plane.
2. Draw a sphere and a plane tangent to the sphere. Draw three lines that are contained in the plane and that pass through the point of tangency. What is true about these lines?
3. Draw two spheres having great circles that are externally tangent circles.

For exercises 4-7, find d$\widehat{XY}$ if A is the center of a sphere, and ∠XAY has the indicated measure. The radius of the sphere is 4 units. Find to the nearest tenth.

4. m∠XAY = 80
5. m∠XAY = 35
6. m∠XAY = 162
7. m∠XAY = 14

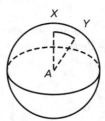

Find the distance between two points H and I on sphere J if ∠HJI and $\overline{IJ}$ have the indicated measures. Find to the nearest tenth.

8. m∠HJI = 30; IJ = 12
9. m∠HJI = 160; IJ = 3
10. m∠HJI = 48; IJ = 16
11. m∠HJI = 90; IJ = 43
12. m∠HJI = 270; IJ = 24

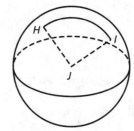

► B. Exercises

Consider the earth to be a sphere. If its diameter is 7900 miles, find the distance between two cities that are the given degree measures apart.

13. 36°
14. 1°
15. 95°

Prove each theorem.

16. Theorem 10.20
17. Theorem 10.21

▶ C. Exercises

Prove each theorem.
18. Theorem 10.22
19. Theorem 10.23
20. Theorem 10.24

■ Cumulative Review

Find each area.

21.

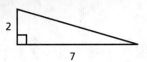

22.

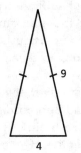

23.

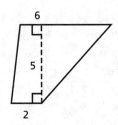

24.

25.

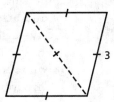

GEORG FRIEDRICH BERNHARD RIEMANN

Georg Bernhard Riemann was born on September 17, 1826, in the village of Breselenz, in Hanover, Germany. His father was a Lutheran pastor. The Lutheran congregation in the small village could not sufficiently support the pastor and his six children, and the Riemann family barely survived. The children were undernourished, and their mother, Charlotte Ebell, died before her children had grown to maturity.

Through the struggle for survival, the Riemann family grew to be a close family, and Bernhard had a very happy childhood. He, however, was a timid boy and remained extremely shy throughout his life. He was often homesick when he was away from his family and made frequent trips home.

Bernhard received his first educational instruction from his father. He loved to learn, and his earliest interest was in history. He began studying arithmetic at the age of six, at which time his mathematical genius became apparent. Bernhard enjoyed solving every problem he could find, and he often made up difficult problems for his brothers and sisters. They found the problems impossible to solve. By the time Bernhard was ten, his father found it necessary to delegate further mathematical instruction to a professional tutor. He began advanced study in arithmetic and geometry but soon surpassed his teacher. His formal education began at fourteen when he entered the Gymnasium at Hanover. Two years later he transferred to the Gymnasium at Lüneburg.

Riemann's father wanted him to study theology, so at the age of nineteen Reimann entered the University of Göttingen to study theology but continued studying mathematics privately. He could not escape his delight in the subject and soon asked his father's permission to leave preparation for the ministry to study mathematics. His father agreed, and Riemann joyously entered the University of Berlin to study math. Throughout his life Riemann remained a faithful Christian.

Riemann finished his doctorate at Göttingen in 1851. Even the greatest mathematician of his day praised his dissertation on the "Foundation for a General Theory of Functions of a Complex Variable." His dissertation showed his ability in pure mathematics but also his interests in applied mathematics. Combining his abilities in pure math and physical science, he excelled in mathematical physics. Investigations in mathematical physics occupied him until he became an assistant professor at the University of Göttingen in 1857 and a full professor in 1859. The professorship improved his financial state, and he married Elise Koch at the age of thirty-six.

Soon after his marriage Riemann became ill and never really recovered. He died of tuberculosis on July 20, 1866. Engraved on his tombstone were the words from Romans 8:28: "All things work together for good to them that love the Lord."

Riemann's major mathematical contributions include his study of functions of both real and complex variables. He also introduced Riemannian geometry—a type of non-Euclidean geometry—in his *Habilitationsshrift*. This essay (and his first lecture as a lecturer in 1854) revolutionized both geometry and physics. Probably his greatest gift was his ability to unify physical and mathematical thoughts connected with some large problem.

He loved to learn, and his earliest interest was in history.

10.7 | Latitude and Longitude

The earth is nearly spherical. The diameter of the earth from the North Pole to the South Pole is 7885 miles, while the diameter at the equator is about 7912 miles. This 27-mile difference is almost nothing compared to the size of the earth. Since we can consider the earth a sphere, the definitions and theorems that you learned about spheres can apply to the earth.

Two important great circles are marked on every globe. The first great circle to consider is the equator. The plane that cuts off the equator passes through the center of the earth and is perpendicular to the diameter that passes through the North and South Poles. The equator divides the earth into two hemispheres called the Northern and Southern Hemispheres. Planes that are parallel

What parts of the Eastern Hemisphere do you recognize on this satellite photo?

to the plane of the equator cut off smaller circles that appear to be parallel to the equator on the globe. These lines are called *latitudinal lines*. These lines help locate cities and other points of interest on the earth in a north-south direction and on most globes are marked off in 10-degree divisions from the equator.

A great circle that passes through Greenwich, England, is also an important reference circle. The semicircle that goes from the North Pole to the South Pole and passes through Greenwich, England, is called the *prime meridian*. The other semicircle of this great circle is called the *International Date Line.*

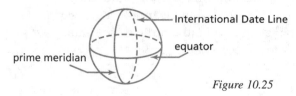

Figure 10.25

The great circle containing the prime meridian and International Date Line divides the earth into two hemispheres called the Eastern and Western Hemispheres. Great circles that pass through the North and South Poles are called *longitudinal lines*.

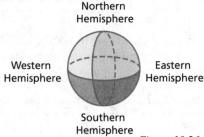

Northern Hemisphere

Western Hemisphere

Eastern Hemisphere

Southern Hemisphere

Figure 10.26

These longitudinal and latitudinal lines form a grid by which places can be located. The means of locating places seems complicated at first but is much like wrapping graph paper around the earth. A place is located according to degree and direction. Latitude is the degree measure of the arc from the equator to a particular location. The degree of latitude is always measured north or south and lies between 0° and 90°. Point *s* is located at 60° north latitude.

Longitude is measured according to degree measure east or west from the prime meridian. This measure ranges from 0° to 180°. Point *x* is located at 110° east longitude. With these longitudinal and latitudinal lines, the location of any place on the earth can be specified.

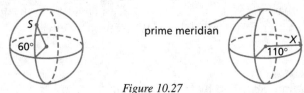

prime meridian

Figure 10.27

EXAMPLE 1   What city is located at 30° north latitude, 90° west longitude?

Answer   1. Start at the intersection of the equator and the prime meridian.
2. Move to 30° north.
3. Move to 90° west along the 30° north latitudinal line.
4. New Orleans is located at the intersection of 30° north latitude and 90° west longitude.

Many times the place you are interested in will not fall directly on an intersection of a longitudinal and a latitudinal line. When this happens, simply estimate the distance between the longitudinal or latitudinal lines.

> **EXAMPLE 2** Give the location of Sydney, Australia.
>
> **Answer** 34° south **1.** Find Sydney, Australia. Move down along the longitudinal line to Sydney, counting the degrees from the equator. Sydney is between 30° and 40° south.
>
> 151° east **2.** Locate the prime meridian and count the degrees east along the equator until you are even with Sydney. It is just beyond 150°.
>
> Sydney, Australia, is located at 34° south, 151° east.

► A. Exercises

Find the city, state, or country at each location. Use globes or atlases.

1. 41° north, 76° west
2. 20° north, 160° west
3. 34° south, 18° east
4. 1° north, 104° east
5. 48° north, 3° east

6. 23° south, 43° west
7. 25° north, 100° west
8. 18° south, 178° east
9. 31° north, 30° east
10. 18° south, 150° west

Find the latitude and longitude of each place indicated below.

11. Manila, Philippines
12. Moscow, Russia
13. Rome, Italy
14. Beirut, Lebanon
15. Alexander Island, Antarctica

16. Madrid, Spain
17. Hong Kong
18. Brasília, Brazil
19. Casablanca, Morocco
20. Nairobi, Kenya

► B. Exercises

Give the location described by the following.
21. Every pair of longitudinal lines intersect at what points?
22. The great circles containing the prime meridian and the equator intersect at what points?

For the following give the latitude, longitude, name of country, and city.
23. Find the location in the Southern Hemisphere corresponding to Honolulu, Hawaii.
24. Find the location opposite Honolulu on a diameter of the earth.
25. Find the location in the Northern Hemisphere on the other side of the world from Honolulu.

► C. Exercises

The Arctic Circle, tropic of Cancer, tropic of Capricorn, and Antarctic Circle divide the earth into five latitudinal zones.

26. Give the latitude of each of the four lines and explain the significance of the numbers.

■ Cumulative Review

27. The average radius of the earth is 3953 miles. Find the surface area to the nearest million square miles.

Give a condition for the congruence of each pair of figures.

28. two circles

29. two parallelograms

30. two segments

31. two cubes

10.8 Spherical Geometry

A farm on the flat plains of Iowa has a rectangle boundary. What shape does it have on the sphere of the earth?

Euclidean geometry results from valid reasoning based on the postulates so far presented. If you replace some of the postulates, you obtain other systems, called *non-Euclidean geometries*. One of these systems is *Riemannian geometry*, named after its developer Bernhard Riemann.

Riemannian geometry is also called spherical geometry because you can use a sphere to represent it. Think about the earth. To go in a straight line from Memphis Tennessee, to Dacca, Bangladesh, you would have to dig a tunnel along the chord of the sphere. An airline pilot considers the shortest distance to be an arc of the great circle of the sphere. In this context it is useful to use great circles as the "lines" of our system and the places on the surface of the earth as points. Thus, the surface of the earth is used to represent the "plane." This system has properties different from Euclidean geometry.

Figure 10.28

Since all great circles intersect, there is no such thing as parallel lines in spherical geometry. For instance, no line (great circle) passes through New Orleans (a given point) parallel to the equator (a given line). In the exercises you will also see that the sum of the measures of the angles of a triangle is never 180° in spherical geometry.

Some scientists believe that space is Riemannian. This is what is meant by "curved space." Just as one can travel around the world without perceiving the curve of the earth's surface, Riemannian scientists conceive that if you could travel through space far enough you could reach your starting point.

How can a Christian explain these contradictory systems? Does truth depend on viewpoint or usefulness? No! Both systems display valid reasoning—but which has true premises? God knows, but as humans we do not. This is an important lesson. Man cannot arrive at truth on his own—not even in math. We cannot know whether apparent parallel lines intersect. We do know that the only source of absolute truth is the Bible. God's Word provides a sure foundation because God wrote it.

These systems also reflect honesty and Christian faith. Each type of geometry clearly states its postulates without double talk; the facts are open to scrutiny. God, likewise, reveals Himself in His Word, which we should study closely. We trust postulate systems as they are used by engineers to build bridges. This faith is reasonable and practical. In the same way, we trust God for salvation. It makes sense to trust God-given math principles in building bridges; but do not put faith in human accomplishment. We must have faith in Jesus Christ: "That your faith should not stand in the wisdom of men, but in the power of God" (I Cor. 2:5).

▶ A. Exercises

1. Imagine walking south along a longitudinal line from the North Pole and turning east at the equator. What angle did you turn?

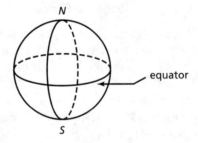

2. Which of the following are "lines" in spherical geometry: the equator, tropic of Cancer, latitudinal circles, longitudinal circles, prime meridian, tropic of Capricorn, Arctic Circle?

3. Minneapolis is at 45° N latitude. The circle of latitude at 45° N is parallel to the equator. Why do we say that there are no parallels in the model of Riemannian geometry?

▶ B. Exercises

Use the system of spherical geometry to check these incidence statements.

4. Does the "plane" of the earth contain at least three noncollinear points?

5. Does every "line" contain at least two points?

6. Are every pair of points on a "line"?

7. Do two intersecting "lines" intersect in exactly one point?

8. Do every pair of points determine exactly one "line"?

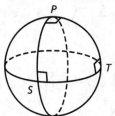

9. How does the sum of the measures of the angles in △PST compare to 180°? Explain.

10. Sketch a triangle with three right angles (an equilateral right triangle).

▶ C. Exercises

Answer the questions using Riemannian geometry.

11. Why are trapezoids and parallelograms impossible?

12. If *ABCD* is a quadrilateral with three right angles, what kind of angle is the fourth angle?

13. How does the measure of an exterior angle of a triangle compare to the sum of the measures of the remote interior angles?

■ Cumulative Review

14. A regular heptagon and a regular octagon are inscribed in congruent circles. Which polygonal region has more area?

15. Is the following argument valid? Sound? What type of argument is it?
    All lizards are reptiles.
    All salamanders are lizards.
    Therefore, all salamanders are reptiles.

16. Sketch and label the altitude, perpendicular bisector, angle bisector, and median to a side in a triangle. They must be different lines and intersect the same side.

17. Draw an illustration for the Angle Addition Postulate and explain it. One angle must be 47°.

18. What definition acts as a Segment Addition Postulate?

Geometry *and* Scripture

The Universe

How big is the universe? Is it Euclidean? Let's look for a Bible answer to each of these two difficult but important questions.

Finite or Infinite?

Some Christians and most of the secular world think the universe is infinite. However, proof would require either infinite travel or omniscience. Other Christians insist that only God is infinite. It may be true that nothing else is infinite, but this cannot be proved from the Bible. Just as man reflects some of God's qualities because he is made in God's image, space may reflect His infinitude. Thus, neither view can be proved.

Evaluate each statement as true or false.

1. Only God can reason.   2. Only God can forgive sin.

> **HIGHER PLANE:** Support your answers to questions 1 and 2 with Scripture.

3. List all Bible references to "infinite." Do any refer to physical things?

The lack of physical examples is the best evidence for a finite universe. This inductive evidence is strong but by no means conclusive. At present many theologians think the universe is finite on this basis.

Euclidean or Non-Euclidean?

You have probably always thought that the universe is Euclidean, but have you ever analyzed your reasons for thinking so?

4. Euclidean geometry teaches that parallel lines never meet. However, we cannot identify parallel lines in our universe. Space may model spherical geometry. What are lines on the earth? Do they always meet?

5. In Euclidean geometry, every triangle's angle measures total 180°. In practice, when we measure three angles and add them up, our measurements always include a margin of error. Suppose the total comes to 179.96° ± 0.05°. Could the true value be less than or greater than 180°? Which would be true spherical geometry?

6. Some argue that only Euclidean geometry has had practical application in physical science; however, scientists have used non-Euclidean geometry in studying the retina of the eye. Moreover, usefulness is not conclusive evidence. Give an ancient view of the earth that was practical for navigation and for predicting seasons but was incorrect.

7. The best evidence for the Euclidean view is in the description of quadrilaterals. What key term in Exodus 27:1 could not exist in non-Euclidean geometry?

This Bible reference to foursquare constitutes weighty evidence but is inconclusive. Sometimes the Bible records events the way they appear to man (using the language of appearance). Match each passage below to the phrase describing the human perspective in the verse.

8. Genesis 32:31	A.	sun sets
9. Psalm 113:3	B.	sun rises
10. Ecclesiastes 9:11	C.	both setting and rising of the sun
11. Luke 4:40	D.	four corners of the earth
12. Revelation 7:1	E.	chance

There is no conclusive evidence that references to foursquare in Exodus 27:1 and elsewhere are using the language of appearance. Such references therefore provide strong evidence for a Euclidean universe.

Line upon Line

AND THOU SHALT make an altar of shittim wood, five cubits long, and five cubits broad; the altar shall be foursquare: and the height thereof shall be three cubits.

EXODUS 27:1

Chapter 10 Review

True/False

1. A dihedral angle separates space into three disjoint sets.
2. Two lines parallel to the same plane are parallel.
3. All cubes are parallelepipeds.
4. All great circles of a sphere are congruent.
5. The faces of a parallelepiped are rectangular regions.

6. Sketch a rectangular prism.
7. How many dihedral angles are determined by your rectangular prism? Name one.
8. Draw a sphere with a tangent plane. Label a radius.

Sketch each figure:

9. two great circles of a sphere that are not perpendicular.
10. a cube in one-point perspective.
11. a cube in two-point perspective.
12. a parallelepiped.

13. How many vertices, edges, and faces does a parallelepiped have?
14. If a convex polyhedron could have forty-three faces and one hundred edges, how many vertices would it have?
15. Find the distance between two points, A and B, on sphere C if $m\angle ACB = 40$ and $AC = 8$ units. Find to the nearest tenth.
16. Show how two planes can separate space into five disjoint sets (including the planes themselves).
17. Show how two planes can separate space into nine disjoint sets (including half-planes and a line).
18. Diagram the possibilities for separating space with three planes. Label each with the number of disjoint sets.
19. Find the city located at 12° south, 78° west.
20. Give the longitude and latitude of Athens, Greece.
21. Give the longitude and latitude for the place that is at the opposite end of a diameter from Athens.
22. If a city is at $x°$ west longitude and $y°$ north latitude, what will the latitude and longitude be at the other end of a diameter of the earth?

Use Riemannian geometry for exercises 23 and 24.

23. Explain the points, lines, and planes in the system.
24. Sketch triangles with one, two, and three right angles.

Prove

25. The radius perpendicular to a chord of a sphere S with center C bisects the chord.

26. Theorem 10.7

27. Theorem 10.13

28. Theorem 10.14

29. *Given:* Planes m, n, and p intersect in $\overleftrightarrow{AB}$, $\overleftrightarrow{CD}$, and $\overleftrightarrow{EF}$ as shown; $\overleftrightarrow{AB} \parallel \overleftrightarrow{CD}$

   Prove: $\overleftrightarrow{EF} \parallel \overleftrightarrow{AB}$ and $\overleftrightarrow{EF} \parallel \overleftrightarrow{CD}$

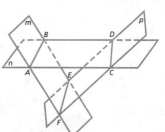

30. Explain the mathematical significance of Exodus 27:1.

▶ Dominion Thru Math

If the latitude and longitude of two cities are known, the distance between them can be approximated using the Pythagorean theorem. The distance between their latitudes and the distance between their longitudes form the sides of a right triangle. The differences can be converted to distances using $d = (x^\circ/360)(2\pi r)$. Estimate each great circle route's distance to the nearest kilometer using the coordinates in parentheses. (The earth has a radius of 6370 km.)

31. Seattle, Washington (47° 36′N; 122° 20′W) to Naples, Italy (40° 50′N; 14° 15′E)

32. Orlando, Florida (28° 33′N; 81° 23′W) to Sydney, Australia (33° 53′S; 151° 10′E)

Trace the routes between the cities on a globe and identify countries crossed.

33. Seattle to Naples

34. Orlando to Sydney

11 Volume

Tomatoes are one of America's favorite fruits. You may not think of them as fruits, but they are, and they are grown from Oregon to sunny Florida.

Who could enjoy spaghetti or pizza without tomato sauce? We eat cherry tomatoes in salads, diced tomatoes in tacos, tomato chunks in shish kabobs, and tomato wedges in stuffed tomatoes; we even drink tomato juice. Many people like tomatoes on hamburgers, and BLT might be one of America's most delicious abbreviations. Yes, this spherical culinary delight is one of the nation's best-loved fruits.

Did you ever consider the geometric problems involved in producing tomato-related products? For example, tomato cartons seek to minimize wasted space, which is not easy with round fruits. Volume measurements are also important in the manufacture of tomato juice. What volume of tomato concentrate is needed to fill a cylindrical can of tomato juice? In this chapter you will calculate the volume of such spheres and cylinders.

Christians should also bear fruit. Like natural fruits, the fruit of the Spirit takes time to mature. John 15 tells some of the things that produce mature fruit when we live as branches of the True Vine. Abiding, purging, depending on God, answered prayer, and steadfast love are all necessary for producing more fruit in our hearts. As the tomato grower who watches over his plants, God is interested in the quality and quantity of our fruit. Remember that fruit is not meant to beautify the plant but to nourish others.

After this chapter you should be able to

1. define volume.
2. state the volume postulates and compare them to the area postulates.
3. state and apply Cavalieri's principle.
4. prove various formulas for volume.
5. state and apply formulas for volumes of prisms, cylinders, pyramids, cones, spheres, and regular polyhedra.
6. explain the construction of some 3-dimensional figures.
7. recognize three classical constructions that are impossible.

11.1 Meaning of Volume

Remember that solid objects consist of both a surface and its interior. In Chapter 8 you learned to compute the surface area of solids. Now you will learn to calculate the volume of the interior of solids. The *volume* of a three-dimensional figure is the number of cubic units needed to fill up the interior. A *cubic unit* is a cube whose side measures one unit—a cubic centimeter, a cubic foot, a cubic yard, a cubic meter, or any cubic unit. Cubic units are different from the square units used for surface area. Keep in mind that volume is the number of these cubes that fill up the solid.

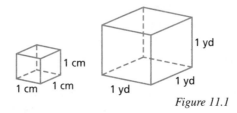

Figure 11.1

Definition

The **volume** of a solid is the number of cubic units needed to fill up the interior completely.

Just as cubic units fill up a three-dimensional object, the love of God should fill up a Christian. Ephesians 3:17-19 says that the Christian can know the love of Christ and that he "might be filled with all the fulness of God." If you are a Christian, you should be demonstrating the love of God to all those around you.

The following postulates form the basis for the study of volume. Compare the volume postulates to the corresponding postulates for area (8.1, 8.2, 8.3, 8.4).

Postulate 11.1
Volume Postulate. Every solid has a volume given by a positive real number.

Postulate 11.2

Congruent Solids Postulate. **Congruent solids have the same volume.**

Postulate 11.3

Volume of Cube Postulate. **The volume of a cube is the cube of the length of one edge: $V = e^3$.**

Cube

Postulate 11.4

Volume Addition Postulate. **If the interiors of two solids do not intersect, then the volume of their union is the sum of the volumes.**

This puzzle contains 27 cubic units of volume.

Finding the volume of simple three-dimensional figures is very easy. Look at figure 11.2 and see how many cubic units are contained in the figure.

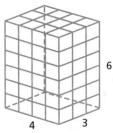

6

4   3

Figure 11.2

In this three-dimensional figure you can count the seventy-two cubes. The area of the base is twelve square units. There are six layers in this figure, each containing twelve cubes. Notice that six layers of twelve make a total of seventy-two cubes. Since the interiors of these cubes do not intersect, the Volume Addition Postulate allows us to add them together for a total of seventy-two cubic units.

The volume of a rectangular prism is found by using the Volume Addition Postulate.

Theorem 11.1

The *volume of a rectangular prism* is the product of its length, width, and height: $V = lwH$.

EXAMPLE   Find the volume of the figure.

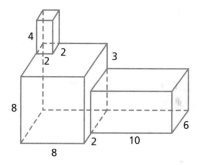

Figure 11.3

Answer   Volume of the central cube = $e^3 = 8^3 = 512$
Volume of the tower = $lwH = 2 \cdot 2 \cdot 4 = 16$
Volume of the rectangular solid = $lwH = 10 \cdot 6 \cdot 5 = 300$

According to the Volume Addition Postulate, add to obtain the total volume.
$V = 512 + 16 + 300 = 828$ cubic units

▶ A. Exercises

Find the volume of each solid.

1.

2.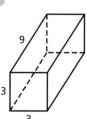

3.

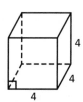

4. A cube edge measures 24 inches
5. A cube with edge x
6. A rectangular prism with a base area of 26 square feet and a height of 16 feet

7. If your bedroom were fifteen feet long by ten feet wide by eight feet high, how many cubic-foot boxes would fit into the room?
8. The Williamses rented a truck for their move that is 24 feet long by 7.5 feet wide. If items can be stored to a height of eight feet, how much can they move per load?

▶ B. Exercises

Find the volume of each solid.

9.

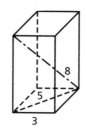

cube

10.

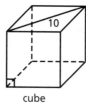

cube

11.

12.

13.

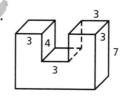

14.

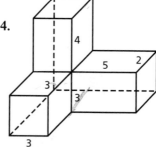

15. Find the volume of a cube with diagonal x.

16. A right rectangular prism has a volume of 3536 cubic feet, and the length of the base of the prism is 9 feet longer than its width. The height of the prism is 26 feet. What are the dimensions of the base of the prism?

17. How many grams of mercury must be poured into a rectangular container with dimensions 12 cm by 18 cm by 10 cm in order to fill it? The density of mercury is $13.6 \frac{g}{cm^3}$.

18. Find the volume of the figure and identify any volume postulates and theorems that you use.

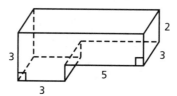

▶ C. Exercises

Prove each volume formula.

19. The volume of a cube having a face with a diagonal x units long is
$$V = x^3 \frac{\sqrt{2}}{4}.$$

20. Prove your formula from exercise 15.

Reread the explanation of the first three area postulates, then explain the following.

21. Which postulates guarantee that areas exist and are meaningful?
22. Which postulate provides a first method for finding an area without counting squares?
23. Find the area of a regular hexagon with a 16-in. side.
24. Find the area of the figure in the diagram.

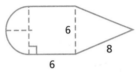

25. Find the surface area of the rectangular prism with dimensions *l*, *w*, and *H*.

11.2 Prisms

You can already find the volumes of rectangular prisms. In order to develop formulas for the volumes of other prisms, you need to understand the idea of a cross section. A *section* is the intersection of a three-dimensional figure and a plane that passes through the figure. A *cross section* is a section that is perpendiular to the altitude.

Look at the two three-dimensional figures in figure 11.4.

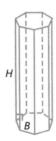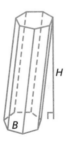

Figure 11.4

The figure on the left is a right prism, while the figure on the right is called an oblique prism. An *oblique prism* is a prism whose slant height is not perpendicular to the base. The bases of these figures are congruent regular heptagons (seven sides) of area B. Every cross section of both figures (*a*) and (*b*) is a regular heptagon with area equal to B. Since the heights (H) are the same and corresponding cross sections have equal areas, the next postulate guarantees that the volumes are equal.

Postulate 11.5

Cavalieri's Principle. **For any two solids, if all planes parallel to a fixed plane form sections having equal areas, then the solids have the same volume.**

Bonaventura Cavalieri, an Italian mathematician, came up with the idea behind this principle. The development of this principle was a steppingstone to the development of integral calculus.

Another helpful theorem in determining the volume of a prism is given below. You will prove a special case of this theorem in the exercises. Notice how it is used in the proof of the volume formula.

Theorem 11.2

A cross section of a prism is congruent to the base of the prism.

Theorem 11.3

The *volume of a prism* is the product of the height and the area of the base: $V = BH$.

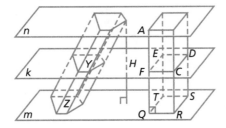

Figure 11.5

STATEMENTS	REASONS
1. Prism with height H, volume V, and base Z having area B	1. Given
2. The bases lie in parallel planes m and n	2. Definition of prism
3. Square $QRST$ exists in plane m with side length $\sqrt{B}$; draw a perpendicular to m at Q, which intersects plane n at a point A; draw other perpendiculars from each vertex of $QRST$ to form a square prism; take any plane k forming cross sections Y of prism P and $CDEF$ of the square prism	3. Auxiliary lines and plane
4. $RS = QR = \sqrt{B}$	4. Definition of square
5. Area $QRST = s^2 = (\sqrt{B})^2 = B$	5. Area of Square Postulate
6. Area Z = Area $QRST$	6. Transitive property of equality
7. $QA = H$	7. Parallel planes are everywhere equidistant
8. $V_{\text{square prism}} = lwH = \sqrt{B}\,\sqrt{B}H = BH$	8. Volume of rectangular prism
9. $Y \cong Z$; $CDEF \cong QRST$	9. Cross section congruent to base of prism
10. Area Y = Area Z; Area $CDEF$ = Area $QRST$	10. Congruent Regions Postulate
11. Area Y = Area $CDEF$ (so corresponding cross sections have equal areas)	11. Substitution (step 10 into 6)
12. $V = V_{\text{square prism}}$	12. Cavalieri's principle
13. $V = BH$	13. Substitution (step 8 into 12)

EXAMPLE Find the volume of the following regular prism.

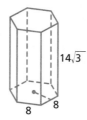

Answer $4^2 + a^2 = 8^2$ **1.** Find the apothem of the base.

$a^2 = 64 - 16 = 48$ A regular hexagon divides into

$a = 4\sqrt{3}$ 6 equilateral triangles.

$B = \frac{1}{2}ap$ **2.** Find the area

$B = \frac{1}{2}(4\sqrt{3})(48)$ of the base.

$B = 96\sqrt{3}$ square units

$V = BH$ **3.** Find the volume.

$V = (96\sqrt{3})(14\sqrt{3})$

$V = 4032$ cubic units

Figure 11.6

▶ A. Exercises

Find the volume of each prism.

1.

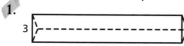

2.

3.

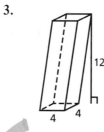

4.

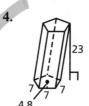

5.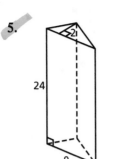

24 21 9

6.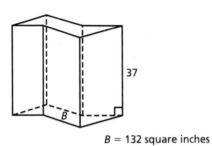

37

B

B = 132 square inches

7.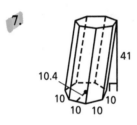

41

10.4

10 10

10 10

8.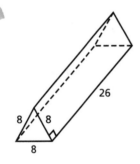

26

8 8

8

9.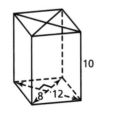

10

8 12

10.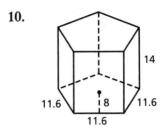

14

11.6 8 11.6

11.6

11.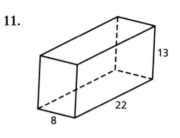

13

22

8

12.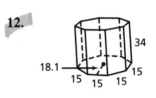

34

18.1

15 15

15 15 15

13.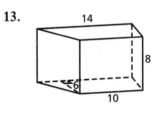

14

8

6

10

14.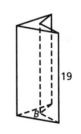

19

8

B = 15 square units

▶ B. Exercises

15. The bases of a prism are rhombi that have diagonals measuring 16 feet and 24 feet. The height of the prism is 3 feet. What is the volume of the prism?

16. A concrete water trough is a rectangular prism. The trough measures 24 by 48 by 96 inches. How many gallons of water will fill the trough if 1 gallon fills 231 cubic inches?

17. If the areas of the bases of a triangular prism and a hexagonal prism are both 52 square centimeters, the heights of both are the same, and the volume of the triangular prism is 468 cubic centimeters, what is the height of each prism and the volume of the hexagonal prism?

18. Consider a square cake pan as a rectangular prism having an 8-by-8-inch base and a height of 2 inches. If 114 cubic inches of batter are poured into the pan, how many cubic inches does the cake need to expand during baking to fill the pan?

19. Find simple formulas for the volume of square prisms and triangular prisms.

▶ C. Exercises

20. Supply reasons to prove Theorem 11.2. A cross section of a prism is congruent to the base.

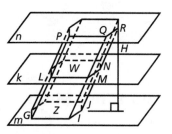

STATEMENTS	REASONS
1. Prism has height H and a base Z with area B; cross section W in plane k with $k \parallel m$	1.
2. The bases of the prism in planes m and n are congruent; $m \parallel n$; each pair of lateral edges are parallel: $\overleftrightarrow{GP} \parallel \overleftrightarrow{IQ}$, $\overleftrightarrow{GP} \parallel \overleftrightarrow{JR}$, $\overleftrightarrow{IQ} \parallel \overleftrightarrow{JR}$, etc.	2.
3. $k \parallel n$	3.

Continued ▶

4. Each pair of lateral edges determines a plane

4.

5. *k* intersects each plane in a line (planes determined by pairs of lateral edges)

5.

6. The sides of *W* are parallel to corresponding sides of *Z*: $\overline{LM} \parallel \overline{GI}$, $\overline{MN} \parallel \overline{IJ}$, etc. (*Z* and *W* have same number of sides). The diagonals of *W* are parallel to corresponding diagonals of *Z*: $\overleftrightarrow{LN} \parallel \overleftrightarrow{GJ}$, etc.

6.

7. Each pair of sides or diagonals of *Z* and *W* determine a parallelogram: *LMIG*, *MNJI*, *LNJG*, etc.

7.

8. Corresponding sides and diagonals of *Z* and *W* are congruent: $\overline{LM} \cong \overline{GI}$, $\overline{MN} \cong \overline{IJ}$, $\overline{LN} \cong \overline{GJ}$

8.

9. Each triangular subdivision of *W* is congruent to a corresponding triangular subdivision of *Z*: $\triangle LMN \cong \triangle GIJ$, etc.

9.

10. $Z \cong W$

10.

▶ Dominion Thru Math

21. An architect designs the foundation for a series of granite columns along a courthouse entrance and needs to know their weight. The cross section of each column is a regular 16-gon with 3.6 inch sides, but every other side is fluted (curved inward) as a semicircle with the same diameter as the flat sides. The columns are 20 feet high, and granite has a density of about 172 lb./ft.^{3} The ratio of a side to the apothem of a regular 16-gon is 0.4. How much does each column weigh?

■ Cumulative Review

State a formula for each.

22. Perimeter of a square

23. Area of a square

24. Surface area of a cube

25. Volume of a cube

26. Surface area of a circular cylinder

11.3 Cylinders

Look at the following diagrams of a prism and a cylinder.

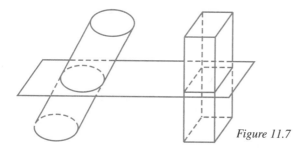

Figure 11.7

Cylindrical oil tanks at a refinery

If the bases of the prism and the cylinder have area *B* and the heights of both of them are the same, what does Cavalieri's principle tell you about the volumes of these three-dimensional figures? Since the areas of the cross sections in every plane parallel to the bases of the figures are the same and the heights are the same, then the volumes are the same. This suggests that the formula for the volume of a cylinder is

$$V = BH,$$

where *B* is the area of the circular base, and *H* is the height of the cylinder.

The proof of this formula uses the fact that cross sections parallel to the base are congruent to the base.

Cylindrical silos on a farm in central Wisconsin

Theorem 11.4

The *volume of a cylinder* is the product of the area of the base and the height: $V = BH$. In particular, for a circular cylinder $V = \pi r^2 H$.

Theorem 11.4 can be proved just like Theorem 11.3 in the previous section. Do you see how to get the alternate formula $V = \pi r^2 H$?

EXAMPLE Find the number of cubic inches of sand needed to fill the following cylinder.

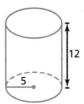

12

5

Figure 11.8

Answer Use the formula to find the volume.

$V = \pi r^2 H$
$V = \pi(5)^2(12)$
$V = 300\pi$
$V \approx 942$ cu. in. of sand

▶ A. Exercises

Find the volume of each cylinder.

1.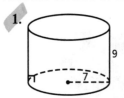
9
7

3.
15
4

5.
19
11

2.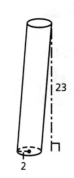
23
2

4.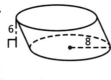
6
8

6.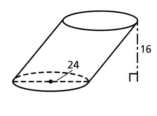
16
24

7. A cylindrical gas tank has a lateral surface area of 300π square feet and is 15 feet high. How many cubic feet of gas will the tank hold?
8. A cylindrical water tower has a volume of $20,000\pi$ cubic feet. What is the height if the radius is 20 feet?

9. A cylinder is lodged inside a square prism similar to the figure shown. How many cubic inches are within the prism but outside the cylinder?

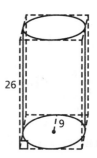

26

9

10. A cube is placed inside a right circular cylinder. How many cubic feet of water will fill the cylinder that has the cube in it?

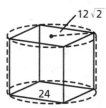

$12\sqrt{2}$

24

▶ B. Exercises

11. The circumference of a cylinder is 86π meters, and the height is 63 meters. Find the surface area and volume of the cylinder.

12. The circumference of a cylinder is 24π inches, and its volume is 4896π cubic inches. What is the height of the cylinder?

13. A cylindrical piece of iron has a diameter of 2 inches and a length of 16 inches. If it is melted down and poured into a rectangular mold that has a 1-by-3-inch base, what will be the length of the rectangular piece of iron?

14. How many gallons of water will fill a circular cylinder that has a radius of 27 feet and a height of 38 feet? (1 cubic foot of water is 7.5 gallons.)

15. How many cubic yards of concrete are needed to pour the patio and sidewalk if the thickness is to be four inches?

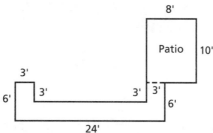

8'

Patio   10'

3'

3'   3' 3'

6'   6'

24'

16. The Leaning Tower of Pisa is about 183 ft. high, but the centers of the bases deviate horizontally by at least 14 ft. The tower is roughly 52 feet in diameter. Since the tower is a right circular cylinder that has begun to lean, approximate the volume of the Leaning Tower of Pisa.

17. Find the volume of a silo that is 50 feet high and 20 feet in diameter.

18. Use $V = BH$ to show that $V = \pi r^2 H$ for a circular cylinder.

▶ C. Exercises

19. What must the length of one side of a square be that has the same area as a circle of radius r?
20. Prove Theorem 11.4 that $V = BH$ for a cylinder. (*Hint*: Mimic proof of Theorem 11.3.)

▶ Dominion Thru Math

A fuel tank is a horizontal cylinder with an inside diameter of 30 inches and a length of 8 feet. The homeowner opens the lid and inserts a yardstick to determine the depth of fuel. There are 7.48 gal./ft.^{3}

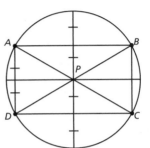

21. Find the capacity of the tank in gallons.
22. How deep is the fuel when the tank is half full?
23. Find the gallons of fuel left if the yardstick reads 22.5 inches. (*Hint*: Find $m\widehat{AB}$ by considering $\triangle PAD$.)
24. Give the percentage of the tank that is full in exercise 23.

■ Cumulative Review

Identify the horizontal cross section for each of the following figures.
25. circular cylinder
26. parallelepiped
27. tetrahedron
28. circular cone
29. sphere

What happens to the volume of each figure?
1. The radius of a cylinder is tripled (without changing the height).
2. The lengths of the edges of a cube are doubled.
3. The area of the faces of a cube are doubled.
4. The diagonals of a cube are multiplied by a factor of 5.

11.4 Pyramids and Cones

The Pyramid at the Louvre in Paris

If cross sections of the pyramid and cone have equal areas, and the height of each three-dimensional figure is H, then according to Cavalieri's principle, the volumes of the two figures must be the same. If we can develop a formula for the volume of one of these figures, then the other formula will follow easily.

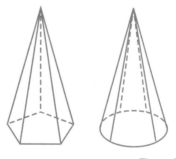

Figure 11.9

First we will develop the formula for the volume of a pyramid. Look at the triangular prism shown in figure 11.10.

If the area of the base of this prism is B, and the height of the prism is H, then $V = BH$. Now the volume of the prism is equal to the total volume of three pyramids as shown in figure 11.11.

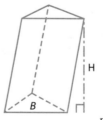

Figure 11.10

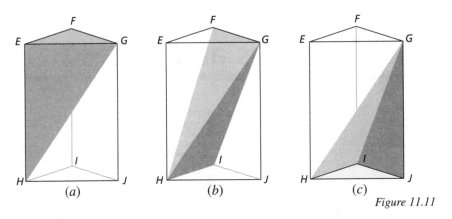

Figure 11.11

Pyramids (*a*) and (*c*) have the same volume. The areas of the bases in pyramids (*a*) and (*c*) are the same since the bases of a prism are congruent. Pyramids (*a*) and (*c*) also have the same height since they share the height of the prism. By Cavalieri's principle, pyramids (*a*) and (*c*) have the same volume.

To show that pyramids (*a*) and (*b*) have the same volume, turn the prism on its side as shown.

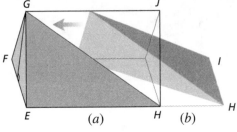

Figure 11.12

Notice that the pyramids have the same height. Also $\triangle EFH$ of pyramid (*a*) and $\triangle HIF$ of pyramid (*b*) form the bases of the pyramids. These bases have the same area since they are each half of the original back left face of the prism. Since the pyramids also have the same height (to point *G*), they must have the same volume. Thus, you can see that the prism is divided into three pyramids with the same volume. So the volume of each pyramid must equal one-third of the volume of the prism. This proves the formula for the volume of a pyramid.

Theorem 11.5

The *volume of a pyramid* is one-third the product of the height and the area of the base: $V = \frac{1}{3}BH$.

As noted earlier, Cavalieri's principle proves that the same formula works for cones. In the case of a circular cone, substitute for *B* using the area of the circular base.

Theorem 11.6

The *volume of a cone* is one-third the product of its height and base area: $V = \frac{1}{3}\pi r^2 H$.

EXAMPLE Find the volume of the cone.

Figure 11.13

Answer $V = \frac{1}{3}BH$

$V = \frac{1}{3}\pi r^2 H$

$V = \frac{1}{3}\pi (4^2)(7)$

$V = \frac{112}{3}\pi$

$V \approx 117.3$ cu. units

▶ A. Exercises

Find the volume of each figure.

1.

3.
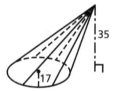

5.
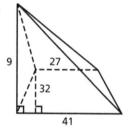

2.
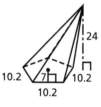

4.
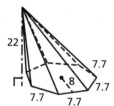

6.
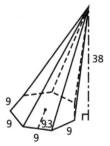

7. The base of a pyramid is a rhombus with diagonals equal to 24 inches and 29 inches. The height of the pyramid is 35 inches. What is the volume of this pyramid?

8. The volume of a circular cone is 1728π cubic feet. If the height is 36 feet, what is the diameter of the base of the cone?

9. Find the volume of a pyramid that has a square base with a side measuring 18 centimeters and an altitude (height) measuring 25 centimeters.

▶ B. Exercises

10. A circular cone is placed inside a rectangular pyramid. The pyramid has a 37-by-43-inch base and a height of 55 inches. The diameter of the cone is 34 inches, and the vertices of both the pyramid and the cone are the same. What is the volume outside the cone but inside the pyramid?

11. A cone sits point down in a cube. The base of the cone is inscribed in the base of the cube. If the cube has a volume of 729 cubic inches, what is the height, radius, and volume of the cone?

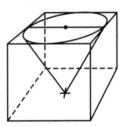

12. A cylinder encloses a cone. The cylinder and cone have the same base area. If the volume of the cone is 324π cubic feet and both have a height of 4 feet, what is the volume and radius of the cylinder?

13. The area of △FED is 27 square centimeters. The area of △ABC is 3 square centimeters. Find the volume of the three-dimensional figure ABCDEF.

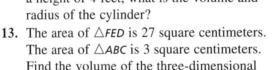

14. Find the volume of the frustum (shaded area) of the cone.

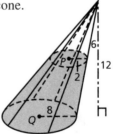

15. The Cheops Pyramid in Egypt is almost a perfect right pyramid. It has a height of 481 feet and a square base with a side measuring 755 feet. What is the volume of this pyramid? What is the surface area of the part exposed to view?

16. Prove Theorem 11.5 by supplying the reasons.

STATEMENTS	REASONS
1. $V_{prism} = V_a + V_b + V_c$ (as in figure 11.11)	1.
2. $V_a = V_b = V_c$ or simply V	2.
3. $V_{prism} = V + V + V$	3.
4. $V_{prism} = 3V$	4.
5. $V_{prism} = BH$	5.
6. $3V = BH$	6.
7. $V = \frac{1}{3}BH$	7.

17. Prove Theorem 11.6.

▶ C. Exercises

18. If a regular tetrahedron has an edge 12 units long, find the volume of the inscribed circular cone.

■ Cumulative Review

19. Any three vertices of a cube determine a right triangle. Is this a true statement?

Find the volume of each.
20. A cube with a 7-meter edge
21. A prism with a height of 4 feet and an 8-by-12-foot rectangular base
22. A cylinder with a height of 9 inches and an oval base of 72 sq. in.
23. If K is a cube, then its vertices can be inscribed in a sphere S. What rule of logic would be needed to prove this conditional statement?

11 Analytic Geometry

Conic Sections

Given a circle and a point *V* not in the plane of the circle, the union of all lines that connect the point *V* to a point of the circle forms a *conical surface*. The connecting lines are the *elements* of the conical surface and the point *V* is the vertex. The circle is the *generating curve* and the line joining *V* to the center of the curve is its *axis*. If the axis is perpendicular to the plane of the curve, it is a right conical surface such as the one shown.

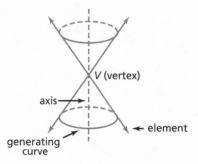

You have already learned to graph three of the figures below. These figures are called *conic sections* because they are cross sections of a conical surface.

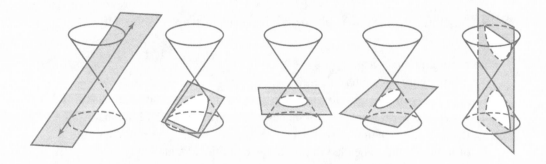

In Chapter 2 you graphed some parabolas. The top or bottom point of a parabola is the *vertex* of the parabola. When the vertex is at the origin, the parabola is in standard position. The equation of a parabola in standard position is $y = ax^2$. You will discover what the sign of the coefficient *a* tells you in the exercises.

▶ Exercises

Identify each conic section and graph it.

1. $x^2 + y^2 = 36$
2. $y = -2x + 5$
3. $y = 2x^2$
4. $y = -2x^2$
5. What does the sign of the coefficient *a* tell you in $y = ax^2$?
   (*Hint:* Compare exercises 3 and 4.)

The Very Large Array near Socorro, New Mexico, opened in 1980. Its 27 radio telescopes linked electronically make it the most powerful radio interferometer in the world. The radio telescopes use parabolas to collect and focus radio waves.

11.5 Polyhedra and Spheres

The goal of this section is to develop the formula for finding the volume of a sphere. We will use Cavalieri's principle to find this formula. Notice how many previously learned principles you need to develop this formula. Figure 11.14 shows a right circular cylinder that has two right circular cones inside it.

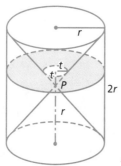

Figure 11.14

The Peachoid at Gaffney, South Carolina, has a volume of over 60,000 gallons. It modifies a typical spherical water tower by adding a 12-foot stem, a nipple at the bottom, a cleft of steel paneling, and a 60-foot leaf that weighs seven tons.

If a plane passes through this figure parallel to the base, then the intersection forms concentric circles. *Concentric circles* are circles that have the same center but radii of different lengths. The region bounded by concentric circles is called an *annulus*. Suppose the plane cuts through the figure at a distance t from the vertex *P*. If you look at this cross section from the top, it forms an annular region. This region is shaded in figure 11.15.

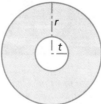

Figure 11.15

How can you find the area of the annulus? Simply subtract the area of the smaller circle from the area of the larger circle.

$$A_{annulus} = A_{lg.\ circle} - A_{sm.\ circle}$$
$$A_{annulus} = \pi r^2 - \pi t^2$$

Now turn your attention to a sphere. In figure 11.16 you see a sphere of radius *r* with a secant plane passing through it at a distance of *t* units from the center. The secant plane intersects the sphere to form ⊙*C* with radius *x*.

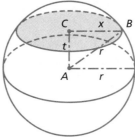

Figure 11.16

Since △*ABC* is a right triangle, the Pythagorean theorem applies to it:

$$t^2 + x^2 = r^2$$

Solving for x^2,

$$x^2 = r^2 - t^2$$

Since the area of this section is a circle, its area is found by

$$A_{section} = \pi x^2$$

Substituting for x^2 above,

$$A_{section} = \pi(r^2 - t^2)$$
$$A_{section} = \pi r^2 - \pi t^2$$

Are you surprised? The area of this circular region is the same as the area of the annulus! The heights of both figures are the same 2*r* (diameter of sphere and height of cylinder). Also, the sections are at the same distance *t* from the centers. Do you see what postulate applies?

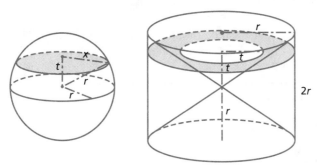

Figure 11.17

Since every horizontal plane cuts the figures into regions of equal areas, Cavalieri's principle applies. Therefore, the volume, *V*, of the sphere is equal to the volume of the solid between the cones and the cylinder.

$$V = V_{\text{cylinder}} - V_{\text{two cones}}$$

$$V = \pi r^2 (2r) - 2\left(\frac{1}{3}\pi r^2\right)(r)$$

$$V = 2\pi r^3 - \frac{2}{3}\pi r^3$$

$$V = \frac{6}{3}\pi r^3 - \frac{2}{3}\pi r^3$$

$$V = \frac{4}{3}\pi r^3$$

That proves the next theorem.

Theorem 11.7

The *volume of a sphere* is four-thirds π times the cube of the radius: $V = \frac{4}{3}\pi r^3$.

EXAMPLE Find the volume of a sphere with a diameter of 10 inches.

Answer Substitute length of the radius into the formula.

$$V = \frac{4}{3}\pi r^3$$

$$V = \frac{4}{3}\pi (5^3)$$

$$V \approx 166.67\pi \text{ cubic inches}$$

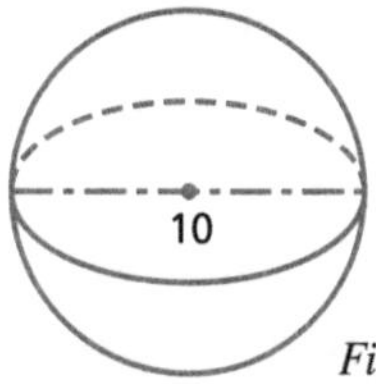

10

Figure 11.18

A regular polyhedron can be inscribed in a sphere. Therefore, the center of the regular polyhedron is equidistant from the vertices; this distance is the radius of the sphere and is also called the radius of the polyhedron. These facts are useful in proving the volume formulas below. In each formula, e is the length of an edge.

Regular Polyhedron	Volume
tetrahedron	$V = \frac{\sqrt{2}}{12}e^3$
cube	$V = e^3$
octahedron	$V = \frac{\sqrt{2}}{3}e^3$
dodecahedron	$V = \left(\frac{15 + 7\sqrt{5}}{3}\right)e^3$
icosahedron	$V = \left(\frac{15 + 5\sqrt{5}}{12}\right)e^3$

The formula for the volume of a cube is a postulate you know already. The formulas for the volumes of a dodecahedron and an icosahedron are extremely difficult to prove. The proofs of the other two formulas are exercises.

▶ A. Exercises

Give the volume of the sphere or regular polyhedron.

1. sphere with a radius of 18 feet
2. sphere with a radius of $\frac{1}{4}$ meter
3. hexahedron with an edge of 3 yd.
4. tetrahedron with an edge of 5 cm
5. octahedron with an edge of $\sqrt{2}$ units
6. sphere with a diameter of $8\sqrt{3}$ inches
7. dodecahedron with an edge of 3 mm
8. tetrahedron with an edge of 2 units
9. sphere that has a great circle with a circumference of 32π centimeters
10. icosahedron with an edge of 6 m
11. A volleyball has a circumference of 27 inches. How many cubic inches of air are needed to inflate the ball?
12. A factory needs to produce ball bearings with 6-cm diameters. How many cubic centimeters of metal are needed to make 3000 ball bearings?

▶ B. Exercises

13. A spherical balloon has 36π cubic inches of air in it. What is the circumference of the balloon?
14. A racquetball is a hollow ball made of rubber. How much rubber is needed to make each ball if the inside hollow sphere has a radius of 2.25 centimeters and the ball has a diameter of 5 centimeters?
15. The volume of a sphere is $16,222.67\pi$ cubic millimeters. What is the radius of the sphere?
16. The circumference of the earth is approximately 40,000 kilometers. If you consider the earth to be a sphere, what is the volume of the earth?
17. A spherical water tower has a diameter of 75 feet. How many gallons of water will it hold? (1 gallon = 0.13398 cubic feet)
18. A ball whose diameter is 8 inches is placed in a cube whose edge measures 8 inches. How many cubic inches of sand will fill the box containing the ball?

19. A metal part is made in the shape of a cylinder with a hemisphere (half of a sphere) on top. Find the volume of the part.

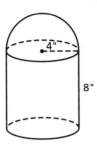

20. An ice-cream cone looks like the following diagram. Approximately how many cubic centimeters of ice cream are used to fill an ice-cream cone like this one?

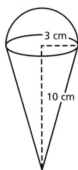

21. Find the number of cubic feet of hot gas needed to fill a spherical balloon that has a radius of 85 feet.

▶ C. Exercises

22. Use the regular tetrahedron shown to derive the formula for its volume in terms of its edge length *e*. (*Hint:* Use formulas for the apothem and altitude of the equilateral triangle [base] to show that the tetrahedron has altitude $H = \dfrac{\sqrt{2}}{\sqrt{3}}e$.)

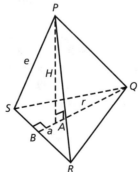

23. Use the regular octahedron shown to derive the formula for its volume in terms of its edge length **e**. Remember that the vertices are equidistant from the center. (*Hint:* Find *XY* in terms of *e* first and consider relation of center to each vertex.)

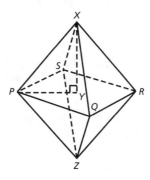

Cumulative Review

Identify each term defined below.

24. A line in the plane of a circle that intersects the circle in exactly one point

25. A triangle with no congruent sides

26. A line that intersects two parallel lines

27. A region of a circle bounded by a chord and the intercepted arc

28. A portion of a sphere determined by intersecting great circles

GEOMETRY AND ENGINEERING

We all enjoy the benefits of technology as we ride in air-conditioned cars, use electric appliances, and listen to recorded music. The people who design all our fascinating gadgets are called engineers. Almost every type of engineer uses geometry in making the drawings of the parts he designs.

Mechanical engineers accomplish little without geometry. They design bearings and cams to make engines and machines. Some mechanical engineers design piping systems to convey liquids, such as water and oil. Pipes are just cylinders connected at turns by elbows and tees. The diameter of a pipe determines its carrying capacity (volume per unit of time). The formula $Q = Av$ says that the capacity (Q) equals the cross-sectional area (A) times the velocity (v). If the engineer wants a pipe to carry 500 gallons per minute at a velocity of 5 feet per second, what must the diameter of his pipe be? Did you get about 6.4 inches?

Circular pools of a water treatment plant

Sanitation engineers provide us with two vital utilities—distribution of water and treatment of wastewater. When the sanitation engineer designs water-treatment plants and wastewater-treatment plants, he sizes and lays out tanks, circular clarifiers, and earthen dikes; he then connects these with various pipes and channels. He incorporates circles, cylinders, trapezoids, and prisms into his designs. Just like the mechanical engineer, the sanitation engineer must know and understand the volume-capacity relationships of all his tanks, basins, and pipes.

Structural engineers design dams, bridges, and buildings that are safe and functional. They must know which shapes best resist bending or breaking. They have to lay out curves and tangents so that forces are distributed properly. The engineers must calculate the volumes of the steel girders in order to plan a foundation that will support the weight of the building and its contents (volume × density = mass, from which weight can be found). Have you ever wondered why dams are shaped the way they are—curved toward the lake and with a long sweeping curve down the face of the dam? The structural engineer must balance the cost of concrete and the mass of concrete needed to hold back the water. The curved shape provides the needed strength but saves millions of dollars by reducing the amount of concrete needed.

Almost every type of engineer uses geometry in making the drawings of the parts he designs.

A construction worker high on the skeleton of a skyscraper

All these engineers must read blueprints and other scale drawings. Engineers and architects draw their diagrams to scale, for example 1 inch to represent 10 feet or $\frac{1}{4}$ inch to represent 1 foot. Thus blueprints convey the engineer's design in a compact geometric form. The men who read these plans and direct the construction of the buildings must be knowledgeable about the geometric principles related to dimensions, diagonals, tangents to circles, and many other measurements. If they did not understand these concepts, builders would erect a poorly constructed building.

11.6 Constructions

Can you construct three-dimensional figures? An architect can build a structure from blueprints that give accurate specifications for each dimension. Likewise, solids are considered constructed if each dimension can be constructed. For example, a sphere is said to be constructed if the radius can be constructed. A right cylinder is considered constructed if its base region and its height can be constructed.

For this reason segment lengths and polygons are important for constructing surfaces and solids.

The cross sections of these basalt columns form hexagons at Giant's Causeway in Northern Ireland.

Construction 16

Segment division

Given: $\overline{AB}$

Construct: Five congruent segments with lengths that total $\overline{AB}$

1. Draw a ray from *A*, forming an acute angle.
2. Place the point of the compass at *A* and, without changing the compass measure, mark off five equal segments on the ray. Label the points *F*, *G*, *H*, *I*, *J*.
3. Draw $\overline{BJ}$.
4. Draw lines parallel to $\overleftrightarrow{BJ}$ through *F*, *G*, *H*, and *I*. You can do this by constructing congruent corresponding angles at each point. Copy ∠*AJB* at vertices *F*, *G*, *H*, and *I*.

These parallel lines cut $\overline{AB}$ into five equal segments.

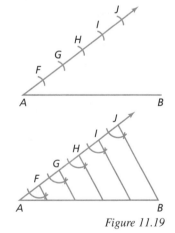

Figure 11.19

Here is a quick way to construct a regular hexagon.

onstruction 17

Regular hexagon

Construct: A regular hexagon

1. Draw a circle.
2. Using the radius of the circle, mark off six consecutive arcs.
3. Connect the arc intersections with segments to form a regular hexagon.

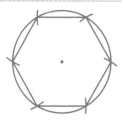

Figure 11.20

The construction works because of the theorem on central angles (Theorem 8.9). What kind of triangles do the central angles determine? This proves that the radius is the same length as the sides of the hexagon.

You should recognize that some constructions are impossible. In ancient Greece there were three famous unsolved construction problems. In modern times they have all been proved impossible as constructions though you could do any of them as drawings with rulers and protractors.

Impossible Construction 1

Squaring the circle

Given: A circle

Construct: A square that has the same area as the circle

Figure 11.21

Impossible Construction 2

Doubling a cube

Given: A cube with edge *x*

Construct: A length *y* that if used as the edge of a cube would form a cube whose volume would be twice the volume of the original cube

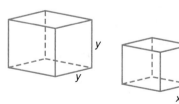

Figure 11.22

Impossible Construction 3

Trisecting an angle

Given: ∠XYZ

Construct: $\overrightarrow{YV}$ and $\overrightarrow{YW}$ such that

∠XYV ≅ ∠VYW ≅ ∠WYZ

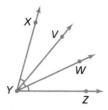

Figure 11.23

▶ A. Exercises

1. To construct a right prism or cone, what must you be able to construct?
2. Construct a regular hexagon.
3. Construct a segment that is one-third of the length of $\overline{XY}$.

4. Divide $\overline{XY}$ into four congruent segments.
5. Divide $\overline{XY}$ into six congruent segments.

Use $\overline{AB}$ (where $AB = p$) for each construction.

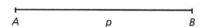

6. Construct a triangle that has an angle measuring 30°, a side congruent to $\overline{AB}$, and another side measuring $\frac{1}{4}p$.
7. Construct a right triangle with a leg congruent to $\overline{AB}$ and the other leg with a measure that is one-third the length of $\overline{AB}$.
8. Construct an equilateral triangle having perimeter p.
9. Construct a square with perimeter p.
10. Construct a regular hexagon with perimeter $2p$.

▶ B. Exercises

Construct each.
11. A regular dodecagon
12. A regular hexagon having an inscribed circle

13. The inscribed and circumscribed circle for the square shown below
14. The inscribed and circumscribed squares for ⊙P shown below
15. Bisector of $\overset{\frown}{AB}$ shown below

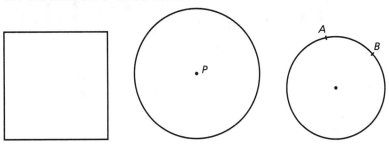

▶ C. Exercises

16. Explain how you solved exercise 15.

▶ Dominion Thru Math

A *net* is a two-dimensional pattern that can be folded into a three-dimensional figure. From another perspective, a net is the surface of a space figure that has been flattened onto a plane. The nets for a right rectangular prism and a right circular cylinder are shown on pages 342 and 344.

17. Construct a net for exercises 6 and 14 on page 468. The dimensions you use are your choice, but the space figure should look like what is shown on that page. Remember, placement of adjoining faces must be chosen so that they do not overlap when laid flat. Fold your net into a space figure as a check.

18. Construct a net for each regular polyhedron shown on page 356. You will want to review how to construct 3-, 4-, and 5-sided regular polygons that will be needed for these nets. (See pages 198, 302, and 406.)

■ Cumulative Review

Copy the figure onto your paper for each construction below. Construct a line through the given point with the given characteristic.

19. point *A* and perpendicular to $\overleftrightarrow{EA}$

20. point *B* and perpendicular to $\overleftrightarrow{EC}$

21. point *A* and parallel to $\overleftrightarrow{EB}$

22. point *D* and parallel to $\overleftrightarrow{EC}$

23. midpoint of $\overline{ED}$ and perpendicular to $\overleftrightarrow{ED}$

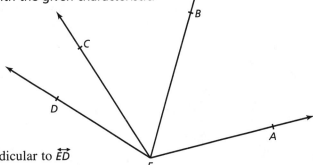

Geometry *and* Scripture

Volume

Recall from your study of area that converting five square feet to square inches is harder than simply multiplying by twelve. Instead, multiply by 144 square inches per square foot. Volume requires even larger conversion factors.

1. How many cubic inches per cubic foot?

2. If we pump helium into a vacuum tube that holds 0.53 cubic feet of gas, how many cubic inches of helium are in the chamber?

The Hebrews used different units for measuring dry goods and fluids, in the same way that we use bushels for corn but gallons for milk. A container just big enough to hold 7.5 gallons of milk would contain one *bath* of milk, according to the Hebrew system.

HIGHER PLANE: Find all the verses that refer to the fluid measurement known as a bath in the Bible.

Read Ezekiel 45:11. In this verse, God set standards of measure for His people.

3. According to God's command, what was the relationship between the size of an ephah and the size of a bath? (An *ephah* was used for measuring dry goods such as corn.)

4. A *homer* could be used for either dry or fluid measure. How many ephahs are in a homer?

The bath and the ephah were the basic units of volume. The homer was used for very large volumes. Four other units helped the Hebrews to measure small volumes.

5. How many *omers* are in an ephah according to Exodus 16:36?

6. Use the conversions above to determine the number of omers in a homer.

The ancient measuring vessels shown are (clockwise from largest) one bath, one seah, one omer, one cab, and one hin. Compare each to the gallon of milk.

The *seah* was the common household unit of dry measure and is usually translated into English as *measure*. A seah is one-third of an ephah. A *cab* is one-sixth of a seah. In fluid measure, a *hin* is one-sixth of a bath.

Remember that a bath (and an ephah) hold about 7.5 gallons. Convert the following Bible measurements to modern units.

7. Numbers 28:14

8. Ruth 2:17

9. I Samuel 25:18

10. II Kings 6:25

11. Which of the four verses above shows that the priests needed to use volume measurements in order to obey God's command concerning sacrifices?

God expected all His people to be honest in their business transactions. No skimping was allowed. This command was explicitly applied to volumes in Ezekiel 45:10.

12. Which units of volume were used to express this warning from God?

Line upon Line

YE SHALL HAVE just balances, and a just ephah, and a just bath. ❧

EZEKIEL 45:10

Chapter 11 Review

Find the volume of each geometric solid.

1.

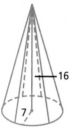

2.

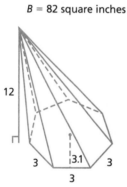

$B = 82$ square inches

3.

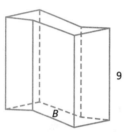

4.

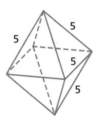

5.

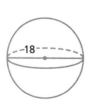

6.

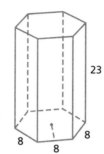

7.

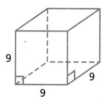

8.

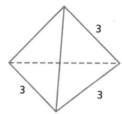

9.

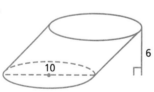

10.

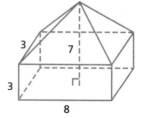

In the following exercises H = height, V = volume, l = slant height, r = radius length, s = side length for base, e = edge length, d = diameter length, a = apothem, and S = surface area. Draw pictures and find the following.

11. Volume of a square prism with $s = 8$ and $H = 22$
12. Volume of an equilateral triangular pyramid with $s = 34$, $H = 38$
13. Volume of a right circular cone with $r = 6$, $H = 10$
14. Volume of a sphere with $r = 14$
15. Volume of a regular tetrahedron with $e = 6$ mm
16. Volume of a regular dodecahedron with a face having a perimeter of 10 m
17. Volume of a regular octahedron with $S = 50\sqrt{3}$ sq. units
18. Volume of a regular icosahedron with $S = 20\sqrt{3}$ sq. ft.
19. Surface area of a dodecahedron with $V = 1{,}993.6$ and $a = 4$
20. Volume of a hemisphere with $d = 12$ in.
21. Give the meaning of *volume*.
22. State Cavalieri's principle in your own words.

Exercises 23-27 involve tori. A *torus* is a surface shaped like a doughnut.

23. Imagine a rubber hose 10 inches long with an inside radius of one inch. Attach the ends to form a torus. Find the volume.
24. How much dough is there in the dough-nut shown if it is one inch thick?
25. How much frosting is needed to cover the doughnut?

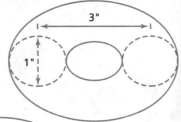

Use a torus with radius r through the dough and radius a across the hole.

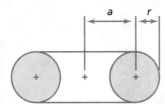

26. Give a formula for the volume of the torus.
27. Give a formula for the surface area of the torus.

Prove the following formulas for a cube with edge e inscribed in a sphere of radius r.

28. $r = \dfrac{\sqrt{3}}{2}e$

29. $V = \dfrac{8\sqrt{3}}{9}r^2$

30. Explain the mathematical significance of Ezekiel 45:10.

12 Transformations and Symmetry

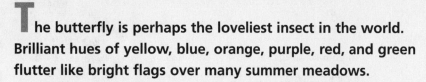

The butterfly is perhaps the loveliest insect in the world. Brilliant hues of yellow, blue, orange, purple, red, and green flutter like bright flags over many summer meadows.

How can a creature that begins life as a homely little caterpillar become such a gorgeous creature? Only God in His wisdom could design it that way.

Few of God's creatures can rival the peacock for loveliness. After finding one feather in 1913, naturalists searched the African plains until 1936 just to find a peacock. Lovers of brilliant hues can understand why those scientists persisted in their search.

The butterfly and the peacock have something more than beautiful colors in common. They are examples of geometric symmetry. The left wing of the butterfly is the mirror image of the right wing. Each peacock feather divides down the center into two perfectly symmetrical halves. Such symmetry contributes to their beauty.

In a similar way we should be a mirror image of God. God made us in His own image (Gen. 1:26), and He wants Christians to reflect that image by imitating His character and actions. Second Corinthians 3:18 tells us that when we look at God's glory in much the same way that we look in a mirror we can be changed from imperfection to perfection.

After this chapter you should be able to

1. identify transformations.

2. perform reflections, rotations, translations, and dilations of given figures.

3. classify transformations as isometries, dilations, or neither.

4. classify isometries as reflections, rotations, translations, or compositions of them.

5. state properties preserved by isometries or dilations.

6. apply isometries to light and to rolling objects.

7. define and identify types of symmetry.

8. define similar figures.

12.1 Transformations

In general usage, a transformation is a major change in the form, appearance, or nature of something. In Acts 9 Saul converted to Christianity (vv. 3-18). This conversion changed his very nature from one of hatefulness (vv. 1-2) to one serving Christ

These zebras display their symmetric stripes at a water hole in Kenya.

(vv. 20-22). Men marveled at the striking transformation that took place in Paul's life after he accepted Jesus Christ as his personal Savior. Every person who comes to know Christ as his personal Savior will experience this same transformation of soul and spirit. Second Corinthians 5:17 says, "Therefore if any man be in Christ, he is a new creature: old things are passed away; behold, all things are become new." Have you been transformed by the shed blood of Jesus Christ?

In mathematics the term *transformation* describes a change in the appearance of points in a plane. It is therefore a correspondence between point locations before and after the change. The geometric figure before a transformation is called the *preimage*. The resulting geometric figure after the transformation is called the *image*.

In figure 12.1 the points of △*ABC* are matched with the points of △*A'B'C'*. △*ABC* is the preimage, △*A'B'C'* is the image, and the correspondence between them is the *transformation*.

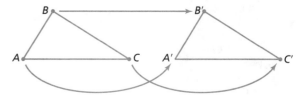

Figure 12.1

Definition

A **transformation** is a one-to-one function from the plane onto the plane.

A transformation can be thought of as a movement of figures. We will study various kinds of transformations in this chapter. Some movements, such as stretching, change the size and shape of a figure, whereas other transformations, such as rotating, do not.

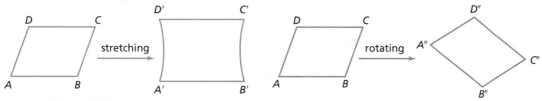

Figure 12.2

The transformation called a reflection is similar to a mirror image. A reflection in a line *l*, is shown in figure 12.3. Line *l* is called the *line of reflection,* or the *mirror.*

In this reflection, $X \rightarrow X'$ (X is mapped to X'), $Y \rightarrow Y'$, and $Z \rightarrow Z'$. If the vertices of the preimage are named in counterclockwise order, $\triangle XYZ$, then the image has clockwise order $\triangle X'Y'Z'$. This order of vertices is called *orientation.* Reversing the orientation is a characteristic of a reflection. Trace figure 12.3 onto your paper and connect each image vertex with its preimage vertex. Measure the distance from the line of reflection to X and to X'. Do this for each pair of

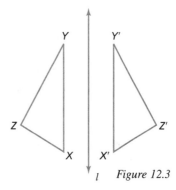

Figure 12.3

vertices. Also measure the angle formed by each segment and *l*. The discoveries you have just made will help you understand the definition of *reflection.*

Definition

A **reflection** in a line, *l*, is a transformation that maps each point A of a plane onto the point A' such that the following conditions are met.

1. If A is on *l*, then $A = A'$.
2. If A is not on *l*, then *l* is the perpendicular bisector of $\overline{AA'}$.

It is easy to find the reflection of any given polygon in a given line.

Wonder Lake captures the reflection of Mt. McKinley at Denali National Park, Alaska.

EXAMPLE   Reflect △LMN in line k.

Answer   1. Draw a line from each vertex
             perpendicular to k.
          2. On each perpendicular line find a point
             on the other side of k that is
             the same distance from k as the
             corresponding vertex.
          3. Connect these points.

Figure 12.4

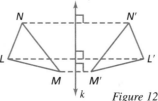

Figure 12.5

An image point for every point of △LMN can be found on △L′M′N′. Choose
any point on △LMN and find its image point on △L′M′N′. What is true about
the orientation of the triangles?

▶ **A. Exercises**

Copy each figure below onto your paper and reflect the preimage across
the given line.

1.

3.

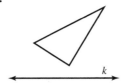

5.

2.

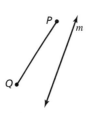

4.

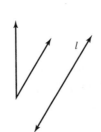

6.

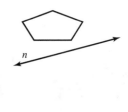

7. Draw a point and a line of reflection. As the point moves closer to the line, what happens to its image?

8. Draw a one-inch segment and a line of reflection. Reflect the segment in the line. Now measure the segment. How long is it?

9. Draw a 45° angle and a line of reflection. Reflect the angle in the line. What is the measure of the reflected angle?

10. Draw a line segment, $\overline{AC}$, with point B between A and C. Draw a line of reflection and reflect $\overline{AC}$ in the line. Is the reflection of B between the reflections of A and C?

▶ B. Exercises

Trace the following figures onto your paper and draw in the lines of reflection. Remember to consider the definition of reflection when doing these problems.

11.

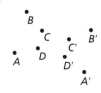

13.

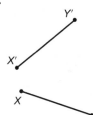

12.

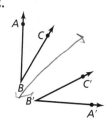

14.

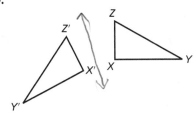

15. Explain how you can find the line of reflection if you are given the preimage and the image of a geometric figure.

▶ C. Exercises

16. *Given:* $\overline{A'B'}$ is a reflection of $\overline{AB}$ in l
    Prove: $\overline{AB} \cong \overline{A'B'}$

17. *Given:* A', B', and C' are reflections of A, B, and C in line l respectively; $\triangle ABC$, $\triangle A'B'C'$
    Prove: $\triangle ABC \cong \triangle A'B'C'$ and $\angle ABC \cong \angle A'B'C'$

18. Name the Platonic solids.

Give the surface area and volume formulas for each figure.

	Figure	**Surface Area**	**Volume**
19.	Sphere		
20.	Cylinder		
21.	Cone		
22.	Prism with regular n-gon as a base		

12.2 Translations and Rotations

The word *translated* is similar in origin to the word *transfer,* which means "to carry across." A translation in geometry has a similar meaning. *Translate* means "to slide into a different position." Figure 12.6 shows a geometric translation.

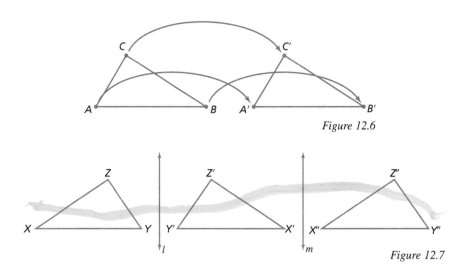

Figure 12.6

Figure 12.7

A translation can be defined in terms of reflections. $\triangle XYZ$ is reflected in line l to form $\triangle X'Y'Z'$, which is then again reflected in line m to obtain $\triangle X''Y''Z''$. Now compare $\triangle XYZ$ with $\triangle X''Y''Z''$. Since you can slide the first onto the second, a translation has occurred. Because a translation involves two reflections, the orientation of the triangle is not changed. The first reflection reverses the orientation, but the second reflection changes it back.

Any time you perform two or more transformations on a geometric figure, you are performing a *composition* of transformations. In figure 12.7 if the first reflection is called R, it is described by $R(\triangle XYZ) = \triangle X'Y'Z'$. The second reflection is called T such that $T(\triangle X'Y'Z') = \triangle X''Y''Z''$. The translation T composed with R is denoted $T \circ R$ or $T(R(\triangle XYZ))$. With this understanding of the composition of transformations, we can now define translation.

Definition

A **translation** is a transformation formed by the composition of two reflections in which the lines of reflection are parallel lines.

The third type of transformation that we will study in this chapter is a rotation. A rotation is also a composition of reflections.

Definition

A **rotation** is a transformation formed by the composition of two reflections in which the lines of reflection intersect.

Figure 12.8 shows an example of a rotation. The image of $\triangle HIJ$ is $\triangle H''I''J''$ by a composition of two reflections. Trace $\triangle HIJ$ onto a piece of paper and then turn the paper, keeping point X fixed. The triangle will rotate to coincide with $\triangle H''I''J''$.

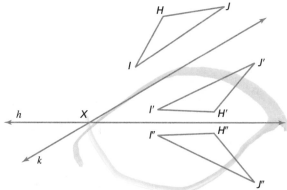

Figure 12.8

The *center of the rotation* in figure 12.8 is the point *X*, the intersection of the two mirrors. The angle that the rotation takes to move to the new position is called the *magnitude* of the rotation. So if $m\angle HXH''$ is 95°, the magnitude of the rotation is 95°. The direction of the rotation is also important. When talking about the magnitude of a rotation, always indicate whether the rotation is clockwise or counterclockwise.

An interesting fact about rotations is that the magnitude is twice the measure of the acute or right angle between the lines of reflection. A 95° rotation would require a 47.5° angle between the lines of reflection. In summary, a rotation can be described by its center, the magnitude of rotation, and the direction of the rotation.

The ferris wheel at Cedar Point, Ohio, rotates ten degrees between cars while loading passengers. Notice also the overlapping red radii and silver kite shapes in the design.

An *identity transformation* is a transformation that maps each point of a geometric figure onto itself. For example, if two 180° rotations have the same center, their composition will map every figure onto itself, thus producing an identity transformation.

MIND OVER MATH

Consider the set of all transformations. Which properties does the operation of composition have?

1. Commutative
2. Associative
3. Identity
4. Inverse

▶ A. Exercises

Trace the following diagrams onto your paper. Reflect the geometric figures in *l* and then in *m* to obtain a composite transformation. Name the types of transformations.

1.

2.

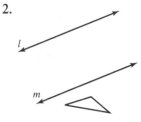

3.

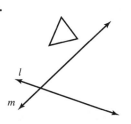

6.

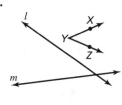

4.

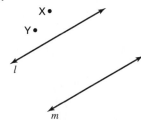

7.

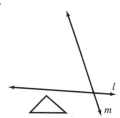

5.

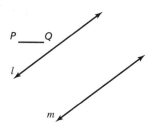

8.

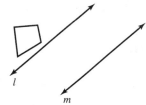

▶ B. Exercises

Translate and rotate each figure below. Indicate whether the size or shape of the geometric figure was changed. Show all your work.

9.

11.

10.

12.

13. If the magnitude of a rotation is 80°, what is the measure of the acute angle between the lines of reflection?

14. If the angle between the lines of reflection is 27°, what is the magnitude of the rotation?

15. Draw an acute triangle and rotate it 70° clockwise about point O. Then rotate the image 70° counterclockwise about point O. What is the composition of these rotations called?

16. Repeat exercise 15, using two different centers. What is the composition?

17. If l and m intersect at point P to form a 40° angle, then what is the composite of the reflections in l and m? Give its center and magnitude.

18. If R is the reflection in l, and T is the reflection in m, does $R \circ T = T \circ R$? (Is the composition commutative?) *Hint:* Answer the question by finding $R(T(X))$ and $T(R(X))$.

▶ C. Exercises

Use the exercises in the previous section to show that rotations and translations share the following two properties.

19. If $\overline{A''B''}$ is a translation or rotation of $\overline{AB}$, then $\overline{AB} \cong \overline{A''B''}$.
    Given: $R(\overline{AB}) = \overline{A'B'}$ and $T(\overline{A'B'}) = \overline{A''B''}$
    Prove: $\overline{AB} \cong \overline{A''B''}$

20. If $\triangle A''B''C''$ is a rotation or translation of $\triangle ABC$, then $\triangle ABC \cong \triangle A''B''C''$.

▶ Dominion Thru Math

21. A men's store installs a mirror so that a customer standing 4 feet away can see from the top of his head to his shoes. Use reflections to find the vertical length v of the mirror and the distance b from the floor to the bottom of the mirror for men of height H and eye height h. If $H - h$ is 5 inches, find v and b for men who are 6'6" and 5'6".

22. Accommodate all customers in this range. Why does it not matter how far away the man stands?

▆ Cumulative Review

23. Decide which numbers are greater than others and put them in increasing order (*Hint:* decimals). $\frac{22}{7}$, 3.14, $\sqrt{10}$, $\sqrt[3]{32}$, $(1.1)^{12}$, π

24. Graph the set on the number line: $\{-2, -\frac{3}{2}, \sqrt{2}, \pi, 4.1\}$

Give the area and perimeter of each figure.

	Figure	Perimeter	Area
25.	Circle		
26.	Rectangle		
27.	Regular polygon		

Translating Conic Sections

Plot $y = 2x^2$ and $y = 2(x - 1)^2$. You should see that the graphs are the same size and shape. The movement of the first graph one unit *to the right* illustrates another application of transformations.

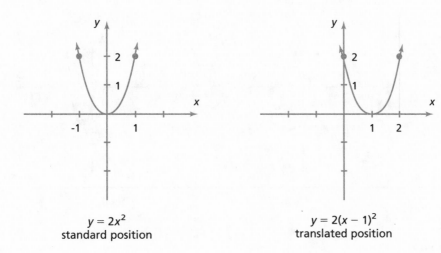

$y = 2x^2$
standard position

$y = 2(x - 1)^2$
translated position

Do you see that replacing x with $x - h$ translates the graph h units horizontally? Similarly, to translate the graph k units vertically, replace y with $y - k$.

	Circle	Parabola
standard position	$x^2 + y^2 = r^2$	$y = ax^2$
translated position	$(x - h)^2 + (y - k)^2 = r^2$ with center (h, k)	$y - k = a(x - h)^2$ or: $y = a(x - h)^2 + k$ with vertex (h, k)

EXAMPLES Graph $(x - 2)^2 + (y + 1)^2 = 9$ and $y = 2(x - 1)^2 - 3$
Find (h, k)

Answers $(2, -1)$ $(1, -3)$
                     $r = 3$ Opens up $2 > 0$

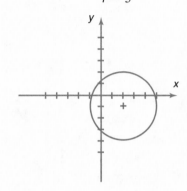   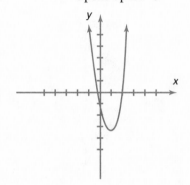

▶ Exercises

Graph.

1. $(x + 2)^2 + y^2 = 4$
2. $y = x^2 + 1$
3. $x^2 + (y - 2)^2 = 1$
4. $y = 2(x + 1)^2 + 4$
5. $(x - 4)^2 + (y - 3)^2 = 25$

12.3 Dilations

When was the last time you went to the eye doctor? Did he put drops into your eyes that caused them to dilate? Dilating the eyes makes the pupils bigger so that the doctor can study them more easily. This enlargement is one type of mathematical dilation. After a few hours the eye drops wear off and the pupils contract to their normal size. Such a contraction represents the other type of mathematical dilation.

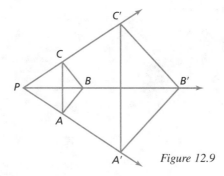

Figure 12.9

Figure 12.9 shows a dilation with center *P*. Notice that △*ABC* and △*A'B'C'* are not congruent.

Definition

A **dilation** is a transformation that expands or contracts the points of the plane in relation to a fixed point, *P*.

Find the ratio of *PA'* to *PA*. Do the same with *PB'* to *PB*, and with *PC'* to *PC*. You should get the same number each time. This constant positive ratio is called the *scale factor* of the dilation and is represented as *k*. Always calculate *k* as a ratio of image to preimage lengths.

So, for any dilation the image of *P* is *P* (the fixed point) and for any other point *R*, the image *R'* is on $\overrightarrow{PR}$ so that *PR'* = *k*(*PR*).

In figure 12.9 the scale factor is greater than 1 and the size of the triangle is expanded. The result of a dilation depends on the scale factor.

The irregular tesselation by M. C. Escher entitled Circle Limit 3 *shows fish shapes successively reduced toward the circular border.*

If $k > 1$, then the dilation is an *enlargement* (or expansion) of the figure.

If $k = 1$, then the dilation does not change the size of the figure; it is the identity transformation.

If $0 < k < 1$, then the dilation results in a *reduction* (or contraction) of the figure.

Values of $k < 0$ can be studied in college.

EXAMPLE   Classify the dilation by its scale factor.

Answer   Measure the corresponding sides of $\triangle XYZ$ and $\triangle X'Y'Z'$. Determine the scale factor of this dilation by finding the ratio $k = X'Y'/XY$. Since $0 < k < 1$, the dilation is a reduction, and the image of $\triangle XYZ$ is a smaller triangle, $\triangle X'Y'Z'$.

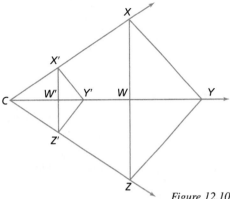

Figure 12.10

Measure the angles of $\triangle XYZ$ in figure 12.10 and compare these angle measurements to the corresponding angle measures of $\triangle X'Y'Z'$. You should find that corresponding angles have the same measure. This means that the triangles have the same shape even though they are not the same size.

When two geometric figures have the same shape but not necessarily the same size, the figures are called *similar figures*. A dilation always results in a similar figure. So in figure 12.10 $\triangle XYZ$ and $\triangle X'Y'Z'$ are similar, designated $\triangle XYZ \sim \triangle X'Y'Z'$. You will study similar figures in the next chapter.

▶ A. Exercises

For each diagram determine the scale factor that would take the figure to its image under a dilation with center O.

1.
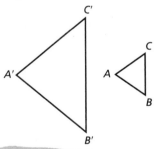

2.
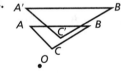

In a dilation, X is the center and A is mapped onto A'; B onto B'; and C onto C'. Find the indicated measure if the scale factor is 5.

3. $AX = 9$ units; find $A'X$
4. $AB = 4$ units; find $A'B'$
5. $XC' = 40$ units; find XC

6. $A'C' = 12$ units; find AC
7. $m\angle ABC = 82$; find $m\angle A'B'C'$
8. $m\angle B'C'A' = 15$; find $m\angle BCA$

Under a certain dilation with center *P*, *L* is mapped onto *L'*, *K* is mapped onto *K'*, *M* is mapped onto *M'*, and *N* is mapped onto *N'*. Find the scale factor for each pair of measurements and classify the type of dilation.

9. *PL* = 10 units; *PL'* = 30 units
10. *PN* = 20 units; *PN'* = 5 units
11. *M'N'* = 8 units; *MN* = 8 units
12. *M'N'* = 14 units; *MN* = 7 units

▶ B. Exercises

Give the scale factor for the dilation that maps *KM* to *K'M'*.

13. *KM* = *x* units; *K'M'* = *y* units

For each diagram determine the scale factor that would take the figure to its image under a dilation with center *O*.

14.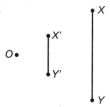

15.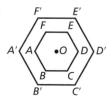

Trace the following figure onto your paper; then using *D* as center, find the image of the given points with the given scale factors.

16. $k = \frac{1}{2}$

17. $k = 3$

18. $k = 2$

19. $k = \frac{1}{3}$

20. $k = 1$

21. $k = \frac{1}{4}$

22. Measure ∠*ABC* in the figure and then measure ∠*A'B'C'* in exercises 16-20. What do you observe?

23. If *X-B-Y* in exercise 14, what can you say about *B'*?

24. Draw $\overleftrightarrow{AB}$ and a point *P* not on the line. Perform a dilation with fixed point *P* and a scale factor of 2. What kind of figure results? Do these figures have any special relation?

▶ C. Exercises

25. A contraction has a scale factor of 0.8. What dilation will return the image to its original size? What is the composition of these two dilations?

26. Prove that two parallel lines are everywhere equidistant.
27. Would exercise 26 be a theorem in Riemannian geometry? Why?

True/False
28. Parallel planes are everywhere equidistant.
29. Skew lines are everywhere equidistant.
30. A plane and a line parallel to the plane are everywhere equidistant.

12.4 Invariance Under Transformations

Invariance means "not varied," or "constant." Invariance in a Christian's spiritual life is important. The main goal of a Christian is to become more Christlike. God is constant and unchanging. James 1:17 states that there is no variableness with God. You need to be constant in obeying biblical truths and displaying godly traits. Consistency and dependability are characteristics that everyone needs to develop fully. Learn to be spiritually constant and true.

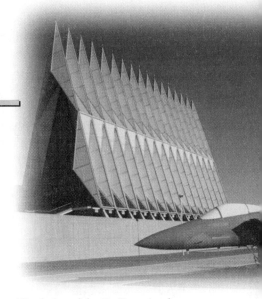

The design of the Air Force Academy chapel in Colorado Springs, Colorado, involves a repeated translation from the front to the back of the building.

Some transformations have invariant qualities. If the preimage and image of a given transformation always share a certain characteristic, the transformation *preserves* that characteristic.

You should remember that reflections, rotations, and translations all preserve distance because the image is always exactly the same size as the preimage.

Definition

An **isometry** is a transformation that preserves distance.

The word *isometry* has a Greek origin. *Isos* means "equal," and *metron* means "measure." So the basic meaning of *isometry* is "equal measure." You have studied the following isometries.

Isometries

Reflection

Translation

Rotation

Identity transformation

Composite of isometries

Six invariant properties are listed below. Remember that each of these invariant properties is left unchanged by the isometries that you have studied.

Properties of Isometries

1. *Distance is preserved.* This property is guaranteed by the definition.
   Example: If $AB = 3$, then after a rotation $A'B' = 3$.
2. *Collinearity of points is preserved.* Example:
   If A, B, and C are collinear, then A', B', and C'
   are collinear after a reflection.

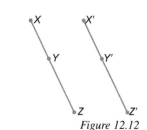

Figure 12.11

3. *Betweenness of points is preserved.*
   Example: If X-Y-Z, then after a translation
   X'-Y'-Z'.

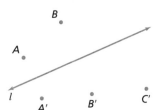

Figure 12.12

4. *Angle measure is preserved.* Example:
   If $\triangle ABC$ is reflected across line l, then
   $m\angle ABC = m\angle A'B'C'$.

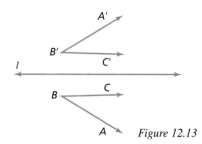

Figure 12.13

5. *Parallelism is preserved.* Example: If $\overleftrightarrow{AB} \parallel \overleftrightarrow{CD}$ and the lines are rotated around a point, then $\overleftrightarrow{A'B'} \parallel \overleftrightarrow{C'D'}$.

6. *Triangle congruence is preserved.* Example: If $\triangle XYZ$ is translated, then $\triangle XYZ \cong \triangle X'Y'Z'$.

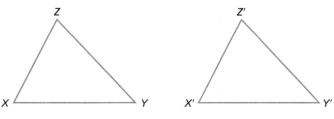

Figure 12.14

Since collinearity, betweenness, and distance are preserved, it follows that subsets of lines are preserved. In other words, the image of a ray is always a ray, and the image of a segment is always a segment, and so on.

An interesting theorem about isometries is given here without a proof.

Theorem 12.1

Isometry Theorem. **Every isometry can be expressed as a composition of at most three reflections.**

Dilations do not preserve distance and so are not isometries. Since dilations preserve angle measure, they preserve shape but may not preserve size. A transformation that preserves shape is a similarity while a transformation that preserves both size and shape is an isometry.

Properties of Dilation

1. A dilation preserves collinearity of points.
2. A dilation preserves betweenness of points.
3. A dilation preserves angle measure.
4. A dilation preserves parallel lines.

Of course similarities and isometries are not the only types of transformations. It is possible to have a transformation with none of these invariant properties. For example, the transformation that doubles the distance from a given line *l* is neither an isometry nor a dilation because it does not preserve angle measure.

▶ A. Exercises

Name the type of transformation illustrated by each pair of figures. Which are isometries?

1.

2.

3.

4.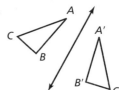

5.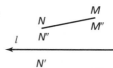

Name the type of transformation illustrated in each figure below.

6. **7.** A **8.** B B

                       A

Trace the following onto your paper and then find the image and identify the type of isometry.

9. Reflect △*ABC* in *m*; then reflect it in *l*.
10. Reflect $\overline{XY}$ in *q*; then reflect it in *k*.

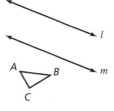

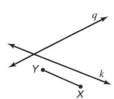

▶ B. Exercises

If each figure below is transformed as indicated, what must the image be and why?
11. A subset of a line by an isometry
12. An angle by a reduction
13. Parallel lines by a reflection
14. Which two properties are invariant for isometries but not for similarities?

Draw the following transformations and, if possible, give the following composite transformations in simpler form.

15. Point *A* rotated 90° around *C* and then rotated another 90° around *C*

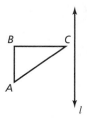

16. △*ABC* reflected in *l* and then rotated 90° clockwise around *B*

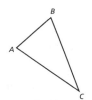

17. $\overline{AB}$ enlarged from point *P* by a scale factor of 6 and then contracted from point *P* by a scale factor of $\frac{1}{2}$

18. Find the center of rotation.

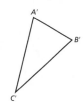

▶ C. Exercises

19. Use Theorem 12.1 to prove that every isometry preserves angle measure (property 4).

20. Prove that every isometry preserves triangle congruence. (*Hint:* Just prove △*ABC* ≅ △*A'''B'''C'''*.)

▥ Cumulative Review

Name a *theorem* that could be used to show that two *segments* are congruent if the segments are related to the figure indicated.

21. quadrilateral
22. triangle
23. circle

24. perpendiculars
25. lines
26. space

DAVID HILBERT

On January 23, 1862, David Hilbert was born to Maria and Otto Hilbert in Königsberg, East Prussia, which is now a part of Russia. Königsberg was close to the Baltic Sea, and David became familiar with boats and the life of a fisherman. His mother was his first teacher, and she introduced him to the constellations and to prime numbers. His father was a very strict judge who strongly believed in living a proper life. He tried to instill virtues such as punctuality, thrift, and discipline into his young son.

At the age of eight, David started his formal education at the Vorschule of the royal Friedrichskolleg. The Hilbert family became close friends with a Jewish family named the Minkowskis. Their children, who were about David's age, were good mathematicians, and they influenced David in this area. David did not do exceptionally well in most subjects, because he had great trouble memorizing material, but mathematics appealed to him because it was easy for him. In 1879 he transferred to Wilhelm Gymnasium, a school that placed much emphasis on the subject. David did excellent work in mathematics there.

In 1880 Hilbert entered the University of Königsberg, where he concentrated on mathematics. He and his friend Herman Minkowski studied there together and shared a deep love for mathematics. On February 7, 1885, Hilbert received his doctor of philosophy degree, and by 1893 he had attained a full professorship at the university. While at Königsberg, Hilbert married Käthe Jerosch. In 1895 the Hilberts moved to the University of Göttingen, where he had a long teaching and research career.

David Hilbert studied number theory and was the first to show that if a geometric contradiction existed, then the corresponding arithmetic of real numbers must also contain contradictions. He also believed in establishing a definite step-by-step procedure for solving problems. In fact, in 1899 his book *Grundlagen der Geometrie* (*Foundations of Geometry*) provided the first systematic treatment of geometry that corrected Euclid's flaws. To him we owe the use of undefined terms (point, line, and plane) as well as the need for more postulates. He used twenty-one postulates, including the axioms of continuity. He also restated the Parallel Postulate in a form similar to the Historic Parallel Postulate.

In Paris in 1900 he gave a lecture to an international math convention in which he challenged mathematicians with twenty-three unsolved problems. Some of these are still unsolved and his challenge has stimulated much of the research of this century.

Hilbert also studied an infinite dimensional geometry called Hilbert Space. In the space, the square of the infinitely many coordinates must have a finite sum. Transformations of many kinds can be studied in Hilbert Space.

Hilbert developed a philosophy of mathematics that abandoned any considerations about the truth of mathematics and was concerned only about the consistency of the mathematical system. He studied proof theory in relation to his philosophy but was never able to establish the truth of the philosophy. Hilbert's philosophy was dangerous because it threw out truth at the beginning. Nothing can be built without certain true building blocks. Therefore, one must be sure that his knowledge is built on a true foundation.

In 1942 Hilbert fell and broke his arm. As a result of complications from the fall, he died on February 4, 1943.

He also believed in establishing a definite step-by-step procedure for solving problems.

12.5 Applications of Isometries

You may often ask (and justifiably so), "When will I ever use this?" You are probably asking that question now about transformations and isometries. One pure mathematical reason for studying isometries is to give a new light to the idea of congruence. But besides the pure mathematical applications of isometries, there are also some practical applications.

The Belle of Louisville *is the oldest of the six remaining sternwheel-style steamboats in the U.S. Its sixteen paddle blades (called bucket planks) are isometric and rotate around the hub, and each blade consists of five panels that are side by side (translated) and are also isometric.*

EXAMPLE 1   A beam will be sent from satellite *A* to satellite *B* by being sent to the earth and reflected by a booster station to satellite *B*. The satellite engineers are trying to place the booster station in the spot where the total distance that the beam travels will be the shortest. If the booster station must be located somewhere along line *h*, what is the best location?

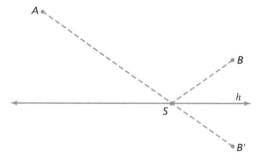

Figure 12.15

Answer   **1.** To find the appropriate location of the booster station, *S*, first reflect *B* in *h* to find *B′*.

2. Connect *A* and *B′*.

3. The point where $\overline{AB'}$ intersects *h* is the appropriate location for the booster station.

You should be asking, "How do I know that $\overline{AS}$ to $\overline{SB}$ represents the shortest path?" Suppose we said some other point on h, called T, was the correct answer.

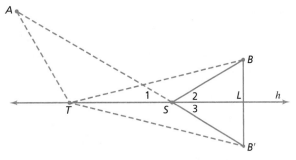

Figure 12.16

If our assumption is true, then $AT + TB < AS + SB$. If we can prove this statement false, S would be the correct position for the booster station. Since reflections preserve distance, $\overline{SB}$ and its reflection $\overline{SB'}$, must be congruent. Likewise, $\overline{TB}$ must be congruent to $\overline{TB'}$. So $AS + SB = AS + SB'$, and $AT + TB = AT + TB'$. Now consider $\triangle ATB'$. Do you remember what the Triangle Inequality states? From this theorem we can conclude that $AT + TB' > AB'$. Because of betweenness of points, $AS + SB' = AB'$. By substitution $AT + TB' > AS + SB'$. Also by substitution $AT + TB > AS + SB$. This contradicts the preliminary assumption, so T cannot be the correct location. Therefore, S must be the correct position for the booster station.

One more example is given before you do some problems on your own.

EXAMPLE 2 Figure 12.17 shows a miniature golf green. Notice that it would be impossible to putt a ball directly from the tee (T) to the hole (H). What spots should you aim for on sides 1 and 2 so that you will make a hole in one?

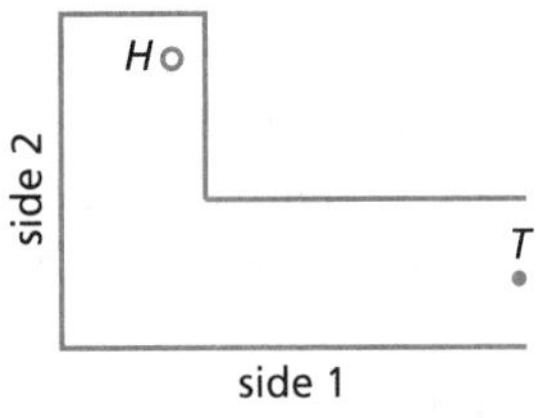

Figure 12.17

Answer
1. Reflect H in side 2.
2. Reflect H' in side 1.
3. Connect H'' and T; the point of intersection of this segment and side 1 is the point you need to aim for on side 1.
4. Connect the point of intersection found in step 3 and H' to find the intersection with side 2.

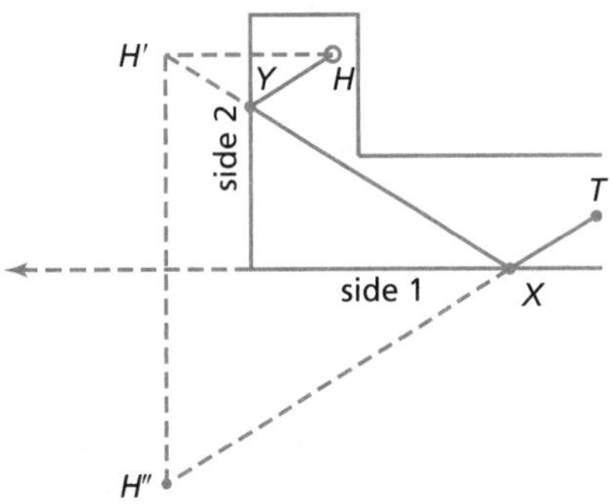

► A. Exercises

Trace the diagram onto your paper and
then follow the steps in exercises 1-5.

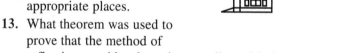

1. Find the point on *k* that marks the
   shortest distance from *A* to *C*.
2. Explain why *E* is not the answer to exercise 1.
3. Find the point *G* on *k* that marks the shortest distance from *A* to *k* to *D*.
4. Compare *AG* + *GD* to *AE* + *ED*.
5. Find the point on *k* that marks the shortest distance from *D* to *k* to *B*.

An electron moves from point *X* to *Y* by bouncing off the sides of a four-
sided enclosure. Find the path it will follow if it bounces off the given side(s).

6. $\overline{DC}$
7. $\overline{AD}$
8. $\overline{AB}$ and then $\overline{BC}$
9. $\overline{AD}$ and then $\overline{DC}$
10. $\overline{AB}$ and then $\overline{DC}$

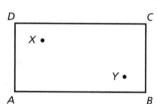

► B. Exercises

The telephone company wants to connect two houses to the main line in
the least expensive way. To reduce expenses, they want to use the least
amount of wire possible. The drawing below illustrates the problem.

11. Find the places on the
    phone line, *l*, where the
    two houses should be
    connected.
12. Explain why these are the
    appropriate places.
13. What theorem was used to
    prove that the method of
    reflection provides the point on a line with the shortest distance to two
    points?
14. Name three types of transformations that preserve distance.
15. What is a transformation called that preserves distance?

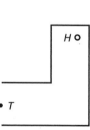

Here is the first hole on a miniature golf course.
Find the appropriate place to aim your putt for a
hole in one under the following conditions.

16. By hitting one side
17. By hitting two sides
18. By hitting three sides
19. By hitting four sides

    20. Putt from the tee to the hole by hitting all five sides.

Cumulative Review

    21. What is the fixed point of a rotation called?

    22. How many fixed points does a translation have?

    23. How many fixed points does a reflection have?

    24. If T is a transformation that is a dilation and X is the fixed point, what is $T(X)$?

    25. If an isometry has three noncollinear fixed points, what can you conclude?

12.6 Symmetry

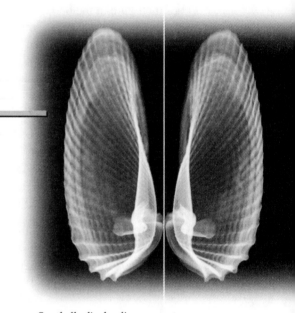

If you reflect the left side of the seashell across line l, it coincides with the right side. The line drawn in the photograph of the shell is called the *axis of symmetry,* and the figure is said to be symmetric.

Seashells display line symmetry.

Definition

A figure has **line symmetry** when each half of the figure is the image of the other half under some reflection in a line.

In figure 12.18 three symmetric geometric figures are shown. Can you find an axis of symmetry for each figure?

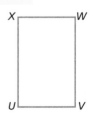

Figure 12.18

Another form of symmetry is called *rotational symmetry.*

Definition

A figure has **rotational symmetry** when the image of the figure coincides with the figure after a rotation. The magnitude of the rotation must be less than 360°.

The left diagram in figure 12.19 has rotational symmetry. Trace it onto your paper and place the traced figure on top of the figure in the book. By keeping the center fixed, rotate the figure until it again coincides with the figure in the book. Since you rotated your paper 120°, the figure has a rotational symmetry of 120°.

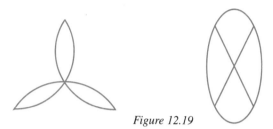

Figure 12.19

Rotational symmetry of 180° is also called *point symmetry.* The right diagram in figure 12.14 provides an example of point symmetry. Notice that you can "reflect" each point of the figure through the center of the rotation to obtain another point of the figure.

The stem is the axis of symmetry for this mimosa.

▶ A. Exercises

How many axes of symmetry does each figure have?

1.

2.

3.

4.

5.

6.

7.

8.

▶ B. Exercises

9. Which figures in exercises 1-8 have rotational symmetry?

10. List the angle of rotation for each rotational symmetry of exercise 9.

11. Which figures in exercises 1-8 have point symmetry?

Draw each figure listed in exercises 12-17 and then draw all lines of symmetry. Identify any figures with rotational symmetry.

12. A square

13. A parallelogram

14. A regular pentagon

15. A concave hexagon

16. An isosceles triangle

17. A circle with one diameter

18. Can a figure have point symmetry without having rotational symmetry? Why?

19. A figure has 90° rotational symmetry. Will it also have point symmetry?

▶ C. Exercises

20. Classify the capital letters of the alphabet according to the number of axes of line symmetry. Identify *any* rotational symmetry.

■ Cumulative Review

Let U = the set of integers.

$A = \{x \mid -3 \le x < 2 \text{ and } x \text{ is an integer}\}$

$B = \{1, 2, 4, 8, 16, \ldots\}$

$C = \{5^0, -\sqrt{9}\}$

$D = \{x \mid x \text{ is a prime number}\}$

21. Write set C in simpler form; then write the correct subset relation for C.

22. Express sets A and D in list form.

23. Express set B in set-builder notation.

24. Find $B \cap D$ and $A \cap D$.

25. Find $A \cup C$ and $B \cup C$.

Geometry and Scripture

Transformations

Remember how God transformed Paul into a man of God? The word *transformed* comes from the idea of a change (*trans*) in form. No Christian is perfect, so all must desire God to change or transform them. God changes both our mind and our body.

1. How does Romans 12:2 tell us to be transformed?

2. Colossians 3:10 says that our new nature in Christ is "renewed in knowledge" of our Creator. Where do we get this knowledge?

3. Who gives us this knowledge and transforms us according to Titus 3:5?

4. How often should we seek this knowledge and let Him transform us (II Cor. 4:16)?

5. Matthew 17:2 and Mark 9:2-3 related a change in form of our Lord Jesus. What happened?

6. Into what does Christ transform our bodies according to Philippians 3:20-21?

7. What kind of transformation is discussed in II Corinthians 11:13-15?

You have just studied three kinds of transformations. God transforms minds to real Christlikeness, he transforms bodies to real Christlikeness, and Satan transforms himself to phony Christlikeness.

This can help you understand transformations in math. Such transformations also represent changes in the position or form of figures.

8. Name three kinds of transformations in math in which the size and shape do not change but the position changes. What is the general term for this?

9. Name two kinds of transformations in which the size changes but not the shape. What is the general term for this?

Did you notice that a translation in math is a transformation that changes the position of the figure? In the Bible, "translation" also emphasizes a change of position.

10. Read Hebrews 11:5. What happened to Enoch?

HIGHER PLANE: Find the Old Testament reference to the translation of Enoch. Was it before or after the Flood?

Hebrews 11:5 refers to Enoch's translation three times. This "translation" clearly conveys the "change in position" idea. The term is used two other times, but in those passages the "change in position" is used more figuratively to describe a change.

11. What change in "position" have Christians already experienced according to Colossians 1:13?

12. The other reference to translation in Scripture is in II Samuel 3:10. Describe the change in this passage.

> ## Line upon Line
>
> By faith Enoch was translated that he should not see death; and was not found, because God had translated him: for before his translation he had this testimony, that he pleased God. ❧
>
> HEBREWS 11:5

Trace the figures below and find the reflections in line *l*.

1.

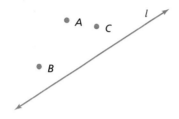

2.

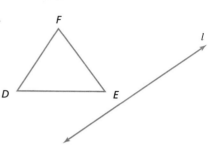

Find the line of reflection for each of these transformations.

3.

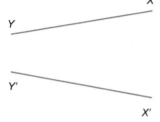

4.

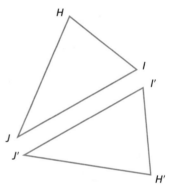

Trace the following figures and then find the composite of two reflections for each exercise. Tell whether the transformations performed result in a reflection, translation, or rotation.

5.

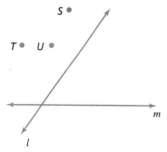

6.

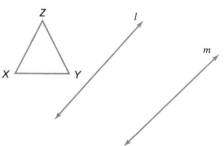

7.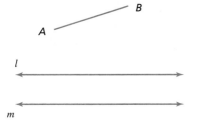

8.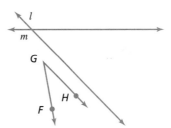

9. If the angle of rotation is 76°, what is the measure of the acute angle between the intersecting lines?

10. Find the correct isometry for the pair of congruent triangles shown.

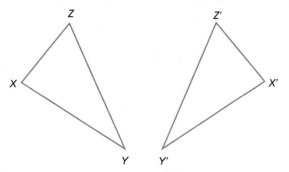

11. An underground cable is to be installed to connect two houses to the main telephone line that runs along the curb of the street. If both lines are to be connected at the same point and the least amount of cable possible is to be used, where should the cable be connected to the main line? The figure below shows the location of the houses and the main line.

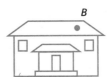

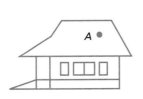

Consider the four figures below.

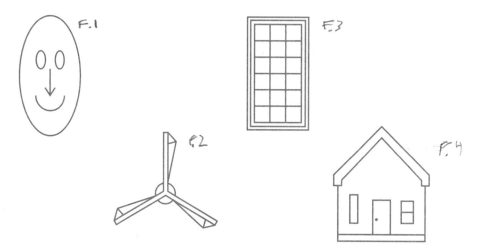

12. Which has both line symmetry and rotational symmetry?
13. Which has neither line symmetry nor rotational symmetry?
14. Which has line symmetry only?
15. Which has rotational symmetry only?
16. Give the magnitude of the rotational symmetry for exercise 15.
17. Which has point symmetry?

Given a scale factor of $k = 3$, answer exercises 18-20.
18. If $MN = 4$, find $M'N'$.
19. If $P'Q' = 51$, find PQ.
20. Using $\triangle ABC$ and point P as shown, find $\triangle A'B'C'$.

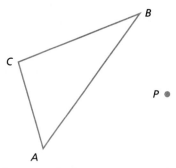

Identify $\triangle ABC$ and its image as being congruent, similar, or neither.
21. Contraction toward a point P
22. Translation composed with a reflection
23. Transformation that triples the distance of points from a given line
24. Rotation around a point P composed with a dilation with center P

Answer the following questions.

25. How would you locate the axis of symmetry for an isosceles right triangle?

26. Find the composition of a 125° clockwise rotation about point *P* and a 235° clockwise rotation about *P*.

27. What is the key characteristic preserved by an isometry?

Assume that dilations preserve angle measure to complete the following proofs.

28. *Given:* $\overrightarrow{OX}$ bisects $\angle ABC$; a dilation with center *O* and scale factor *k* maps *A* onto *D*, *B* onto *E*, and *C* onto F

    Prove: $\overrightarrow{OE}$ bisects $\angle DEF$

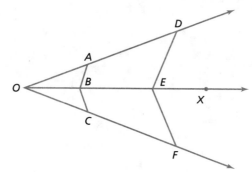

29. *Given:* $\overline{KI} \perp \overline{HJ}$; a dilation with scale factor *k* and center *O* maps *H* onto *H'*, *I* onto *I'*, *J* onto *J'*, *K* onto *K'*

    Prove: $\overline{K'I'} \perp \overline{H'J'}$

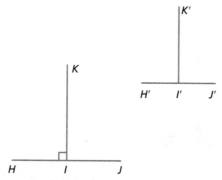

30. Explain the mathematical significance of Hebrews 11:5.

13 Similarity

What do a submarine, a poet named Oliver Wendell Holmes, the ocean floor, and geometry all have in common?

The chambered nautilus, of course! Did you know that the first nuclear submarine was named the U.S.S. *Nautilus*? Holmes wrote a famous poem on the subject: "The Chambered Nautilus." And the ocean floor is the home of the lovely shellfish named the nautilus.

But what about geometry? Well, the nautilus is one of nature's geometric masterpieces. As the nautilus grows and builds new chambers, its size increases, but its shape remains the same. Thus some geometry teachers use the nautilus to illustrate proportion and ratio. The spiral shape of the nautilus also demonstrates the geometric concept of the golden rectangle and the golden spiral. The ancient Greeks thought that this geometric shape enhanced the beauty of an object. The nautilus also illustrates two other geometric ideas—rotation and dilation.

Not just the nautilus but all God's creation shows His wonderful design. You might feel inferior, but Psalm 139 says that you are "fearfully and wonderfully made." God's work is "marvellous." God did more, however, than just design your body. He died on the cross to pay the penalty for your sin. Have you trusted Christ as Savior? If so, God has forgiven you, adopted you as His child, and provided an eternal home in heaven.

After this chapter you should be able to

1. define similar polygons.
2. solve proportion problems.
3. state and apply similarity criteria.
4. define and find geometric means.
5. prove and apply theorems on proportions.
6. solve word problems using similar triangles.
7. prove and apply theorems on lengths of segments determined by secants or tangents of circles.
8. define and recognize applications of the Golden Ratio.

13.1 Similar Figures

Recall that two figures are similar if they have the same shape but not necessarily the same size. A more formal definition is given here.

The scale model of Jerusalem showing the city at the time of Christ is a favorite tourist stop in Israel.

Definition

Similar polygons are polygons having corresponding angles that are congruent and corresponding sides that are proportional. If △*ABC* and △*DEF* are similar, the proper notation is △*ABC* ~ △*DEF*.

Do you remember what a proportion is? A ratio is the comparison of two numbers, usually integers, using division. If *a* and *b* are integers and $b \neq 0$, then $\frac{a}{b}$ is their ratio. $\frac{1}{4}$ and $\frac{9}{1}$ are examples of ratios. A proportion is an equation with two equal ratios. For example, $\frac{1}{2} = \frac{3}{6}$ is a proportion.

You may remember a process called cross multiplication, which helps you check or solve proportion problems.

If $\frac{1}{2} = \frac{3}{6}$

then $1(6) = 2(3)$.

After cross multiplying, you get the same answer on each side; thus the ratios are equal, and the equation is a proportion. You can solve certain first-degree equations by using cross multiplication.

EXAMPLE Solve $\frac{x}{5} = \frac{9}{15}$.

Answer $\frac{x}{5} = \frac{9}{15}$

$15x = 45$ 1. Cross multiply.

$x = 3$ 2. Solve.

So $\frac{3}{5}$ and $\frac{9}{15}$ are equivalent ratios.

The lengths of the sides of similar figures are proportional. Look at figure 13.1.

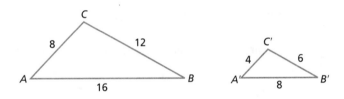

Figure 13.1

Look at the ratios of the corresponding sides:

$\frac{A'B'}{AB}$, $\frac{B'C'}{BC}$, and $\frac{A'C'}{AC}$

What number do you get in each case? Thus if $\triangle ABC \sim \triangle A'B'C'$,

$\frac{A'B'}{AB} = \frac{B'C'}{BC} = \frac{A'C'}{AC}$

If the corresponding angles of two polygons are congruent and the sides are proportional, then you know that the two figures are similar.

In the dilations that you studied earlier, the scale factor of the dilation equals the common ratio of the corresponding sides. It may help to do the cumulative review questions on dilations first.

▶ A. Exercises

Solve each proportion.

1. $\dfrac{x}{100} = \dfrac{1}{2}$

4. $\dfrac{5}{y} = \dfrac{55}{121}$

2. $\dfrac{2}{3} = \dfrac{y}{18}$

5. $\dfrac{1}{8} = \dfrac{x}{2}$

3. $\dfrac{6}{x} = \dfrac{54}{63}$

6. $\dfrac{9}{5} = \dfrac{18}{y}$

Find the ratio of the lengths in the right figure (image) to those in the left figure (preimage) for each pair of similar figures.

7.

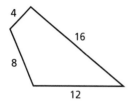

8.

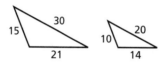

Given that the figures are similar in each problem, find the length of the indicated sides.

9.

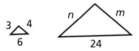

11.

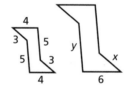

10.

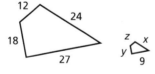

12.

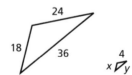

▶ B. Exercises

13. If $\triangle LPQ \sim \triangle RST$, what angles are congruent, and what sides are proportional?
14. Is every square similar to every other square? Why or why not?
15. Are congruent triangles also similar?
16. What is the common ratio or scale factor in exercise 15?
17. *Prove:* If two triangles are congruent, then they are also similar.
18. *Prove:* Similarity of triangles is reflexive.

19. If $\triangle ABC \sim \triangle XYZ$, find the perimeter of $\triangle ABC$ and $\triangle XYZ$. Are the perimeters of the triangles in any particular ratio?

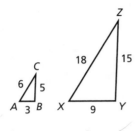

▶ C. Exercises

20. *Prove:* The ratio of the perimeters of two similar triangles is equal to the ratio of the lengths of any pair of corresponding sides.

■ Cumulative Review

21. Find the center of the dilation.
22. Give the scale factor.

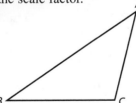

23. If the image of a dilation is congruent to the preimage, then what is the scale factor?
24. Classify three types of dilations based on scale factors.
25. Find $\triangle A'B'C'$, if P is the center of a dilation with scale factor $\frac{4}{3}$.

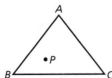

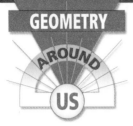
GEOMETRY AND ART

Math is rational—figures, theorems, formulas, strict rules. Art is creative—visual effects, interpretations, feelings. So art and math have nothing in common, right? Wrong. Of course artists are not mathematicians; they often lay out their drawings or sculptures freehand without following specific rules or formulas. But there are some mathematical principles that artists follow to make their creations beautiful. Still doubtful? Come along.

Egyptian painting

Have you ever seen Egyptian art? To most of us it looks a little crude: the people stand in awkward positions, and sizes seem to be all mixed up. What makes this art look peculiar? Well, the Egyptians did not follow the laws of perspective. They felt that important people should be drawn large, and unimportant people should be drawn smaller, no matter where they were in the picture. Due to studies by artists during the Renaissance, modern artists know that a drawing without perspective, like Egyptian art, will not look realistic.

Since the Renaissance, artists have tended to use the geometric principle of proportion. First the artist establishes a vanishing point somewhere on (or even slightly off) the canvas. All the objects in the picture are drawn in relation to this point. The artist draws people in the foreground larger than people in the background. Thus a person standing ten feet from the viewer appears twice as large as a person standing twenty feet away. Artists use this method to make the painting more realistic. The use of proportions to make similar figures enables the painting to portray depth.

Photographers also deal with similarity every day. When developing and enlarging pictures, photographers rely on equipment that uses the principles of similarity. For example, when a picture is enlarged or reduced, the final print is similar to the negative—only the size changes. In like manner, a slide projector produces an image on the screen similar to but much larger than the image on the slide.

Proportion also determines what forms we consider beautiful. Some artists use geometric shapes to design pottery and beautiful vases. Artists who draw or sculpt the human body also recognize the proportions in God's design. Leonardo da Vinci wrote down proportions for artists to use that he thought reflected God's perfect design.

Leonardo da Vinci (1452–1519). The Last Supper. *1498. Post-restoration. S. Maria delle Grazie, Milan, Italy.*

There are some mathematical principles that artists follow to make their creations beautiful.

13.2 Similar Triangles

Scale model airplane

Let's review the characteristics
of similar triangles mentioned
in Section 13.1. Look at figure 13.2.

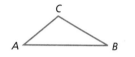

Figure 13.2

$\triangle ABC \sim \triangle XYZ$. This means that $\angle A \cong \angle X$, $\angle B \cong \angle Y$, $\angle C \cong \angle Z$, and
$$\frac{XY}{AB} = \frac{YZ}{BC} = \frac{XZ}{AC}.$$

In this lesson you will be learning ways of proving that two triangles are similar. Many of the similarity theorems resemble the congruence theorems that you learned in Chapter 6. The theorems that will be developed are based on the AA (Angle-Angle) Postulate.

Postulate 13.1
AA Similarity Postulate. **If two angles of one triangle are congruent to two angles of another triangle, then the two triangles are similar.**

Why can we say that two triangles are similar if only two pairs of corresponding angles are congruent? Recall the earlier theorem: If two angles of a triangle are congruent to two angles of another triangle, the third angles are congruent.

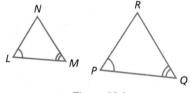

Figure 13.3

Since two pairs of angles are congruent in the diagram, $\angle N$ and $\angle R$ are also congruent. Furthermore, according to the AA Similarity Postulate, $\triangle LMN \sim \triangle PQR$. Let's experiment to discover some similarity theorems.

Experiment

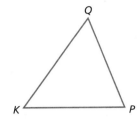

Figure 13.4

1. Copy △*KPQ* onto your paper.
2. With a compass, measure the length of $\overline{KP}$. Construct a segment that is twice as long as $\overline{KP}$; call the segment $K'P'$.
3. Measure $\overline{PQ}$ and construct a segment twice as long as $\overline{PQ}$. Using P' as an endpoint, mark an arc with radius equal to 2*PQ*.
4. Construct a segment that equals 2*KQ*. Using K' as center and using a radius equal to 2*KQ*, form an arc that intersects the arc you made in step 3. Call the intersection point Q'.
5. Connect Q', K', and P' to form a triangle.

Look at △*K'P'Q'* and △*KPQ*. Measuring the corresponding angles with a protractor will convince you that although the triangles are not congruent, they are similar since the corresponding angles are congruent. What is the ratio of the side lengths of the two triangles? These measurements suggest the next theorem.

Theorem 13.1
SSS Similarity Theorem. If the three sides of one triangle are proportional to the corresponding three sides of another triangle, then the triangles are similar.

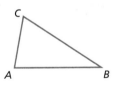 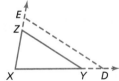

Figure 13.5

Given: In △*ABC* and △*XYZ*,
$$\frac{XY}{AB} = \frac{YZ}{BC} = \frac{XZ}{AC}$$

Prove: △*ABC* ~ △*XYZ*

STATEMENTS	REASONS
1. $\frac{XY}{AB} = \frac{YZ}{BC} = \frac{XZ}{AC}$	1. Given
2. Draw a segment congruent to $\overline{AB}$ by extending $\overline{XY}$ and call it $\overline{XD}$; $\overline{AB} \cong \overline{XD}$	2. Auxiliary lines
3. $AB = XD$	3. Definition of congruent segments
4. $\frac{XY}{XD} = \frac{YZ}{BC}$	4. Substitution (step 3 into 1)
5. Construct $\overleftrightarrow{DE}$ parallel to $\overleftrightarrow{YZ}$	5. Auxiliary line
6. $\angle XYZ \cong \angle XDE$; $\angle XZY \cong \angle XED$	6. Corresponding Angle Theorem
7. $\triangle XDE \sim \triangle XYZ$	7. AA
8. $\frac{XY}{XD} = \frac{YZ}{DE} = \frac{XZ}{XE}$	8. Definition of similar triangles
9. $\frac{YZ}{DE} = \frac{YZ}{BC}$	9. Transitive property of equality (steps 4 and 8)
10. $\frac{XZ}{XE} = \frac{XZ}{AC}$	10. Substitution (steps 8 and 1 into 9)
11. $(YZ)(BC) = (YZ)(DE)$ $(XZ)(AC) = (XZ)(XE)$	11. Multiplication property of equality (cross multiplication; see steps 9, 10)
12. $BC = DE$, $AC = XE$	12. Multiplication property of equality
13. $\overline{BC} \cong \overline{DE}$, $\overline{AC} \cong \overline{XE}$	13. Definition of congruent segments
14. $\triangle ABC \cong \triangle XDE$	14. SSS
15. $\angle B \cong \angle XDE$; $\angle C \cong \angle XED$	15. Definition of congruent triangles
16. $\angle B \cong \angle XYZ$; $\angle C \cong \angle XZY$	16. Transitive property of congruent angles (see step 6)
17. $\triangle ABC \sim \triangle XYZ$	17. AA
18. If the three sides of one triangle are proportional to the corresponding three sides of another triangle, then the triangles are similar	18. Law of Deduction

You should know two other theorems on similar triangles.

Theorem 13.2
SAS Similarity Theorem. If two sides of a triangle are proportional to the corresponding two sides of another triangle and the included angles between the sides are congruent, then the triangles are similar.

Theorem 13.3
Similarity of triangles is an equivalence relation.

The table summarizes three quick ways to prove that two triangles are similar.

	Given	Result
AA	$\angle A \cong \angle X, \angle B \cong \angle Y$	$\triangle ABC \sim \triangle XYZ$
SSS	$\dfrac{XY}{AB} = \dfrac{YZ}{BC} = \dfrac{XZ}{AC}$	$\triangle ABC \sim \triangle XYZ$
SAS	$\dfrac{XY}{AB} = \dfrac{YZ}{BC}; \angle B \cong \angle Y$	$\triangle ABC \sim \triangle XYZ$

▶ A. Exercises

State whether the lengths (in units) below could be sides of similar triangles.

1. 3, 8, 7, and 12, 32, 28
2. 2, 6, 7 and 10, 30, 42
3. 12, 15, 9 and 4, 5, 3

4. 2, 3, 4 and 3, 4.5, 6
5. 7, 4, 9 and 14, 10, 18
6. $\dfrac{1}{5}, \dfrac{4}{9}, \dfrac{1}{3}$ and $\dfrac{1}{10}, \dfrac{2}{9}, \dfrac{1}{6}$

Which pairs of triangles are similar? Why?

7.

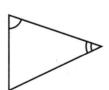

8.

9.

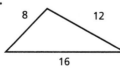

10.

11.

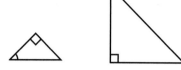

12.

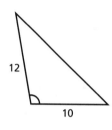

▶ B. Exercises

Prove the following statements.

13. *Given:* WXYZ is a parallelogram
Prove: △WXY ~ △YZW

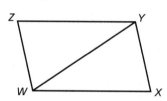

Use the figure below for the proofs in exercises 14-15.

14. *Given:* $\overleftrightarrow{MN} \parallel \overleftrightarrow{OQ}$
Prove: △MNP ~ △QOP

15. *Given:* △MNP ~ △QOP
Prove: $\overleftrightarrow{MN} \parallel \overleftrightarrow{QO}$

16. *Prove:* Similarity of triangles is transitive (If △ABC ~ △LMN, and △LMN ~ △PQR, then △ABC ~ △PQR.)

17. *Prove:* Similarity of triangles is symmetric

Use the following figure for the proofs in exercises 18-19.

18. *Given:* $\overleftrightarrow{DB} \perp \overleftrightarrow{DE}; \overleftrightarrow{DB} \perp \overleftrightarrow{AB}$

    Prove: $\triangle ABC \sim \triangle EDC$

19. *Given:* $\overleftrightarrow{DB} \perp \overleftrightarrow{DE}; \overleftrightarrow{DB} \perp \overleftrightarrow{AB}$

    Prove: $\dfrac{DE}{AB} = \dfrac{CE}{AC}$

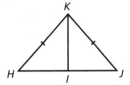

▶ C. Exercises

20. *Given:* $\overleftrightarrow{KI} \perp \overleftrightarrow{HJ}; \triangle HJK$ is an isosceles triangle

    Prove: $(JI)(HK) = (HI)(JK)$

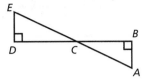

■ Cumulative Review

21. What is a relation that is reflexive, symmetric, and transitive?
22. List three symbols that represent equivalence relations.
23. Does the set of rotations with a given center P form an equivalence relation?
24. Suppose two regions are related if they have the same area. Is this an equivalence relation? Why?
25. Suppose two solids with the same volume are related. Is this an equivalence relation? Why?

Use the ideas in this chapter to show that $x = \sqrt{a}$. Make a construction that represents the square root of 5.

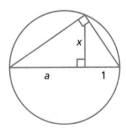

Figure 13.6

13.3 Similar Right Triangles

Right triangles have some peculiar characteristics: they satisfy the Pythagorean theorem and also form the basis of trigonometric ratios (in Chapter 14). Right triangles also have some special characteristics related to similarity.

A sloop, such as this one on Chesapeake Bay, Virginia, has one mast with two sails (mainsail and jib) that are both right triangles.

Theorem 13.4

An altitude drawn from the right angle to the hypotenuse of a right triangle separates the original triangle into two similar triangles, each of which is similar to the original triangle.

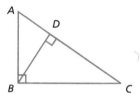

Figure 13.7

Theorem 13.4 means that if $\overline{BD}$ is an altitude of $\triangle ABC$, then 1) $\triangle BCD \sim \triangle ACB$, 2) $\triangle ABD \sim \triangle ACB$, and 3) $\triangle ABD \sim \triangle BCD$. The proof of the second case is shown here, and the other two cases are exercises.

Given: $\overline{BD}$ is the altitude of right $\triangle ABC$
Prove: $\triangle ABD \sim \triangle ACB$

STATEMENTS	REASONS
1. $\overline{BD}$ is the altitude of right $\triangle ABC$	1. Given
2. $\overline{BD} \perp \overline{AC}$	2. Definition of altitude
3. $\angle BDA$ is a right angle	3. Definition of perpendicular
4. $\angle ABC \cong \angle BDA$	4. All right angles are congruent
5. $\angle A \cong \angle A$	5. Reflexive property of congruent angles
6. $\triangle ABD \sim \triangle ACB$	6. AA

In the proportion $\frac{a}{x} = \frac{x}{b}$, notice that the denominator of one ratio is the same as the numerator of the other ratio. When this happens, x is called the *geometric mean*. So 8 is the geometric mean between 16 and 4 because

$$\frac{16}{8} = \frac{8}{4}.$$

Solving for a geometric mean involves the use of a quadratic equation, which you studied in algebra.

EXAMPLE 1 Find the geometric mean between 3 and 27.

Answer

$\frac{3}{x} = \frac{x}{27}$ 1. Set up a proportion; let x represent the geometric mean.

$x^2 = 81$ 2. Cross multiply.

$\sqrt{x^2} = \pm\sqrt{81}$ 3. Solve by finding the square root of both sides.

$x = 9$ 4. Use the principal square root.

Although there are two answers ($\pm$), select the positive root (or principal root) because the context demands a result between 3 and 27.

EXAMPLE 2   Find the geometric mean between 6 and 9.

   Answer $\frac{6}{x} = \frac{x}{9}$ **1.** Write the proportion.

   $x^2 = 54$ **2.** Cross multiply.

   $x = \pm\sqrt{54}$ **3.** Solve.

   $x = 3\sqrt{6}$ **4.** Simplify your answer.

The geometric mean is used in the next theorem.

Theorem 13.5
In a right triangle, the altitude to the hypotenuse cuts the hypotenuse into two segments. The length of the altitude is the geometric mean between the lengths of the two segments of the hypotenuse.

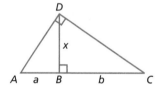

Figure 13.8

Given: Right $\triangle ACD$; $\overline{DB}$ is an altitude of $\triangle ACD$
Prove: $\frac{a}{x} = \frac{x}{b}$, or $\frac{AB}{DB} = \frac{DB}{BC}$

STATEMENTS	REASONS
1. Right $\triangle ACD$; $\overline{DB}$ is an altitude of $\triangle ACD$	**1.** Given
2. $\triangle ABD \sim \triangle DBC$	**2.** The altitude divides a right triangle into two similar triangles
3. $\frac{AB}{DB} = \frac{DB}{BC}$	**3.** Definition of similar triangles

Another theorem about right triangles and similarity involves a geometric mean.

Theorem 13.6

In a right triangle, the altitude to the hypotenuse divides the hypotenuse into two segments such that the length of a leg is the geometric mean between the hypotenuse and the segment of the hypotenuse adjacent to the leg.

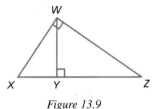

Figure 13.9

$$\frac{XZ}{WX} = \frac{WX}{XY}$$

$$\frac{XZ}{WZ} = \frac{WZ}{YZ}$$

You will prove this theorem as an exercise.

EXAMPLE 3 Given the measurements in $\triangle HIJ$, find x, y, and z.

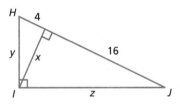

Figure 13.10

Answer $\dfrac{4}{x} = \dfrac{x}{16}$

$x^2 = 64$

$x = 8$

1. Solve the equation for the length of the altitude according to Theorem 13.5.

$\dfrac{4}{y} = \dfrac{y}{20}$ and $\dfrac{16}{z} = \dfrac{z}{20}$

$y^2 = 80$ and $z^2 = 320$

$y = 4\sqrt{5}$ and $z = 8\sqrt{5}$

2. Solve equations for the legs according to Theorem 13.6.

▶ A. Exercises

Solve each proportion; assume that x is positive.

1. $\dfrac{x}{6} = \dfrac{17}{3}$

2. $\dfrac{x}{4} = \dfrac{3}{11}$

3. $\dfrac{x}{9} = \dfrac{4}{x}$

4. $\dfrac{x-1}{3} = \dfrac{4}{1}$

5. $\dfrac{x-3}{2} = \dfrac{2}{x}$

▶ B. Exercises

Given that $\triangle ABC$ is a right triangle and $\overline{DC}$ is an altitude to the hypotenuse, $\overline{AB}$, find the length of the indicated sides.

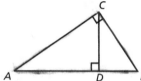

6. $AD = 15$ units; $DB = 5$ units; find CD
7. $AD = 15$ units; $DB = 5$ units; find $\overline{AC}$
8. $AD = 15$ units; $DB = 5$ units; find BC
9. $AB = 32$ units; $DB = 6$ units; find CD
10. $AB = 32$ units; $DB = 6$ units; find BC
11. $AD = 6$ units; $AB = 10$ units; find CD
12. $AD = 8$ units; $DB = 3$ units; find BC
13. $AD = 11$ units; $DB = 5$ units; find AC
14. $AD = x$ units; $DB = y$ units; find CD
15. $AD = 12$ units; $AB = 18$ units; find CB

Prove exercises 16-17 using the figure.

16. *Given:* Right $\triangle ABC$ with altitude $\overline{CD}$; $\overline{DE} \perp \overline{AC}$
    Prove: $\triangle AED \sim \triangle ACB$
17. *Given:* Right $\triangle ABC$ with altitude $\overline{CD}$
    Prove: $(CD)^2 = (AD)(BD)$
18. *Given:* Right $\triangle LMN$ with altitude $\overline{NO}$; $(LM)(NO) = (NM)^2$
    Prove: $MO = NO$

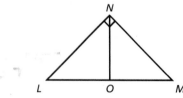

▶ C. Exercises

Prove the other two cases of Theorem 13.4.
19. $\triangle BCD \sim \triangle ACB$ (Case I)
20. $\triangle ABD \sim \triangle BCD$ (Case III)

▶ Dominion Thru Math

Photocopiers apply concepts of similarity in their settings to reduce or enlarge an image. The settings are given as a percent and express the scale factor for the linear measures of the image and pre-image.

21. Express the area for a reduction if the original has area A and the copier reduces at $r\%$.

22. Stan makes many copies of a flier but must reduce the total amount of paper by 35%. Provided smaller fliers are acceptable, what reduction setting should he use on the copier?

■ Cumulative Review

Which pairs of figures are similar? For each pair of similar figures, give the scale factor, k.
23. Two circles with radii 3 and 6
24. Two rectangles: 6 by 9 and 8 by 12
25. Two rectangles: 6 by 8 and 16 by 18
26. Two regular tetrahedra with sides of length 9 and 6 respectively
27. Two squares with sides of length s and t respectively

Slopes of Perpendicular Lines

You can use similar triangles to show that the product of the slopes of perpendicular lines is -1.

Theorem

If two distinct nonvertical lines are perpendicular, then their slopes are negative reciprocals.

Given: $l_1 \perp l_2$
Prove: $m_1 \cdot m_2 = -1$

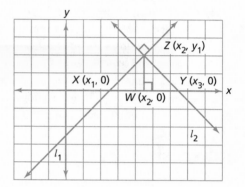

STATEMENTS	REASONS
1. $l_1 \perp l_2$	**1.** Given
2. Draw the altitude $\overline{ZW}$ of right $\triangle XYZ$	**2.** Auxiliary line
3. $\dfrac{XW}{ZW} = \dfrac{ZW}{WY}$	**3.** The altitude is the geometric mean between the two segments of the hypotenuse
4. $XW = x_2 - x_1 \quad ZW = y_1 \quad WY = x_3 - x_2$	**4.** Distance formula for number lines
5. $\dfrac{x_2 - x_1}{y_1} = \dfrac{y_1}{x_3 - x_2}$	**5.** Substitution (step 4 into 3)
6. $m_1 = \dfrac{y_1}{x_2 - x_1}$ $m_2 = \dfrac{-y_1}{x_3 - x_2} = -\left(\dfrac{y_1}{x_3 - x_2}\right)$	**6.** Definition of slope
7. $m_1 \cdot m_2 = \dfrac{y_1}{x_2 - x_1} \cdot -\left(\dfrac{y_1}{x_3 - x_2}\right)$	**7.** Multiplication property of equality
8. $m_1 \cdot m_2 = \dfrac{y_1}{x_2 - x_1} \cdot -\left(\dfrac{x_2 - x_1}{y_1}\right)$	**8.** Substitution (step 5 into 7)
9. $m_1 \cdot m_2 = -1$	**9.** Inverse property of multiplication

Therefore if the slope of one line is $\frac{3}{7}$, the slope of a perpendicular line is $-\frac{7}{3}$. You often use this relationship to find equations of lines in algebra.

▶ Exercises

Give the equation of the line perpendicular to the line described and satisfying the given conditions.

1. $y = -\frac{4}{3}x + 5$ with y-intercept $(0, -8)$

2. $y = 2x - 1$ and passing through $(1, 4)$

3. the line containing $(2, 5)$ and $(3, 4)$ at the first point

4. $y = \frac{1}{2}x + 5$, if their point of intersection occurs when $x = 2$

5. *Prove:* The diagonals of a rhombus are perpendicular to each other. (*Hint*: Find a in terms of b and c.)

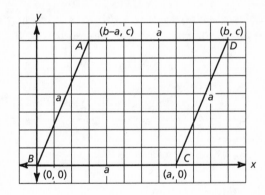

13.4 Similar Triangles and Proportions

This section presents some of the many proportions associated with similar triangles. Theorem 13.10 was proved in exercise 20 of Section 13.1, and the rest are exercises in this section.

St. Louis Cathedral in New Orleans, Louisiana, has similar pyramids for steeples. Built in 1794, it is the oldest cathedral in the United States.

Theorem 13.7

In similar triangles the lengths of the altitudes extending to corresponding sides are in the same ratio as the lengths of the corresponding sides.

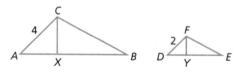

Figure 13.11

In figure 13.11, if $\triangle ABC \sim \triangle DEF$ and $\frac{AC}{DF} = 2$, then $\frac{CX}{FY} = 2$. If $CX = 3$ units, then $FY = \frac{3}{2}$ or 1.5 units.

Theorem 13.8

In similar triangles the lengths of the medians extending to corresponding sides are in the same ratio as the lengths of the corresponding sides.

Do you remember that a median extends from a vertex to the midpoint of the opposite side?

EXAMPLE   Since $\triangle ABC \sim \triangle XYZ$, find AD.

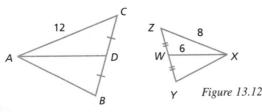

Figure 13.12

Answer

$\dfrac{AD}{WX} = \dfrac{AC}{XZ}$ 1. Apply Theorem 13.8.

$\dfrac{AD}{6} = \dfrac{12}{8}$

$\dfrac{AD}{6} = \dfrac{3}{2}$ 2. Reduce the fraction.

$AD = \dfrac{3}{2}(6) = 9$ 3. Solve the proportion.

Draw a diagram to illustrate Theorem 13.9.

Theorem 13.9

In similar triangles the lengths of the corresponding angle bisectors from the vertices to the points where they intersect the opposite sides are in the same ratio as the lengths of the corresponding sides of the triangles.

Theorem 13.10

In similar triangles the perimeters of the triangles are in the same ratio as the lengths of the corresponding sides.

So far, you have seen that altitudes, medians, angle bisectors, and perimeters are all proportional to the sides. What do you predict for the ratio of the areas? Does the next theorem surprise you?

Theorem 13.11

In similar triangles the ratio of the areas of the triangles is equal to the square of the ratio of the lengths of corresponding sides.

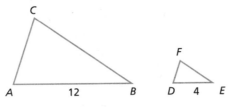

Figure 13.13

If $\triangle ABC \sim \triangle DEF$ and the sides have lengths as indicated, then the ratio of the areas of the triangles is $\left(\dfrac{12}{4}\right)^2 = 3^2 = 9$ or 9 to 1.

If the area of $\triangle DEF$ is 6 square units, then the area of $\triangle ABC$ is 54 square units.

▶ A. Exercises

Find the indicated lengths in exercises 1-10. Assume that $\triangle ABC \sim \triangle MNO$.

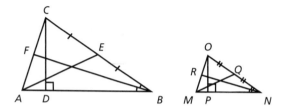

1. $AC = 8$ units; $MO = 4$ units; $CD = 6$ units; find OP
2. $AB = 15$ units; $MN = 5$ units; $NR = 6$ units; find BF
3. $BC = 24$ units; $NO = 12$ units; $AE = 10$ units; find MQ
4. $FB = 16$ units; $RN = 12$ units; $NO = 15$ units; find BC
5. $CD = 9$ units; $OP = 3$ units; $AE = 12$ units; find MQ
6. $MO = 9$ units; $AC = 18$ units; perimeter of $\triangle ABC = 70$ units; find the perimeter of $\triangle MNO$
7. $FB = 10$ units; $RN = 7$ units; perimeter of $\triangle MNO = 28$ units; find the perimeter of $\triangle ABC$
8. $AC = 8$ units; $MO = 4$ units; area of $\triangle ABC = 36$ square units; find the area of $\triangle MNO$
9. $BC = 9$ units; $NO = 6$ units; area of $\triangle MNO = 20$ square units; find the area of $\triangle ABC$
10. Area of $\triangle ABC = 81$ square units; area of $\triangle MNO = 9$ square units; $AB = 18$ units; find MN

▶ B. Exercises

Use the same triangles in exercises 1-10 to answer exercises 11-15.
11. State a proportion involving the medians in the triangles.
12. State a proportion involving the altitudes in the triangles.
13. State a proportion involving the angle bisectors in the triangles.
14. State a proportion involving the area of the triangles.
15. State a proportion involving the perimeters of the triangles.

Prove each theorem.

16. Theorem 13.7
17. Theorem 13.8
18. Theorem 13.9
19. Theorem 13.11

▶ C. Exercises

20. Prove that if two regular tetrahedra are similar, then the ratio of the volumes is equal to the cube of the ratio of the sides.

▶ Dominion Thru Math

Join three points at the vertices of a scalene right triangle with the minimum length. Connecting pipes, sidewalks, and electrical wires all use such minimizations. One path follows the legs, but a second path follows the hypotenuse and the altitude to it.

21. Three machines at points *A*, *B*, and *C* use water and lie at the vertices of a right triangle with *B* at the right angle. *AB* is 60 ft., and *BC* is 45 ft. Find the shortest path for a water pipe to connect all three machines.

22. Show that exercise 21 works for all right triangles. Use the TI-83 graphing calculator. If you use the TI-83, let the two legs be y_1 and y_2, the hypotenuse's two parts be *a* and *b*, and the altitude from the right angle be *h*. Let y_1 be *x*, the smaller leg; let y_2 be *x* + *D*, and choose different values of *D*. Compare the graphs of $y = y_1 + y_2$ and $y = a + b + h$. (*Hint:* Use proportional sides of similar right triangles to express *a*, *b*, and *h* in terms of *x*.)

▪ Cumulative Review

Give the name of each shaded figure. Classify each as convex or concave.

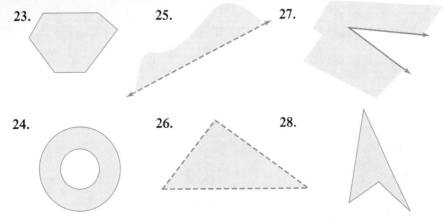

23.

25.

27.

24.

26.

28.

13.5 Similar Triangles and Problem Solving

You should apply new material that you learn whenever possible. A Christian should also glean biblical principles and learn to apply them to his life (Prov. 2:1-5; 23:12). In this lesson you will see several applications of similar triangles.

EXAMPLE 1   Janet stands in front of the Capitol in Washington, D.C. She wants to know the building's height above the street level but cannot measure it directly. How can she find the height of the building?

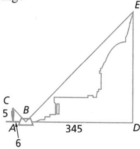

Figure 13.14

Answer $\angle ABC \cong \angle DBE$

1. Janet puts a mirror on the ground 345 feet from the center of the rotunda in the Capitol. She stands 6 feet from the mirror and sees the top of the Capitol reflected in it. The angle at which the image reflects is the same in both directions.

$\triangle ABC \sim \triangle DBE$

2. Since the heights form right angles with the ground, the triangles are similar by AA.

$\dfrac{DE}{AC} = \dfrac{DB}{AB}$

$\dfrac{DE}{5} = \dfrac{345}{6}$

3. Corresponding sides are proportional. Janet's eyes are 5 ft. above ground.

$6DE = 5(345)$

$6DE = 1725$

$DE = 287.5 \text{ ft.}$

You can work most of these problems in more than one way by using different similar triangles. Can you find a different way to solve example 1?

EXAMPLE 2 A state park is building a boardwalk trail for visitors through a wetland. The plan calls for the trail to cross part of the pond as it winds through the surrounding marsh and bog. How long must the bridge be to cross the pond at the points indicated?

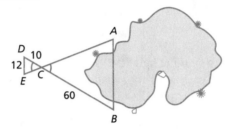

Figure 13.15

Answer $\triangle ABC \sim \triangle EDC$

1. To avoid wading in the pond, the rangers measure angles at *D* and *E* congruent to angles at *B* and *A*. This makes $\overleftrightarrow{DE} \parallel \overleftrightarrow{AB}$. The triangles are therefore similar by AA.

$$\frac{AB}{ED} = \frac{BC}{DC}$$

$$\frac{AB}{12} = \frac{60}{10} = 6$$

2. Write a proportion and substitute the known values.

$AB = 6(12)$

$AB = 72$ ft.

3. Solve the proportion.

▶ A. Exercises

The scale on a map shows 5 miles represented by 2 inches. Make a ratio as a scale factor and then answer the questions below.

1. What does one inch represent?
2. How far apart are two places if the distance on the map is twelve inches?
3. What distance corresponds to seven inches on the map?
4. What length represents twenty miles?
5. How does the map represent three miles?

▶ B. Exercises

6. Karen wants to find the height of the flagpole in front of the school. An 8-foot pole casts a 10-foot shadow at the same time the flagpole casts a 150-foot shadow. How high is the flagpole?

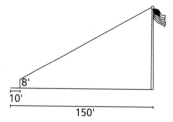

7. Al is an archaeologist and is trying to find the height of a pyramid that he is studying. If he uses the mirror technique described in example 1, he must stand 8 feet from the mirror that is placed 640 feet from the center of the pyramid. Al's eyes are approximately 6 feet from the ground. How high is the pyramid?

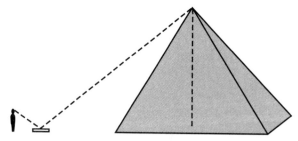

8. Find the length of the shadow that a 30-foot telephone pole casts when a 2-foot pole casts a shadow of 3 feet.

9. Dan is standing at an observation deck on top of Pine Mountain, which is 5,872 feet high. From this viewpoint Dan can see a river on the far side of Buck Ridge (as shown). If Buck Ridge is 2,936 feet high and 4,875 feet from the river, how far from the river is Pine Mountain?

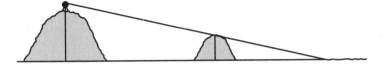

10. John and his friends have a tree house by a stream. They want to place a log across the stream as a bridge. How long must the log be in order to reach across the stream? See the figure.

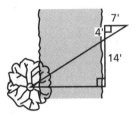

11. How tall is the Washington Monument if it casts a 185-foot shadow at the same time that a 3-foot pole casts a 1-foot shadow?

Washington Monument in Washington, D.C.

12. A tent rope goes from the top of the tent to the ground and is connected to two poles, one at the top and one at the edge of the tent. How long must the rope be if the distance from the stake to the first pole is 3 feet, the distance between the two poles is 8 feet, and the distance from the stake to the top of the first pole is 5 feet?

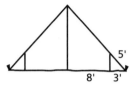

13. A flashlight shines directly on a strip of metal. How long is the strip of metal if its shadow on the wall is 16 inches long and the flashlight is 8 inches from the metal and 24 inches from the wall?

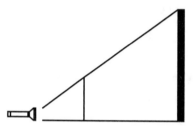

14. How tall is a tree if it casts an 84-foot shadow at the same time that a 10-foot pole casts a 4-foot shadow?

15. Find the depth of a rock quarry, using the information in the figure.

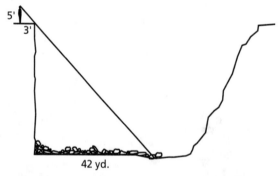

16. Mrs. Brown has a small vegetable garden. The garden forms a right triangle with legs of five and seven feet respectively. She plans to enlarge the garden to have nine times more area; how long will the legs be for the new garden?

17. Farmer Smith has a triangular pasture that is fenced. The longest side is 120 feet, and he used 300 feet of fencing. He plans to expand the pasture. The new pasture will be the same shape, and the longest side will be 180 feet. How much fencing will he use to fence it? How much more must he buy if he can still use all the old fencing?

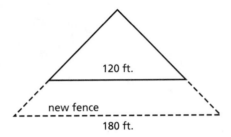

120 ft.

new fence

180 ft.

18. The scale of a map shows that one inch represents forty miles. A traveling salesman drives to Grand Island and Fargo and then returns home to Sioux City. According to the map, the perimeter of the triangular circuit was $21\frac{1}{4}$ inches. How far did he drive?

▶ C. Exercises

19. Mr. Cramer owns a large triangular lot. Because a river bounds two sides, he will fence only the longest side. Using 30 feet of fence, he partitioned off an area covering 80 square feet. How much fencing would he need to partition off an area covering 180 square feet?

20. Suppose Mr. Cramer's triangular lot has a perimeter of 200 feet. If a portion with an 80-foot perimeter requires a fence 30 feet long on the long side, how much fence is needed on the long side of the entire lot?

▦ Cumulative Review

Match the shape of each symbol (numeral or dash) to its classification.

21. 6
22. –
23. 0
24. 30
25. 8

A. Simple Closed Curve
B. Simple Curve (not closed)
C. Closed Curve (not simple)
D. Curve (neither simple nor closed)
E. Not a curve

13.6 Circles and Proportions

The Fuglevad windmill in the Netherlands has secants that intersect at the center. Note also the pyramidal base.

In Chapter 9 you studied chords, secants, and tangents. You learned how to compute the angles formed by intersections of these lines. When such lines intersect, it is also possible to find the lengths of the segments formed by using proportions.

Similar triangles provide a means of proving these proportions. Some of these relationships and proportions are basic to trigonometry, as you will see in the next chapter.

The first theorem describes a relationship between intersecting chords of a circle.

> **Theorem 13.12**
> If two chords intersect in the interior of a circle, then the product of the lengths of the segments of one chord is equal to the product of the lengths of the segments of the other chord.

Figure 13.16

This theorem states that since $\overline{AC}$ and $\overline{BD}$ intersect in the interior of $\odot O$, then $(AE)(EC) = (DE)(EB)$. Here is a proof of this theorem.

Given: $\odot O$ with chords $\overline{AC}$ and $\overline{BD}$ that intersect at E

Prove: $(AE)(EC) = (DE)(EB)$

Figure 13.17

STATEMENTS	REASONS
1. $\odot O$ with chords $\overline{AC}$ and $\overline{BD}$ that intersect at E	1. Given
2. Draw $\overline{AD}$ and $\overline{BC}$	2. Line Postulate
3. $\angle AED \cong \angle BEC$	3. Vertical Angle Theorem
4. $\angle ADB \cong \angle BCA$	4. Angles inscribed in the same arc are congruent
5. $\triangle AED \sim \triangle BEC$	5. AA
6. $\frac{EB}{AE} = \frac{EC}{DE}$	6. Definition of similar triangles
7. $(DE)(EB) = (AE)(EC)$	7. Multiplication property of equality
8. If $\overline{AC}$ and $\overline{BD}$ are chords intersecting at E, then $(AE)(EC) = (DE)(EB)$	8. Law of Deduction

The next theorem applies to problems in which the secants intersect in the exterior of the circle instead of the interior.

A *secant segment,* such as $\overline{AC}$ in figure 13.18, includes the chord of the circle and has as one endpoint a point in the exterior of the circle. The portion in the exterior of the circle, $\overline{BC}$, is called an *external secant segment.*

Theorem 13.13

If two secants intersect in the exterior of a circle, then the product of the lengths of one secant segment and its external secant segment is equal to the product of the lengths of the other secant segment and its external secant segment.

Given: Secants $\overleftrightarrow{AB}$ and $\overleftrightarrow{ED}$ intersect at point C in the exterior of $\odot O$

Prove: $(AC)(BC) = (EC)(DC)$

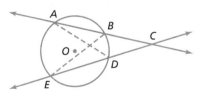

Figure 13.18

STATEMENTS	REASONS
1. Secants $\overleftrightarrow{AB}$ and $\overleftrightarrow{ED}$ intersect at point C in the exterior of $\odot O$	1. Given
2. Draw $\overline{BE}$ and $\overline{AD}$	2. Line Postulate
3. $\angle CAD \cong \angle CEB$	3. Inscribed angles intercepting the same arc are congruent
4. $\angle ACE \cong \angle ACE$	4. Reflexive property of congruent angles
5. $\triangle DAC \sim \triangle BEC$	5. AA
6. $\frac{BC}{DC} = \frac{EC}{AC}$	6. Definition of similar triangles
7. $(AC)(BC) = (EC)(DC)$	7. Multiplication property of equality
8. If secants $\overleftrightarrow{AB}$ and $\overleftrightarrow{ED}$ intersect at C in the exterior of $\odot O$, then $(AC)(BC) = (EC)(DC)$	8. Law of Deduction

▶ **A. Exercises**

Find each length using the figure shown.

1. If $LN = 18$ units, $YN = 8$ units, and $MY = 12$ units, find YO.
2. If $LY = 4$ units, $YN = 6$ units, and $MY = 8$ units, find YO.
3. If $LY = 8$, and YN is half of MY, find YO.
4. If OY is one-third of OM, and the product of LY and YN is 8 units, find OM.

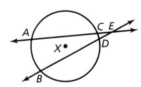

Using the diagram, find each length.
5. If $AE = 20$ units, $CE = 5$ units, and $EB = 25$ units, find ED.

6. If $EB = 32$ units, $DB = 14$ units, and $AE = 36$ units, find AC.

7. If $AE = 42$ units, $CE = 12$ units, and $DE = 8$ units, find BE.

8. If $BE = 24$ units, $DE = 6$ units, and $AE = 36$ units, find CE.

Find the indicated lengths.

9.

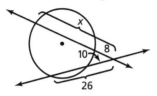

11.

10.

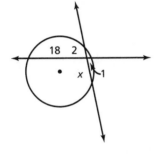

▶ B. Exercises

Find the indicated lengths.

12.

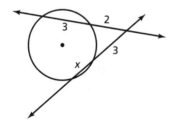

14.

13.

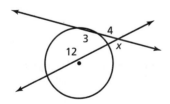

15. Find the radius.

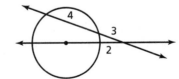

Exercises 16-19 refer to the following information about the figure shown.

Given: $\overrightarrow{AB}$ is tangent to $\odot P$ at A

Draw: Auxiliary $\overleftrightarrow{PB}$ intersects $\odot P$ at M, and $\overline{MN} \perp \overline{AP}$ at N

Label: $AB = t$, $MN = s$, $PN = c$, $PB = v$, and r is the radius of $\odot P$

16. Prove $\triangle MPN \sim \triangle BPA$.

Use the similarity in problem 16 for exercises 17-19.

17. Write the proportion for the sides.
18. Express t in terms of s and c if the radius is 1.
19. Express v in terms of s or c if $r = 1$.

▶ C. Exercises

20. *Prove:* If a secant and a tangent intersect in the exterior of a circle, then the square of the length of the tangent segment equals the product of the lengths of the secant segment and the external secant segment.

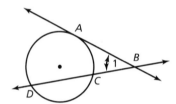

▪ Cumulative Review

On a flat map, which curves below are simple? Which are closed?
21. The boundary of Kansas
22. The path traced by the life of a man born in Phoenix who lived for a while in Detroit and then Atlanta and died in Seattle
23. The route of Columbus's first roundtrip voyage to the New World
24. The flight path of an airplane pilot that flies from Denver via New York and Miami and back to Denver
25. The international boundary between the United States and Mexico

13.7 The Golden Ratio and Other Applications of Similarity

An interesting phenomenon appears often in the art and sculpture of the ancient world. In many of Michelangelo's and Leonardo da Vinci's paintings a special rectangular ratio occurs. This ratio is called the *Golden Ratio*. The Golden Ratio is the ratio of the length to the width of a golden rectangle. A *golden rectangle* is a rectangle with the following characteristic: if a square unit is cut from one end of the rectangle, then the resulting rectangle has the same length-to-width ratio as the original rectangle. Figure 13.19 illustrates an approximation of the golden rectangle.

Leonardo da Vinci (1452–1519). Mona Lisa. *Oil on wood, 77 × 53 cm. Photo: Lewandowski/LeMage/Gattelet.*

If you compare the lengths of the rectangles to the widths, you get a ratio of 1.619. If a 21-by-21 unit square is cut from this rectangle, a 21-by-13 unit rectangle results. The length-to-width ratio is now $\frac{21}{13} = 1.615$. Find the ratio of the length to the width of each of the resulting smaller rectangles in figure 13.19. Each time the ratio should be approximately 1.618. We can find the exact value by using algebra.

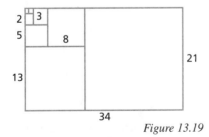

Figure 13.19

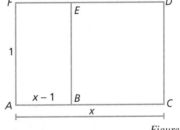

Figure 13.20

Figure 13.20 shows a general golden rectangle whose length is *x* units and whose width is 1 unit. Since the Golden Ratio is the ratio of the length to the width, the value of *x* is the value of the Golden Ratio. Because the ratio of the length to the width is the same in both the small and large rectangles, *ABEF* and *ACDF*, we can set up the proportion as follows.

$\dfrac{AC}{AF} = \dfrac{AF}{AB}$	1. Golden Ratios are equal.
$\dfrac{x}{1} = \dfrac{1}{x-1}$	
$x^2 - x = 1$	2. Cross multiply.
$x^2 - x - 1 = 0$	3. Solve this quadratic
$x = \dfrac{1 \pm \sqrt{1+4}}{2}$	equation, using the quadratic formula.
$x = \dfrac{1 \pm \sqrt{5}}{2}$ units	

The golden spiral is an important application of similar figures. Look at the golden spiral shown. The spiral continues to form smaller and smaller rectangles, but in the portion that is shown you should see five golden rectangles. Because their sides are proportional, all these golden rectangles are similar. What is the common ratio that shows that all the rectangles are in proportion?

God included the Golden Ratio in the design of His creation. You can observe the Golden Ratio in the spiral of a nautilus shell (page 534) and the whorls of a pineapple.

There are many other applications of similarity too. You have already seen how surveyors can use similar triangles to determine tree and building heights. Likewise similar figures form when a photo is enlarged or reduced. Craft books often provide patterns that you must enlarge before using.

Figure 13.21

This pineapple from Hawaii displays a spiral pattern.

▶ A. Exercises

For each pair of photographs, find the scale factor of the dilation that would produce the second photograph.

1.

USS Constitution *at Boston, Massachusetts*

2.

USS Alabama *at Mobile, Alabama*

3.

USS Eisenhower *at sea*

4.

USS Pintado *off the California coast*

Look at the series of rectangles below. Find the ratio of length to width for each. It will help to measure in millimeters.

5.

7.

9.

6.

8.

10.

▶ B. Exercises

11. If the ratio is within 0.02 of 1.618, then we call it a Golden Ratio. Which of the rectangles in exercises 5-10 are golden rectangles?

12. What kind of transformation takes place when a photograph is enlarged or reduced?

13. Measure the rectangles on the *Mona Lisa*. Which are golden rectangles?

Leonardo da Vinci (1452–1519). Mona Lisa *(detail).*

Study the architecture of the Parthenon to answer exercises 14-18. The triangular part at the top is called a pediment. Find the indicated ratios. Which are golden?

Parthenon at Athens

14. Length to width of either half of the colonnade
15. Length to width of any rectangle determined by consecutive columns
16. Length to width of any small rectangle supporting the pediment
17. Base to leg of pediment
18. Length to altitude of entire building

▶ C. Exercises

19. Study the pattern in the dimensions of this series of golden rectangles: 5 by 8, 8 by 13, 13 by 21. What would be the dimensions of the next three golden rectangles?
20. If you start with a 55 by 89 golden rectangle and remove a square from one end, what size rectangle will result? Is the resulting rectangle a golden rectangle? What size is the next smaller golden rectangle of this series?

■ Cumulative Review

21. Which two conditions are theorems for proving triangles similar: SSS, ASA, SSA, SAA, SAS, AAA?
22. Three of the other conditions in exercise 21 guarantee triangle similarity. Which three?
23. Why are the three similarity theorems in exercise 22 not needed?
24. Which of the six conditions does not guarantee similarity?
25. Prove (by counterexample) your answer to exercise 24.

Geometry and Scripture

Art

Artists use similar figures when they paint objects with the same shape but with different sizes to make one appear more distant. In the Bible, some people made figures similar to real creatures.

1. For what purpose was the fashioning of likenesses to living creatures prohibited (Exod. 20:3-5)?

2. How did Aaron break this law (Exod. 32:4)?

At other times, God commanded people to make such figures—although, of course, never to be worshiped.

3. What colors and likenesses did God want on the curtains of His tabernacle?

4. Who carried out this command (Exod. 36:2, 8; 38:21-23)?

5. Hiram (the artist for the temple) made cherubim that were similar. For what three different places did he design cherubim (II Chron. 3:7-14)?

God's restriction on art makes some people think that all art is wrong. God's commands to create images make others think that all art is good. Neither extreme is correct.

God was the first "artist" when He created the universe. He made art for good, but it can be misused. God required artistic figures (similar to real creatures) for beauty but prohibits idolizing the art.

For instance, God commanded Moses to make a brazen (bronze) statue in the likeness of a serpent (Num. 21:8-9). Later men worshiped it, and it had to be destroyed (II Kings 18:4).

> **Higher Plane:** Find the verse in John in which Jesus taught the similarity between eternal salvation by trusting His death on the cross and the physical healing found in looking on the brazen serpent.

Read Genesis 1:26-31, which teaches us about similarity, art, math, and God. God, as the first designer, made man *similar* to Himself.

6. Give the phrases in the passage that refer to this.

God made us like Himself with the ability to reason. Art, language, and math all involve reasoning since they permit us to express ideas and experiences. All three skills are essential to architecture, physics, aviation, chemistry, and medicine. Such skills help man exercise dominion over the earth as God commanded (vv. 26, 28).

7. What does *dominion* mean?

Identify the reason from Genesis 1:26-31 that shows why we should learn math. Use each once.

8. Math was created good.

9. Math reflects God's creativity and reasoning.

10. Math is a tool for fulfilling God-given responsibility.

11. Similar figures were part of God's design.

A. "let them have dominion" (v. 26)

B. "it was very good" (v. 31)

C. "Let us make man . . . after our likeness" (v. 26)

D. "God created man in his own image" (v. 27)

Art, language, and even math also require creativity. As you learn new theorems and formulas, you creatively apply them to problems you've never seen before. It also takes great creativity to design new technology like Archimedes did, prove new theorems like Euclid and Heron did, or open new fields of knowledge like Hilbert and Riemann did.

Line upon Line

AND GOD SAID, Let us make man in our image, after our likeness: and let them have dominion over the fish of the sea, and over the fowl of the air, and over the cattle, and over all the earth, and over every creeping thing that creepeth upon the earth. ❧

GENESIS 1:26

Use the diagram and perform the indicated dilation, given the scale factor. Use *O* as the center of dilation.

1. $\frac{1}{2}$

2. 4

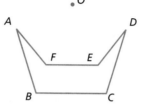

In a given dilation, *O* is the center, *A* is mapped onto *A'*, *B* is mapped onto *B'*, and *C* is mapped onto *C'*. Find the indicated measures using a factor of $\frac{1}{4}$.

3. If *AO* = 16 units, find *A'O*

4. If *AB* = 12 units, find *A'B'*

5. If *OC* = 60 units, find *OC'*

Solve each proportion.

6. $\frac{x}{9} = \frac{18}{54}$

7. $\frac{15}{9} = \frac{x}{2}$

8. $\frac{1}{9} = \frac{12}{x}$

State whether the following pairs are similar figures. Explain why they are similar or why they are not similar.

9.

10.

11.

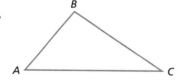

12. *Given:* △*AED* ~ △*CEB*

   Prove: *ABCD* is a trapezoid

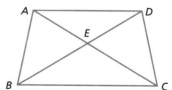

13. *Given:* $\triangle ABF \sim \triangle EBC$;
   $\overline{GB} \perp \overline{AF}$; $\overline{BD} \perp \overline{CE}$

   Prove: $\dfrac{FB}{CB} = \dfrac{GB}{DB}$

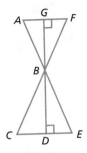

14. If $\triangle ABC \sim \triangle XYZ$, then what are the lengths of the missing sides?

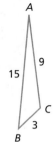

Can the lengths (in units) given here be possible side lengths of similar triangles?

15. 4, 9, 12 and 16, 36, 48
16. 2, 5, 6 and 1, $\frac{5}{2}$, 3
17. 8, 9, 11 and 4, 6, $\frac{11}{2}$
18. State three ways to prove that two triangles are similar.

If $\overline{CX}$ is the altitude to the hypotenuse of right $\triangle ABC$, find the lengths indicated.
19. $AX = 12$ units; $XB = 4$ units; find CX
20. $AX = 27$ units; $XB = 3$ units; find CX
21. $AX = 8$ units; $AB = 10$ units; find AC

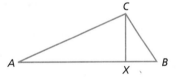

$\triangle HIJ \sim \triangle ZPQ$, $\overline{JK}$ and $\overline{QR}$ are altitudes, $\overline{HL}$ and $\overline{ZS}$ are medians, and $\overline{IM}$ and $\overline{PT}$ are angle bisectors. Find the indicated lengths.

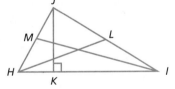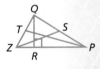

22. $JI = 8$ units; $QP = 4$ units; $JK = 6$ units; find QR
23. $HJ = 15$ units; $ZQ = 3$ units; $HL = 20$ units; find ZS
24. $HI = 25$ units; $ZP = 15$ units; $MI = 20$ units; find TP
25. Perimeter of $\triangle HIJ = 26$ units; $JK = 10$ units; $QR = 5$ units; find the perimeter of $\triangle ZPQ$
26. $HL = 18$ units; $ZS = 8$ units; Area $\triangle HIJ = 42$ square units; find Area $\triangle ZPQ$
27. If a telephone pole casts a 110-foot shadow at the same time that a 4-foot pole casts a 10-foot shadow, how tall is the telephone pole?
28. Explain what the Golden Ratio and the golden rectangle are.
29. Name three occupations that use the idea of similar figures.
30. Explain the mathematical significance of Genesis 1:26.

14 Trigonometry

This fire tower stands on Rich Mountain in Pisgah National Forest near Hot Springs, North Carolina. During daylight hours in the summer, men stay in such towers looking for fires. If a watcher spots one, he finds its bearing and estimates its distance.

The person who spots the fire does not try to put it out. Instead, he stays at his post and calls the firefighters. How will they know where to go? One way to locate the fire is to find the Rich Mountain fire tower on their map and follow the bearing reported. They know that the fire is on that ray. However, they have only an estimate of the distance.

They can be more exact if they can get another fire tower to confirm the sighting and report its bearing from there. The bearings enable them to find the two angles at the towers, and they can find the distance between the towers on the map. By ASA, the triangle formed by the two towers and the fire is determined. You already know how to find the other angle of this triangle, but how can you find the distance of the fire from the towers? Finding the other sides of this triangle requires trigonometry.

After this chapter you should be able to

1. define six trigonometric ratios.

2. evaluate trigonometric ratios.

3. state and apply exact trigonometric ratios for 30°, 45°, and 60° angles.

4. use tables or calculators to approximate trigonometric ratios.

5. solve right triangles.

6. prove trigonometric identities.

7. apply trigonometry to surveying and other word problems.

8. derive and apply formulas for regular polygons using trigonometry.

14.1 Trigonometric Ratios

A ratio is a comparison of two numbers. When you think of a comparison, you probably think of the similarities and differences between two things. In our spiritual lives we are to make sure that our spiritual conditions measure up to the standards of the Bible. First Corinthians 2:13 states, "Which things also we speak, not in the words which man's wisdom teacheth, but which the Holy Ghost teacheth; comparing spiritual things with spiritual." Make sure that you compare your spiritual condition to the truths found in the Bible and not to the condition of your friends or other people. Keep your eyes on Jesus Christ.

In Chapter 13, while studying similar figures and triangles, you learned that there are many ratios and proportions associated with these figures. In this chapter we will study some special ratios of right triangles. These ratios are called *trigonometric ratios*. Trigonometry is the name given to the field of mathematics that deals with triangles, their angles, and the ratios of their side measurements. The word *trigonometry* is derived from the Greek words *trigonon* and *metrica* and means "triangle measure." So trigonometry is the study of triangle measurement.

Look at $\triangle ABC$ and $\triangle XYZ$.

Since $\triangle ABC$ and $\triangle XYZ$ are similar, the sides are proportional.

$$\frac{XY}{AB} = \frac{YZ}{BC} = \frac{XZ}{AC}$$

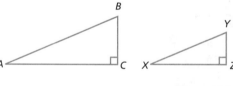

Figure 14.1

From this you can easily obtain the following proportions.

$$\frac{BC}{AB} = \frac{YZ}{XY}, \frac{AC}{AB} = \frac{XZ}{XY}, \frac{BC}{AC} = \frac{YZ}{XZ}$$

As an example, to derive the first proportion, cross multiply and then divide both sides by $(XY)(AB)$.

$$\frac{XY}{AB} = \frac{YZ}{BC}$$
$$(BC)(XY) = (AB)(YZ)$$
$$\frac{BC}{AB} = \frac{YZ}{XY}$$

The proportions above show that for similar triangles, the ratio of any two sides of one triangle is the same as the ratio of the corresponding sides of the other triangle. In other words, the ratio will be the same regardless of the size of the triangles. This is important as the basis of the entire field of trigonometry.

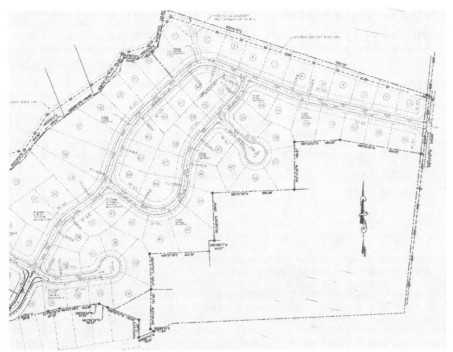

Gray Engineering used trigonometry to prepare this layout for a subdivision in South Carolina.

Notice that if the corresponding acute angles of the two right triangles are congruent, the triangles will always be similar by AA. Therefore, the measures of the acute angles determine the right triangle and thus the ratios of the triangle. If you can find the ratio of particular sides of a right triangle that is determined by a given acute angle, then the ratio remains constant for all triangles determined by this acute angle. Since the ratios remain constant, mathematicians have given them names.

In $\triangle ABC$ the side opposite the right angle is called the hypotenuse. The sides are labeled with lowercase letters corresponding to their opposite angles. The side opposite $\angle A$ is a.

Figure 14.2

The ratio of the side opposite $\angle A$ to the hypotenuse is $\frac{a}{c}$. This ratio is called the *sine* of $\angle A$. The sine of $\angle A$ is abbreviated sin A. So sin A = opposite side over hypotenuse = $\frac{a}{c}$.

Another common ratio is the *cosine* ratio. The cosine of an angle is the ratio of the adjacent side to the hypotenuse. The cosine of $\angle A$ is $\frac{b}{c}$. In trigonometry this relationship is symbolized by cos A = adjacent side over hypotenuse = $\frac{b}{c}$.

The last ratio that we will look at in this brief introduction to trigonometry is the *tangent* ratio. The tangent of $\angle A$, abbreviated tan A, is the ratio of the opposite side to the adjacent side. These three basic trigonometric ratios are summarized in the following table. You should remember that the lowercase letters represent lengths of the sides of the triangle, so the ratios are always numbers.

Trigonometric Ratios			
Name	Abbreviation	Meaning	Side ratio
sine of $\angle A$	sin A	opposite over hypotenuse	$\frac{a}{c}$
cosine of $\angle A$	cos A	adjacent over hypotenuse	$\frac{b}{c}$
tangent of $\angle A$	tan A	opposite over adjacent	$\frac{a}{b}$

Examples are given below to help you find these special ratios.

EXAMPLE 1   Find the sine of $\angle A$ and the sine of $\angle B$.

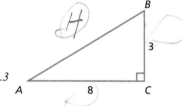

Figure 14.3

Answer

$a^2 + b^2 = c^2$

$3^2 + 8^2 = c^2$

$9 + 64 = c^2$

$73 = c^2$

$\sqrt{73} = c$

1. Find the length of the hypotenuse by applying the Pythagorean theorem.

$\sin A = \dfrac{\text{opposite}}{\text{hypotenuse}}$

$= \dfrac{3}{\sqrt{73}}$

$= \dfrac{3}{\sqrt{73}} \cdot \dfrac{\sqrt{73}}{\sqrt{73}}$

$= \dfrac{3\sqrt{73}}{73}$

$\sin B = \dfrac{\text{opposite}}{\text{hypotenuse}}$

$= \dfrac{8}{\sqrt{73}}$

$= \dfrac{8}{\sqrt{73}} \cdot \dfrac{\sqrt{73}}{\sqrt{73}}$

$= \dfrac{8\sqrt{73}}{73}$

2. Find the required ratios, using the lengths of the sides of the triangle. Remember from algebra that you should not leave a radical in the denominator of a fraction. Rationalize the denominator.

EXAMPLE 2   Find the tangent of $\angle A$.

Answer $\tan A = \dfrac{\text{opposite}}{\text{adjacent}}$

$= \dfrac{a}{b}$

$= \dfrac{4}{5}$

Figure 14.4

▶ A. Exercises

Find the indicated trigonometric ratio for each triangle. Apply the Pythagorean theorem if needed.

1. $\cos A$

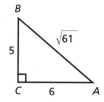

4. $\tan B$

2. $\tan A$

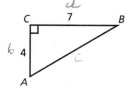

5. $\sin B$

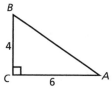

3. $\sin A$

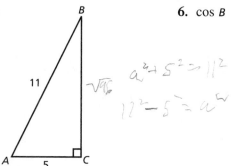

6. $\cos B$

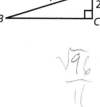

Sketch a right triangle that has the given trigonometric ratio.

7. $\cos A = \frac{2}{5}$

8. $\tan A = \frac{4}{3}$

9. $\sin B = \frac{1}{2}$

10. $\tan B = 6$

▶ B. Exercises

Find the sine, cosine, and tangent of each acute angle of the right triangles below.

11.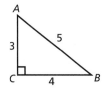

13.

12.

14.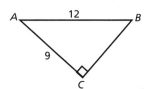

In each problem below, a trigonometric ratio for a certain angle in right $\triangle ABC$ is given. Find the other two trigonometric ratios for the angle.

15. $\cos A = \frac{1}{3}$

16. $\tan B = \frac{4}{5}$

17. $\sin A = \frac{2}{3}$

18. $\cos A = \frac{5}{11}$

Find the indicated trigonometric ratios. (*Hint:* Draw a picture of a triangle that would correspond to the given ratio in each problem.)

19. If $\sin A = \frac{3}{4}$, find $\cos A$.

20. If $\tan B = \frac{4}{5}$, find $\sin B$.

21. If $\sin A = \frac{2}{5}$, find $\tan B$.

22. If $\cos B = \frac{7}{9}$, find $\cos A$.

▶ C. Exercises

23. Use the Pythagorean theorem to prove $\sin^2 A + \cos^2 A = 1$ in right $\triangle ABC$, where $\angle C$ is the right angle.

▶ Dominion Thru Math

If two fire towers give bearings on the same fire, the fire fighters can locate the fire. A bearing is the number of degrees in the arc of a circle measured clockwise from due north. For example, 90° is east. Find all the angles in each triangle.

24. Two towers at the same latitude are 5 miles apart. The bearing to the fire from tower 1 is 135° and from tower 2 is 210°.

25. Two towers are 6 miles apart. The bearing from tower 1 to tower 2 is 65°, and the bearing from tower 1 to the fire is 110°. The bearing from tower 2 to the fire is 145°.

■ Cumulative Review

26. Find x if $\triangle ABC \sim \triangle DEF$.

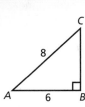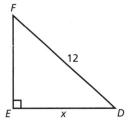

Use the diagrams below for exercises 27-29. Angles A and B have the same measure, and $\angle C$ and $\angle D$ are right angles.

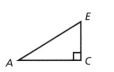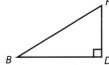

27. If $m\angle A = m\angle B$, prove that $\triangle ACE \sim \triangle BDF$.
28. Prove $\sin A = \sin B$.
29. Using exercise 27, show that $\cos A = \cos B$ and $\tan A = \tan B$. Write a proportion; do not write a proof.
30. Why do you always get the same number for the trigonometric ratios no matter how far you extend the angle?

14.2 Special Triangles

Two special triangles are used frequently in trigonometry. These two triangles are singled out as "distinct among the set of triangles." Therefore, *special* triangles are both "distinctive" and "useful." Christians are special in the eyes of God. They should also be useful in the hands of God. If you are a Christian, you should be willing to be used as God directs you. If you trust the Lord and let Him lead you, He will give you wisdom and understanding (Prov. 3:1-13).

All the property lines in this subdivision had to be surveyed and mapped before construction could begin.

The first special triangle is the 45-45 right triangle. Since the sides opposite congruent angles in a triangle are congruent, this triangle is an isosceles triangle.

Suppose that $CB = 1$ unit; then $AC = 1$ unit. To find AB, use the Pythagorean theorem.

$$(AC)^2 + (CB)^2 = (AB)^2$$
$$1^2 + 1^2 = (AB)^2$$
$$2 = (AB)^2$$
$$\sqrt{2} = AB$$

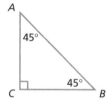

Figure 14.5

If $CB = 2$ units, then $AC = 2$ units.

$$(AC)^2 + (CB)^2 = (AB)^2$$
$$2^2 + 2^2 = (AB)^2$$
$$8 = (AB)^2$$
$$2\sqrt{2} = AB$$

Do you see the pattern? Repeat the process with x.

If $CB = x$, then $AC = x$ units

$$(AC)^2 + (CB)^2 = (AB)^2$$
$$x^2 + x^2 = (AB)^2$$
$$2x^2 = (AB)^2$$
$$x\sqrt{2} = AB$$

This proves the following theorem.

Theorem 14.1

If the length of a leg of an isosceles right triangle (45-45 right triangle) is x, then the length of the hypotenuse is $x\sqrt{2}$.

From this theorem you can find the trigonometric ratios and the lengths of all the sides of a 45-45 right triangle if you are given one side of the triangle. Notice that the three sides are always in the ratio of $1:1:\sqrt{2}$.

EXAMPLE 1   Find the three trigonometric ratios for a 45° angle and the lengths of the sides of the triangle.

Answer $b = 1$ $a = 1$ $c = \sqrt{2}$ **1.** Use Theorem 14.1 to find each side length. See Figure 14.6.

$$\sin 45° = \frac{\text{opposite}}{\text{hypotenuse}}$$

$$= \frac{1}{\sqrt{2}} = \frac{\sqrt{2}}{2}$$

2. Find the trigonometric ratios of the 45° angle.

$$\cos 45° = \frac{\text{adjacent}}{\text{hypotenuse}}$$

$$= \frac{1}{\sqrt{2}} = \frac{\sqrt{2}}{2}$$

$$\tan 45° = \frac{\text{opposite}}{\text{adjacent}}$$

$$= \frac{1}{1} = 1$$

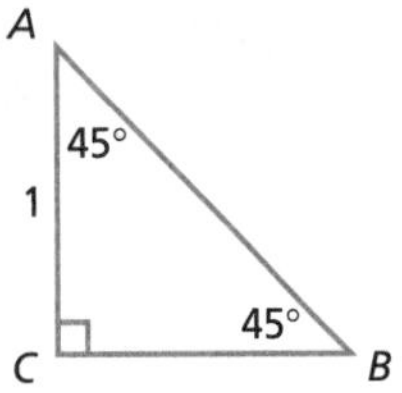

Figure 14.6

Consider the equilateral triangle in figure 14.7. Since an equilateral triangle is equiangular, all of its angles measure 60°. The altitude bisects the vertex angle into two angles that measure 30° each. Thus the altitude divides the equilateral triangle into two 30-60 right triangles. The 30-60 right-triangle is the other special triangle.

Thus, $\triangle ABC$ is a 30-60 right triangle. If BC is 1 unit, then DB is 2 units because the altitude of an equilateral triangle intersects the opposite side at the midpoint. Since $\triangle ABD$ is an equilateral triangle, AB is 2 units also. Now that we know $BC = 1$ unit and $AB = 2$ units, we can find AC by using the Pythagorean theorem.

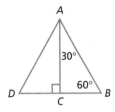

Figure 14.7

$$(BC)^2 + (AC)^2 = (AB)^2$$
$$1^2 + (AC)^2 = 2^2$$
$$1 + (AC)^2 = 4$$
$$(AC)^2 = 3$$
$$AC = \sqrt{3} \text{ units}$$

Once you know the length of each side of a right triangle, you can find the trigonometric ratios for the angles. Now we will derive the pattern for the 30-60 right triangle as a theorem.

If $BC = x$ units, then $DB = 2x$ units and $AB = 2x$ units. By the Pythagorean theorem,

$$(BC)^2 + (AC)^2 = (AB)^2$$
$$x^2 + (AC)^2 = (2x)^2$$
$$(AC)^2 = 4x^2 - x^2$$
$$(AC)^2 = 3x^2$$
$$AC = x\sqrt{3} \text{ units}$$

Theorem 14.2 summarizes these findings about the 30-60 right triangle.

Theorem 14.2

If the length of the leg opposite the 30° angle of a 30-60 right triangle is x, then the length of the leg opposite the 60° angle is $x\sqrt{3}$, and the length of the hypotenuse is $2x$.

Notice the sides are always in the ratio $1:\sqrt{3}:2$.

EXAMPLE 2   Find the side lengths and the trigono-metric ratios for the acute angles of the following 30-60 right triangle.

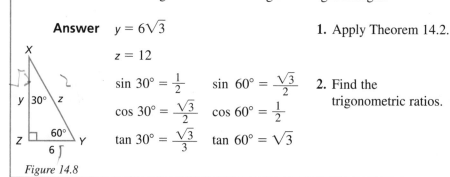

Answer $y = 6\sqrt{3}$ 1. Apply Theorem 14.2.

$z = 12$

$\sin 30° = \frac{1}{2}$ $\sin 60° = \frac{\sqrt{3}}{2}$ 2. Find the trigonometric ratios.

$\cos 30° = \frac{\sqrt{3}}{2}$ $\cos 60° = \frac{1}{2}$

$\tan 30° = \frac{\sqrt{3}}{3}$ $\tan 60° = \sqrt{3}$

Figure 14.8

▶ A. Exercises

For each of the given measures, find the lengths of the sides of the triangles.

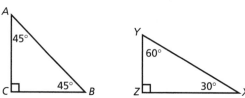

	AB	AC	BC
1.			18 units
2.		9 units	
3.			12 units
4.	5 units		
5.	4 units		

	XY	XZ	YZ
6.			5 units
7.	12 units		
8.		24 units	
9.			2 units
10.		15 units	

▶ B. Exercises

Given the following side lengths, determine which of these triangles are special and give the name of each special right triangle.

11. $6, 6, 6\sqrt{2}$

12. $1, 1, \sqrt{2}$

13. $7, 7\sqrt{3}, 14$

14. $8, 8\sqrt{2}, 16$

15. $12, 12, 12\sqrt{3}$

16. $4, 4\sqrt{3}, 8$

Give the following trigonometric ratios. Draw pictures if necessary.

17. $\sin 45°$

18. $\cos 30°$

19. $\tan 60°$

20. $\tan 45°$

21. $\cos 60°$

22. $\sin 30°$

▶ C. Exercises

23. For acute $\triangle ABC$, show that Area $\triangle ABC = \frac{1}{2}bc \sin A$. (*Hint:* Let $\overline{CD}$ be the altitude from vertex C and use the resulting right triangles.)

■ Cumulative Review

Give the dimensions of the figure with the following areas.

24. A square with area of 20 square units

25. An equilateral triangle with an area of $100\sqrt{3}$ square units

26. A regular hexagon with an area of $50\sqrt{3}$ square units

27. A circle with an area of 5π square units

28. A cube with a surface area of $\frac{8}{3}$ square units

SURVEYING

While driving by a highway construction site, most of us have seen surveyors at work. One man holds an orange pole upright while his partner peers through an object resembling a telescope and waves his arms back and forth. Although we can describe their actions, we might have trouble explaining what they are doing. But surveyors would explain that they are measuring plots of land using trigonometry.

Engineers often need to know the topography of an area of land to design roads, buildings, ditches, and dikes. Surveyors can determine vertical changes in the property by standing a calibrated pole on the ground and observing it through the telescope of a level instrument. The surveyor uses a standard elevation as a reference point for the measurements so that he can draw a contour (topographic) map. Also called a benchmark, the standard elevation was established by the United States Geological Survey (USGS) and is relative to

The surveyor uses a standard elevation as a reference point for the measurements so that he can draw a contour map.

Surveyors sight on ruled rods to determine vertical distance above a base point.

mean sea level. The surveyor shows the elevations on a drawing by labeling a contour line with a number, thus indicating that any point on that line is at a certain elevation above or below the mean sea level.

All of the land in the United States has been mapped on topographic maps. The maps are so detailed that mapping North Carolina requires over 900 maps. These maps were made with steel tape and transits. The tapes (called chains) were used to measure horizontal distance. When steep slopes prohibited the surveyor from holding the tape horizontally, he used trigonometry. By measuring the distance along the sloped ground and using the transit to find the angle (A) of the incline, he could use the cosine to find the horizontal distance:

$$\cos A = \frac{\text{horizontal distance}}{\text{distance along the slope}}$$

Surveyors once measured angles by turning the telescope on their transit from one side of the angle to the other. They then read the angle measure from the base. Surveyors call this method "turning an angle."

Modern technology has greatly improved the methods of measuring distances and turning angles. The transit and steel tape as well as the compass and data notebook have all been combined into one instrument called a *total station*. An electronic distance meter sends a laser beam to a reflector at the other end of the distance to be measured. This beam reflects back to the meter, which converts the time of travel into decimal units of length and stores the information electronically. Measuring angles is also done electronically. The surveyor later downloads the data into a computer at his office, which performs the calculations and drawings.

The USGS topographic map of Mt. Whitney, California, uses contours representing 20 meters (65.62 ft.).

Electronic total station for surveying

14.3 Solving Right Triangles

In the last section you used a method for finding trigonometric ratios for the more common angle measures, such as 30, 45, and 60. But what do you do when you need the trigonometric ratio for angles other than these special angles? You could draw a triangle that has an acute angle congruent to the angle you are interested in. By measuring the sides and finding the appropriate ratios, you could find the trigonometric ratios for any acute angle. But this method is both time consuming and imprecise.

The height of a lighthouse, such as this one at Portland, Maine, is known to the keeper. By sighting a ship and measuring the angle below the horizontal, the keeper can solve the triangle to determine the distance of a distressed ship from the shore.

Georg Rhaeticus (1514-76), a German mathematical astronomer, spent twelve years of his life developing two very accurate trigonometric tables, which were published after his death in 1613. These tables made computations using trigonometric ratios consistent. These tables also saved much time for later mathematicians. Rhaeticus was the first mathematician to relate trigonometric ratios to the sides of a right triangle.

Calculators have made finding trigonometric ratios even easier. You may use a calculator or the trigonometric table in the back of this book to solve triangles. However, you must learn to recognize the special angles and ratios. Whenever a special angle is involved, the exact answer is expected. Calculators and trigonometric tables provide only approximate answers, and an approximate answer will not be accepted for the special angles.

Solving a right triangle means finding all the angle measures and all the side lengths of the triangle from the information given. By using trigonometric ratios, the Pythagorean theorem, and the sum of the angles of a triangle, you should be able to solve any right triangle. If you take a more advanced class in trigonometry, you will study methods for solving acute and obtuse triangles. Study the following examples.

EXAMPLE 1   Given right $\triangle ABC$, find the measures of each side and angle.

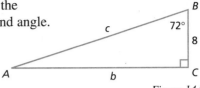

Figure 14.9

Answer $\tan 72° = \frac{b}{8}$

1. Since the tangent is the ratio of the opposite side to the adjacent side, write an equation for *b*.

$3.0777 \approx \frac{b}{8}$
$(3.0777)(8) \approx b$
$25 \approx b$

2. Find $\tan 72° = 3.0777$ using a calculator or the table on page 616. Solve for *b*.

$m\angle A = 90° - m\angle B$
$\quad\quad = 90° - 72° = 18°$

3. $\angle A$ must be the complement of $\angle B$ since this is a right triangle.

$a^2 + b^2 = c^2 \quad \cos 72° = \frac{8}{c}$
$8^2 + (25)^2 = c^2 \quad 0.3090 \approx \frac{8}{c}$
$64 + 625 = c^2 \quad 0.3090c \approx 8$
$689 = c^2 \quad\quad c \approx \frac{8}{0.3090}$
$26 \approx c \quad\quad\quad c \approx 26$

4. Find *c* by using either the Pythagorean theorem or another trigonometric ratio.

The complete solution to this triangle follows.

$m\angle A = 18°$	$a = 8$ units
$m\angle B = 72°$	$b \approx 25$ units
$m\angle C = 90°$	$c \approx 26$ units

Often there is more than one way to solve a right triangle. Use the method that seems easiest for you.

EXAMPLE 2 Solve right △*ABC* if ∠*C* is
the right angle, *m*∠*A* = 38°,
and *c* = 26 units.

Answer Draw a picture of △*ABC*.
A picture will often help
you figure out the rest of
the triangle.

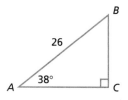

Figure 14.10

$$\cos 38° = \frac{b}{26}$$

1. To find *b*, use the
   cosine ratio.

$$0.7880 \approx \frac{b}{26}$$
$$20.5 \approx b \text{ (nearest tenth)}$$

2. Find cos 38°.

$$\sin 38 = \frac{a}{26}$$
$$0.6157 \approx \frac{a}{26}$$
$$16.0 \approx a$$

3. To find *a*, use the sine
   ratio.

$$m\angle B = 90° - 38° = 52°$$

4. To find *m*∠*B*, find the
   complement of ∠*A*.

EXAMPLE 3 Solve right △*ABC*.

Answer We know that *m*∠*C* = 90°.

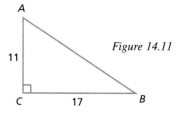

Figure 14.11

$$\tan A = \frac{17}{11}$$

1. To find *m*∠*A*, use
   the tangent ratio.

$$\tan A \approx 1.5455$$
$$m\angle A \approx 57°$$

2. Find the closest
   number to 1.5455
   in the tangent
   column in the
   table. Identify
   the angle.

$$m\angle B = 90° - m\angle A \quad \tan B = \frac{11}{17}$$
$$m\angle B \approx 90° - 57° \quad \tan B \approx 0.6471$$
$$m\angle B \approx 33° \quad m\angle B \approx 33°$$

3. To find *m*∠*B*, find
   the complement of
   ∠*A* or use the
   tangent ratio again.

Continued ▶

$$a^2 + b^2 = c^2$$
$$11^2 + 17^2 = c^2$$
$$410 = c^2$$
$$20 \approx c$$

$$\sin 57° = \frac{17}{c}$$
$$0.8387 \approx \frac{17}{c}$$
$$c \approx \frac{17}{0.8387}$$
$$c \approx 20$$

4. To find c, use the Pythagorean theorem or a trigonometric ratio.

▶ A. Exercises

Find the indicated trigonometric ratios. See the table on page 616.

1. sin 41°
2. cos 12°
3. tan 82°
4. tan 16°
5. cos 59°

Find $m\angle A$, given the following trigonometric ratios. See the table on page 616. Find the angle to the nearest degree.

6. tan A = 9.514
7. cos A = 0.8746
8. sin A = 0.9925
9. sin A = 0.2079
10. tan A = 4.000

▶ B. Exercises

Use the triangles shown. Name the ratio or theorem that you would use to find the indicated measurement and then calculate it.

11. *DF*
12. *GH*
13. *EF*
14. *m∠G*
15. *m∠E*

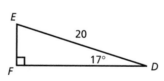

Solve each right triangle. Round your answers to the nearest tenth or to the nearest degree.

16.

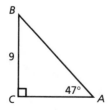

17.

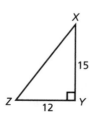

18.

19.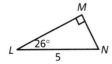

20.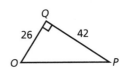

Solve right △ABC if ∠C is the right angle. Round your answers to the nearest tenth or to the nearest degree.

21. $a = 15$ units; $c = 34$ units

22. $m\angle B = 82°$; $a = 34$ units

23. $m\angle A = 47°$; $b = 18$ units

24. $a = 33$ units; $b = 17$ units

25. $c = 28$ units; $m\angle B = 78°$

▶ C. Exercises

26. Find *PS* using the diagram.

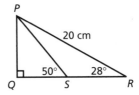

■ Cumulative Review

Which are congruent, similar, or neither? Why?

27.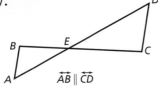

$\overleftrightarrow{AB} \parallel \overleftrightarrow{CD}$

30.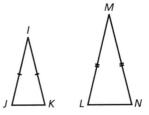

28.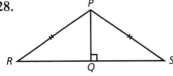

31.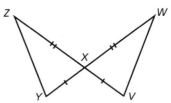

29.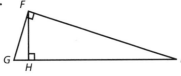

In Chapter 11 you learned that trisecting an angle is an impossible construction. You can prove this using trigonometry. However, trisecting an angle is possible as a drawing. Trisect an angle with a protractor, and then study the mechanical device for trisecting angles called the Conchoid of Nicomedes.

Measurement

A surveyor needs to use a variety of calculation methods because direct measurements may be hindered by trees or cliffs. For distances he may need the distance formula, the Pythagorean theorem, or a trigonometric ratio. For finding areas of plots of land he may need $A = \frac{1}{2}ab \sin C$, or Heron's formula. Using such calculations would be foolish if they had not been proved correct. Analytic geometry provides the simplest means for proving many of the useful formulas. The following example proves a method used by builders to check that a foundation is precisely rectangular.

EXAMPLE Prove that the diagonals of a rectangle are congruent (Theorem 7.28) using analytic geometry.

Answer Locate the rectangle on a Cartesian graph in an appropriate position. Place one vertex at the origin and label the other vertices with arbitrary ordered pairs.

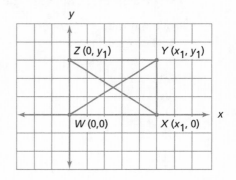

Given: Quadrilateral $WXYZ$ is a rectangle

Prove: $\overline{ZX} \cong \overline{WY}$

STATEMENTS	REASONS
1. Quadrilateral $WXYZ$ is a rectangle	**1.** Given
2. $XZ = \sqrt{(x_1-0)^2 + (0-y_1)^2} = \sqrt{x_1^2 + y_1^2}$ $WY = \sqrt{(x_1-0)^2 + (y_1-0)^2} = \sqrt{x_1^2 + y_1^2}$	**2.** Distance formula applied to $\overline{XZ}$ and $\overline{WY}$
3. $WY = XZ$	**3.** Transitive property of equality
4. $\overline{WY} \cong \overline{XZ}$	**4.** Definition of congruent segments

Use the figure for exercises 1-2.

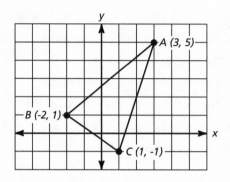

1. Find the perimeter of the triangle.
2. Find the area of the triangle.

Prove the following theorems.
3. The segment joining the midpoints of two sides of a triangle is parallel to the third side.
4. The length of the segment joining the midpoints of two sides of a triangle is one-half the length of the third side.

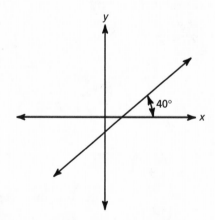

5. Use trigonometry to find the slope of the line.

14.4 Trigonometric Identities

Three *reciprocal ratios* are multiplicative inverses of the sine, cosine, and tangent ratios respectively. The abbreviations for the reciprocal ratios are csc A, meaning *cosecant* of $\angle A$; sec A, meaning *secant* of $\angle A$; and cot A, meaning *cotangent* of $\angle A$. So in all there are six trigonometric ratios, and they are related in pairs. The definitions of the three reciprocal ratios follow.

$$\csc A = \frac{1}{\sin A} \qquad \sec A = \frac{1}{\cos A} \qquad \cot A = \frac{1}{\tan A}$$

The most important relationship between two trigonometric ratios is proved below.

Theorem 14.3
Sum of Squares Identity. For any angle x,
$$\sin^2 x + \cos^2 x = 1.$$

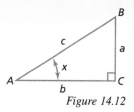

Figure 14.12

Notice the form of notation used for the square of sine x. In trigonometry we write $\sin^2 x$ rather than $(\sin x)^2$ to save writing parentheses. By contrast, $\sin x^2$ means that you square the angle measure before finding the sine.

STATEMENTS	REASONS
1. $m\angle A = x$ and $\triangle ABC$ is right	**1.** Given
2. $\sin x = \frac{a}{c}$	**2.** Definition of sine
3. $\cos x = \frac{b}{c}$	**3.** Definition of cosine
4. $a = c \sin x$ and $b = c \cos x$	**4.** Multiplication property of equality
5. $a^2 + b^2 = c^2$	**5.** Pythagorean theorem
6. $(c \sin x)^2 + (c \cos x)^2 = c^2$	**6.** Substitution (step 4 into 5)
7. $c^2 \sin^2 x + c^2 \cos^2 x = c^2$	**7.** Power of product property
8. $\sin^2 x + \cos^2 x = 1$	**8.** Multiplication property of equality

A trigonometric property that is true for all values of the variable is called a *trigonometric identity*. Like theorems, identities must be proved. When doing so, you must start with one side of the equation and transform it to the other side without working on both sides of the equation. If you work on both sides of the equation, you have assumed the truth of the statement you want to prove. Remember that this is the fallacy called circular argument.

To avoid this error, you must begin with a known truth. Even though you do not have to prove identities in two-column format, you should still have a reason in your mind for each step. One good way to get started is to select one side to work on and then apply the reflexive property to this expression. Study the examples below.

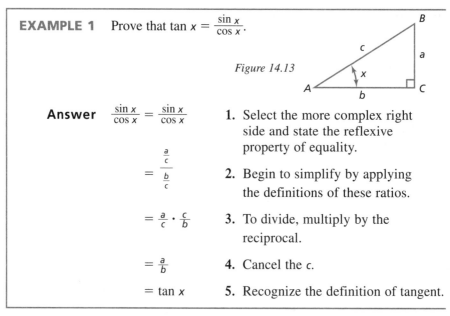

EXAMPLE 1   Prove that $\tan x = \frac{\sin x}{\cos x}$.

Figure 14.13

Answer $\frac{\sin x}{\cos x} = \frac{\sin x}{\cos x}$ 1. Select the more complex right side and state the reflexive property of equality.

$= \frac{\frac{a}{c}}{\frac{b}{c}}$ 2. Begin to simplify by applying the definitions of these ratios.

$= \frac{a}{c} \cdot \frac{c}{b}$ 3. To divide, multiply by the reciprocal.

$= \frac{a}{b}$ 4. Cancel the c.

$= \tan x$ 5. Recognize the definition of tangent.

The identity is now proved. Notice that by the transitive property of equality the top and bottom lines in the proof are equal: $\frac{\sin x}{\cos x} = \tan x$.

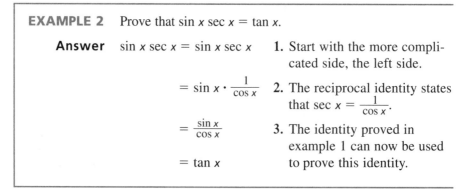

EXAMPLE 2   Prove that $\sin x \sec x = \tan x$.

Answer $\sin x \sec x = \sin x \sec x$ 1. Start with the more complicated side, the left side.

$= \sin x \cdot \frac{1}{\cos x}$ 2. The reciprocal identity states that $\sec x = \frac{1}{\cos x}$.

$= \frac{\sin x}{\cos x}$ 3. The identity proved in example 1 can now be used

$= \tan x$ to prove this identity.

God's intricate order provides many interesting relationships among these ratios. The following example shows how to use the Sum of Squares Identity (Theorem 14.3) in proofs. By subtracting from both sides, $\sin^2 x + \cos^2 x = 1$ has two other forms: $\sin^2 x = 1 - \cos^2 x$ and $\cos^2 x = 1 - \sin^2 x$.

EXAMPLE 3   Prove that $(1 + \cos x)(1 - \cos x) = \sin^2 x$.

Answer   Begin with the left side. Simplify using the FOIL method. The last step uses the Sum of Squares Identity.

$$(1 + \cos x)(1 - \cos x) = (1 + \cos x)(1 - \cos x)$$

$$= 1 + \cos x - \cos x - \cos^2 x$$

$$= 1 - \cos^2 x$$

$$= \sin^2 x$$

Notice that this identity could also have been proved by substituting $1 - \cos^2 x$ for $\sin^2 x$ on the right side and factoring the difference of two squares.

Finally, remember that $\sec^2 x$ means $(\sec x)^2$. This enables you to substitute reciprocal ratios even when a quantity is squared. For instance, $\sec^2 x = \frac{1}{\cos^2 x}$ since $\sec x = \frac{1}{\cos x}$.

▶ **A. Exercises**

Prove each identity.

1. $\tan x = \frac{\sec x}{\csc x}$

2. $1 + \tan^2 x = \sec^2 x$

3. $\cot x = \frac{\cos x}{\sin x}$

4. $\sec x \cot x = \csc x$

5. $\cos x \csc x = \cot x$

▶ **B. Exercises**

6. $\csc x \cos^2 x + \sin x = \csc x$
7. $\cot^2 x + 1 = \csc^2 x$
8. $\csc^2 x (1 - \cos^2 x) = 1$
9. $\cot x + \tan x = \csc x \sec x$
10. $\cos^2 x(1 + \tan^2 x) = 1$
11. $\sin x \sec x \cot x = 1$
12. $(\sin x + \cos x)^2 = 1 + 2\sin x \cos x$

13. $1 - \sin x \cos x \tan x = \cos^2 x$
14. $\sec x - \sin x \tan x = \cos x$
15. $\csc x - \cos x \cot x = \sin x$
16. $\csc x - \sin x = \cos x \cot x$
17. $\sec^2 x (1 - \sin^2 x) = 1$
18. $(\cos x + \sin x)^2 +$ $(\cos x - \sin x)^2 = 2$

▶ **C. Exercises**

19. $\frac{1 + \tan x}{1 + \cot x} = \tan x$

20. $\sin x \cos x - \sec x \sin x = -\sin^2 x \tan x$

Match to each description the letter of the most specific term that applies.

21. Any composite of two reflections across concurrent lines

22. Any mapping with a scale factor greater than one

23. Any one-to-one correspondence from the plane onto the plane

24. Any mapping that preserves distance

25. Any mapping that preserves shape

A. translation

B. enlargement

C. isometry

D. reduction

E. reflection

F. rotation

G. similarity

H. transformation

14.5 Applications of Trigonometry

Builders translate mathematically made blueprints into architecture such as the state capitol at Providence, Rhode Island.

One definition of *apply* is "to put into practice." It is good to know and understand different areas of study, but if you do not put your knowledge into practice, it does not benefit you. It is important, then, that you not only learn material but that you also apply it. Likewise, a Christian needs to search the Word of God and apply its principles to his life. "Be ye doers of the word and not hearers only" (James 1:22). Do you apply the principles you learn from math and from Scripture?

In this lesson you will see some simple applications of trigonometry. There are many other advanced applications of trigonometry in surveying work, engineering, navigation, and astronomy. But in this lesson we will concentrate on problems such as finding the height of a mountain without climbing it or the distance across a deep river. You must understand the next two terms before you can solve the problems.

Definitions

The **angle of elevation** is the angle formed by a horizontal line and the line of sight toward an object that is above the horizontal. The measure of this angle is the *inclination*.

The **angle of depression** is the angle formed by a horizontal line and the line of sight toward an object that is below the horizontal. The measure of this angle is the *declination*.

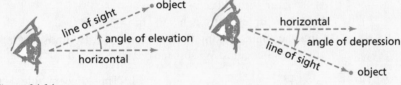

Figure 14.14

EXAMPLE 1 The KVLY-TV antenna near Blanchard, North Dakota, is the tallest structure in the world. Standing a distance of 1006 feet from the base of the antenna, you find the measure of the angle of elevation to the top of the antenna to be 64°. Find the height of the antenna.

Answer

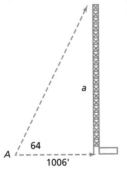

Figure 14.15

1. Draw a diagram and label the unknown height a.

$\tan 64° = \frac{a}{1006}$

$2.0503 \approx \frac{a}{1006}$

2063 ft. $\approx a$

2. Use a trigonometric ratio to write an equation involving a. Solve it.

There are some general rules to follow when solving these kinds of problems.

1. Read the problem carefully.
2. Plan your solution by sketching the problem, labeling it correctly, and identifying the appropriate trigonometric ratio.
3. Solve an equation using the trigonometric ratio.
4. Check to see that you answered the question and that your answer is reasonable.

EXAMPLE 2 A surveyor must find the distance between two particular points that are on opposite sides of a river. He is unable to measure it directly. He measures a distance of 180 feet from point B to another point A on the same side of the river, directly across the river from C. With his instrument at B, he turns the angle, looking first at A, then at C. He finds that this angle is 43 degrees. What is the distance from B to C?

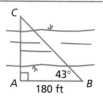

Figure 14.16

Answer $\cos 43° = \frac{180}{a}$

1. The adjacent side of the given angle is known and the hypotenuse to be found. Select the cosine ratio since it involves these two sides.

$$0.7314 \approx \frac{180}{a}$$
$$0.7314a \approx 180$$
$$a \approx \frac{180}{0.7314}$$
$$a \approx 246 \text{ ft.}$$

2. Solve the resulting equation.

▶ A. Exercises

1. A 40-foot guy wire is connected to a telephone pole, forming a 21° angle with the pole. How far from the pole is the guy wire anchored to the ground?

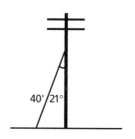

2. An observer from a lighthouse sees a boat that is at an angle of depression of 8°. If the lighthouse is 140 feet above sea level, how far is the boat from the lighthouse?

▶ B. Exercises

Round your answers to the nearest tenth or to the nearest degree.

3. Dan finds the angle of depression from his location at the top of a mountain to a stream below to be 78°. The mountain is 4830 feet above the stream. Find the horizontal distance from Dan to the stream.

4. Scientists can estimate the height of objects on different planets and on the moon by measuring the shadows of the objects. If a crater rim casts a 127-yard shadow when the sun shines on it at a 46° angle of elevation from the moon's surface to the sun, what is the height of the crater rim?

5. Karen has an exotic tropical bird that escaped from its cage. The bird is now on a tree branch 13 feet above the ground. If Karen places a 15-foot ladder against the tree branch, what angle does the ladder make with the ground?

6. An airplane takes off from Harrisburg Airport at an angle of elevation of 4°. If the plane is traveling at 215 miles per hour and maintains a constant speed and rate of climb, how high is the plane after 6 minutes?

7. The slope of a line is defined by the ratio of the vertical change to the horizontal change. If the slope of a line is $\frac{5}{3}$, what angle does the line make with the horizontal?

8. A hot-air balloon rises and moves away from you. The total distance it travels is 1382 feet. The angle of elevation from you to the balloon is 29°. How high is the balloon?

9. Donna needs to find the distance across a pond from point A to point B (see illustration). The distance from A to C is 500 feet, and $m\angle C = 37°$. What is the distance from A to B?

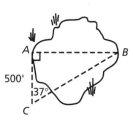

10. A submarine makes a dive after being on the surface of the ocean. The angle of the dive with the surface is 12°. If the submarine levels off at a depth of 348 feet, how far does the submarine travel to get to this depth?

11. Rob is installing a 15-foot TV antenna on the flat roof of an apartment building. The guy wires will be attached eight feet from the base of the antenna. How long should the guy wires be?

12. Happytown is 60 miles due west of Parkville. Welcome is due south of Happytown. Two planes depart from Parkville with an angle of 46° between their flight paths, one for Happytown and one for Welcome. How far is it from Welcome to Happytown?

13. A balloonist floats 800 feet above Barker Park. He observes his house at a declination of 35°. How far is his house from the park?

14. You measure an inclination of 25° to a cloud 700 feet above ground. How far is the cloud from you?

15. Mauna Kea, Hawaii, is 13,796 feet above sea level. A westbound ship first sights the summit at an angle of elevation of 5°. How far from land is the ship?

16. An air survey team is flying at 7000 feet above sea level. They measure a declination to the summit of Mt. Mitchell at $7\frac{1}{2}°$. After flying another 2400 feet, they are directly above the summit. How high is Mt. Mitchell?

17. Ribbon Falls, Colorado, flows 140 feet along a slope inclined 23° above the horizontal. How far does it drop vertically?

18. If Sliding Rock, North Carolina, drops 50 feet in a horizontal distance of 130 feet, how steep is the slide?

19. Chuck found an old cabin far from the trails in the woods. His map shows that he is six miles from Tadmor and five miles south of a pond that is due west of Tadmor. How many degrees east of north should he face on his compass to head for Tadmor?

▶ C. Exercises

20. A city water tower is to be constructed in the shape of a prism with a regular hexagon as base. How long should an edge of the base be if the height is to be 30 feet and the tower must hold 9000 cubic feet of water?

■ Cumulative Review

21. Give the measure of a central angle of a regular nonagon.
22. Give the perimeter of a regular heptagon with a side of length 12.
23. Give the area of a regular triangle with side of length 5.
24. Give the area of a regular hexagon with side of length 5.
25. Give the area of a regular octagon with side of length 6 and apothem of length 4.1.

14.6 Trigonometry and Regular Polygons

Consider the regular pentagon below.

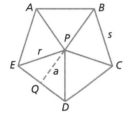

An aerial view of the Pentagon (U.S. military headquarters) reveals the regular polygon.

Figure 14.17

Recall that the 5 central angles, being congruent, measure $\frac{360}{5} = 72°$ each. Since the radii are congruent, the triangles are isosceles. This means that each apothem ($\overline{PQ}$) bisects the central angles into two 36° angles. Also, it is the perpendicular bisector of the side to which it is drawn. This means that we can focus on the right triangle thus formed.

Figure 14.18

Using trigonometry, $\tan 36° = \dfrac{\frac{1}{2}s}{a}$ and $\sin 36° = \dfrac{\frac{1}{2}s}{r}$. If the length of a side is known, $a = \dfrac{s}{2 \tan 36°}$ and $r = \dfrac{s}{2 \sin 36°}$.

EXAMPLE 1 Find the apothem and radius of a regular pentagon with a side of length 12.

Answer $a = \dfrac{12}{2 \tan 36°} = 6 \cot 36° \approx 8.3$

$r = \dfrac{12}{2 \sin 36°} = 6 \csc 36° \approx 10.2$

Remember that to find reciprocal ratios on a calculator, you will probably need to use the definition. For instance, to find cot 36°, use the definition $\cot 36° = \dfrac{1}{\tan 36°}$. Find tan 36° and then press the reciprocal $\left(\dfrac{1}{x}\right)$ button.

EXAMPLE 2   If the radius of a circle is 10 cm, find the length of the side of an inscribed regular pentagon.

Answer $r = \dfrac{s}{2 \sin 36°}$

$10 = \dfrac{s}{2 \sin 36°}$

$s = 20 \sin 36° \approx 11.8$ cm

You should be able to derive similar formulas for other regular polygons.

▶ A. Exercises

Refer to the regular pentagon to find the following.

1. $m\angle CPD$ and $m\angle CPM$

2. $m\angle PCM$

3. PM and PD

4. area of the pentagon

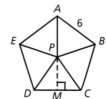

▶ B. Exercises

Using a regular decagon of side s, find
5. the measure of a central angle and its bisected angle.
6. the apothem in terms of the side.
7. the radius in terms of the side.
8. the apothem and radius if the side is 10 mm.
9. the side if the circumscribed circle has a radius of 12 m.
10. the area if the side is 7 feet.

Use a regular octagon of side s. Find formulas for the
11. apothem and radius.
12. area.
13. If the apothem is 7 in., how long is the side?

Part of a regular n-gon with side s units long is shown.
14. Express AM, $m\angle APM$, and the
perimeter in terms of n and s.

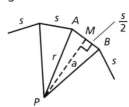

▶ C. Exercises

Find a formula for each part of the *n*-gon.

15. apothem

16. radius

17. area

18. Prove your formula for the apothem.

Find a formula for the volume of the solid with height *H* and a regular *n*-gon of side *s* as a base.

19. prism   **20.** pyramid

▶ Dominion Thru Math

Randy works after school for Sports Outlet. His employer wants an entrance display of balls stacked to form a regular tetrahedron. He wants the information based on the radius *r* so that different kinds of balls can be displayed during different seasons.

21. Find a pattern for the number of balls in each layer starting at just 1 ball per side of the base triangle, then 2, 3, 4, 5 and finally *n*.

22. Find the total number of balls beginning with 12 on a side.

23. If *n* balls are on each side of the base triangle, what is the length of each side of the wooden frame that holds the bottom layer in place?

24. Find the height of a pyramid with 12 basketballs on a side. The circumference of a basketball is 30 inches.

■ Cumulative Review

Find the dimensions of each.

25. Cube with a volume of 64 cu. in.

26. Sphere with a volume of 36π cu. ft.

27. Regular octahedron with a volume of $72\sqrt{2}$ cubic inches

Find the volume of each.

28. A rectangular solid has a square base with edges 5 cm long. The lateral faces are golden rectangles.

29. A pyramid is 11 units high and has a regular polygon as its base. The polygon has 20 sides each 4 units long. (*Hint:* Use trigonometry to find the apothem first.)

Geometry and Scripture

Angle Measure

Trigonometry means triangle measure. You know that each trigonometric function is a ratio of lengths that depend on the size of the given angle. Angles are basic to trigonometry. Let's look at angles mentioned in Scripture.

1. What Line upon Line verses contained references to the concept of right angles?

You know that the angle of inclination of the sun enables us to tell time and also to find elevations using shadows.

2. According to Genesis 1, why did God make the sun?

In Chapter 10 we saw that sunrise and sunset are figures of speech that draw on the language of appearance. Give the apparent angle of elevation of the sun in each of these verses.

3. Mark 16:2   5. Judges 19:14
4. Acts 26:13

The day from sunrise to sunset is divided into twelve hours. Give the time of day and determine the approximate angle of elevation of the sun for each event in the Crucifixion.

6. Jesus is crucified (Mark 15:25).

7. God darkens the sun (Luke 23:44).

8. Jesus' death (Luke 23:44-47)

For the Hebrews in the Old Testament, the night was divided into three watches, or guard duties.

9. Judges 7:19, I Samuel 11:11, and Lamentations 2:19 refer to these three watches. Give the names of the watches in order from dusk to dawn.

Romans divided the night into four watches. These watches could be referred to in different ways. Sometimes they were just numbered first watch, second watch, third watch, and fourth watch.

> **HIGHER PLANE:** Find three verses in the Gospels that refer to the night watches by number.

The watches could also be referred to by name: even (dusk until 9 P.M.); midnight (until 12 A.M.); cockcrowing (until 3 A.M.); morning (until dawn).

10. Which of these four watches are mentioned by name in Mark 13:35?

The third, sixth, and ninth hours split the day into four parts based on the angle of inclination of the sun. Shadows could be measured for greater accuracy. Two major types of dials were used: a traditional sundial and a series of steps. Either way, the dial measured an angle: the angle of elevation of the sun.

11. The dial of Ahaz in II Kings 20 probably consisted of steps down which the shadow would fall. How did God use this dial to reverse His created order (v. 11)?

> ### Line upon Line
>
> AND ISAIAH THE PROPHET cried unto the Lord: and he brought the shadow ten degrees backward, by which it had gone down in the dial of Ahaz. 🙠
>
> **II KINGS 20:11**

Chapter 14 Review

Find the three basic trigonometric ratios for each acute angle of each right triangle below.

1.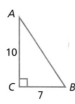

2.

3. Find the other two trigonometric ratios for $\angle A$ if $\cos A = \dfrac{7\sqrt{193}}{193}$.

Given that $\triangle ABC$ is a 45-45 right triangle, and $\triangle MNO$ is a 30-60 right triangle, find the measures of the other two sides.

4. $AC = 9$ units

5. $AB = 12$ units

6. $NO = 7$ units

7. $MN = 26$ units

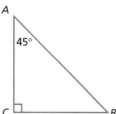
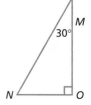

Give the following trigonometric ratios.

8. $\cos 30°$ **10.** $\sin 39°$ **12.** $\sin 11°$
9. $\tan 45°$ **11.** $\cos 86°$ **13.** $\tan 72°$

Find the measure of $\angle A$.

14. $\sin A = 0.7314$ **15.** $\cos A = 0.8828$ **16.** $\tan A = 2.246$

Solve each right triangle.

17.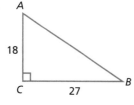

18.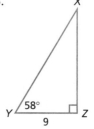

Prove the following trigonometric identities.

19. $\sec x - \cos x = \sin x \tan x$

20. $\sin x (\csc x - \sin x) = \cos^2 x$

21. Meteorologists can find the height of a cloud ceiling by shining a light vertically and then finding the angle of elevation to the spot where the light meets the clouds. Find the height of the cloud ceiling by the information given in the illustration.

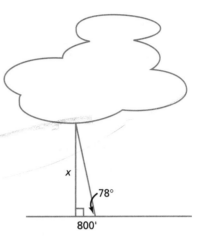

Write the three reciprocal identities.

22. $\sec x$

23. $\cot x$

24. $\csc x$

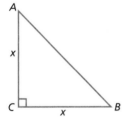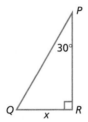

25. Solve the isosceles right triangle shown.

26. Solve the special triangle shown.

27. Give the three basic trigonometric ratios for angle A in exercise 25.

28. Give the three basic trigonometric ratios for angles P and Q in exercise 26.

29. If $\triangle ABC \sim \triangle DEF$ and $\cos B = 0.3657$, what is $\cos E$?

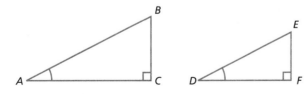

30. Explain the mathematical significance of II Kings 20:11.

Trigonometric Table

DEGREES	sin	cos	tan	cot	sec	csc	
0	.0000	1.0000	.0000		1.000		90
1	.0175	.9998	.0175	57.290	1.000	57.299	89
2	.0349	.9994	.0349	28.636	1.001	28.654	88
3	.0523	.9986	.0524	19.081	1.001	19.107	87
4	.0698	.9976	.0699	14.301	1.002	14.336	86
5	.0872	.9962	.0875	11.430	1.004	11.474	85
6	.1045	.9945	.1051	9.514	1.006	9.567	84
7	.1219	.9925	.1228	8.144	1.008	8.206	83
8	.1392	.9903	.1405	7.115	1.010	7.185	82
9	.1564	.9877	.1584	6.314	1.012	6.392	81
10	.1736	.9848	.1763	5.671	1.015	5.759	80
11	.1908	.9816	.1944	5.145	1.019	5.241	79
12	.2079	.9781	.2126	4.705	1.022	4.810	78
13	.2250	.9744	.2309	4.331	1.026	4.445	77
14	.2419	.9703	.2493	4.011	1.031	4.134	76
15	.2588	.9659	.2679	3.732	1.035	3.864	75
16	.2756	.9613	.2867	3.487	1.040	3.628	74
17	.2924	.9563	.3057	3.271	1.046	3.420	73
18	.3090	.9511	.3249	3.078	1.051	3.236	72
19	.3256	.9455	.3443	2.904	1.058	3.072	71
20	.3420	.9397	.3640	2.747	1.064	2.924	70
21	.3584	.9336	.3839	2.605	1.071	2.790	69
22	.3746	.9272	.4040	2.475	1.079	2.669	68
23	.3907	.9205	.4245	2.356	1.086	2.559	67
24	.4067	.9135	.4452	2.246	1.095	2.459	66
25	.4226	.9063	.4663	2.145	1.103	2.366	65
26	.4384	.8988	.4877	2.050	1.113	2.281	64
27	.4540	.8910	.5095	1.963	1.122	2.203	63
28	.4695	.8829	.5317	1.881	1.133	2.130	62
29	.4848	.8746	.5543	1.804	1.143	2.063	61
30	.5000	.8660	.5774	1.732	1.155	2.000	60
31	.5150	.8572	.6009	1.664	1.167	1.942	59
32	.5299	.8480	.6249	1.600	1.179	1.887	58
33	.5446	.8387	.6494	1.540	1.192	1.836	57
34	.5592	.8290	.6745	1.483	1.206	1.788	56
35	.5736	.8192	.7002	1.428	1.221	1.743	55
36	.5878	.8090	.7265	1.376	1.236	1.701	54
37	.6018	.7986	.7536	1.327	1.252	1.662	53
38	.6157	.7880	.7813	1.280	1.269	1.624	52
39	.6293	.7771	.8098	1.235	1.287	1.589	51
40	.6428	.7660	.8391	1.192	1.305	1.556	50
41	.6561	.7547	.8693	1.150	1.325	1.524	49
42	.6691	.7431	.9004	1.111	1.346	1.494	48
43	.6820	.7314	.9325	1.072	1.367	1.466	47
44	.6947	.7193	.9657	1.036	1.390	1.440	46
45	.7071	.7071	1.0000	1.000	1.414	1.414	45
	cos	sin	cot	tan	csc	sec	DEGREES

Symbols

{ }	set braces or empty set	$\overrightarrow{AB}$	vector AB	$\parallel$	is parallel to
$\{x \mid x\}$	set builder notation	$\angle ABC$	angle ABC	$\perp$	is perpendicular to
$\in$	is an element of	$m\angle ABC$	measure of angle ABC	$\vee$	or
$\notin$	is not an element of	$\triangle ABC$	triangle ABC	$\wedge$	and
$\subseteq$	is a subset of	$\angle A$-BC-D	dihedral angle with edge BC	$\sim$	not or is similar to
$\nsubseteq$	is not a subset of	$\odot P$	circle P	$p \rightarrow q$	p implies q
$\subset$	is a proper subset of	$m\overset{\frown}{AB}$	measure of arc AB	$p \leftrightarrow q$	p if and only if q
$\varnothing$	empty set	$\mid x \mid$	absolute value of x	$\forall$	for all
U	universal set	$=$	is equal to	$\exists$	there exists
$\mathbb{R}$	set of real numbers	$\neq$	is not equal to	$\cos A$	cosine of angle A
$\cup$	union	$\approx$	is approximately equal to	$\sin A$	sine of angle A
$\cap$	intersection	$>$	is greater than	$\tan A$	tangent of angle A
AB	distance between A and B	$<$	is less than	π	pi
$\overline{AB}$	segment AB	$\geq$	is greater than or equal to	$\circ$	degree
$\overleftrightarrow{AB}$	line AB	$\leq$	is less than or equal to	$\sqrt{}$	square root
$\overrightarrow{AB}$	ray AB	$\cong$	is congruent to		
$\overset{\circ}{\longleftrightarrow}{AB}$	half-line AB	$\ncong$	is not congruent to		

Postulates and Theorems

Postulate 1.1 *Expansion Postulate.* A line contains at least two points. A plane contains at least three noncollinear points. Space contains at least four noncoplanar points.

Postulate 1.2 *Line Postulate.* Any two points in space lie in exactly one line.

Postulate 1.3 *Plane Postulate.* Three distinct noncollinear points lie in exactly one plane.

Postulate 1.4 *Flat Plane Postulate.* If two points lie in a plane, then the line containing these two points lies in the same plane.

Postulate 1.5 *Plane Intersection Postulate.* If two planes intersect, then their intersection is exactly one line.

Theorem 1.1 If two distinct lines intersect, they intersect in one and only one point.

Theorem 1.2 A line and a point not on that line are contained in one and only one plane.

Theorem 1.3 Two intersecting lines are contained in one and only one plane.

Theorem 1.4 Two parallel lines are contained in one and only one plane.

Postulate 2.1 *Line Separation Postulate.* Every point divides any line through that point into three disjoint sets: the point and two half-lines.

Postulate 2.2 *Plane Separation Postulate.* Every line divides any plane containing the line into three disjoint sets: the line and two half-planes.

Theorem 2.1 *Jordan Curve Theorem.* Any simple closed curve divides a plane into three disjoint sets: the curve itself, its interior, and its exterior.

Postulate 3.1 *Ruler Postulate.* Every point of a line can be placed in correspondence with a real number.

Postulate 3.2 *Completeness Postulate.* Given a ray, $\overrightarrow{AB}$, and any positive real number r, there is exactly one point C on the ray so that $AC = r$.

Distance Formula The distance, d, between two points $A(x_1, y_1)$ and $B(x_2, y_2)$ is $d = \sqrt{(x_1 - x_2)^2 + (y_1 - y_2)^2}$.

Theorem 3.1 *Midpoint Theorem.* If M is the midpoint of $\overline{AB}$, then $AM = \frac{1}{2}AB$.

Theorem 3.2 The perimeter of a regular n-gon with sides of length s is $n \cdot s$.

Postulate 4.1 *Protractor Postulate.* For every angle A there corresponds a positive real number less than or equal to 180. This is symbolized $0 < m\angle A \leq 180$.

Postulate 4.2 *Continuity Postulate.* If k is a half-plane determined by $\overleftrightarrow{AC}$, then for every real number, $0 < x \leq 180$, there is exactly one ray, $\overrightarrow{AB}$, that lies in k such that $m\angle BAC = x$.

Postulate 4.3 *Angle Addition Postulate.* If K lies in the interior of $\angle MNP$, then $m\angle MNP = m\angle MNK + m\angle KNP$.

Theorem 4.1 All right angles are congruent.

Theorem 4.2 If two angles are adjacent and supplementary, then they form a linear pair.

Theorem 4.3 Angles that form a linear pair are supplementary.

Theorem 4.4 If one angle of a linear pair is a right angle, then the other angle is also a right angle.

Theorem 4.5 *Vertical Angle Theorem.* Vertical angles are congruent.

Theorem 4.6 Congruent supplementary angles are right angles.

Theorem 4.7 *Angle Bisector Theorem.* If $\overrightarrow{AB}$ bisects $\angle CAD$, then $m\angle CAB = \frac{1}{2}m\angle CAD$.

Theorem 5.1 The conditional $p \to q$ is equivalent to the disjunction $\sim p \vee q$.

Theorem 5.2 *Contrapositive Rule.* A conditional statement is equivalent to its contrapositive. In other words, $p \to q$ is equivalent to $\sim q \to \sim p$.

Postulate 6.1 *Parallel Postulate.* Two lines intersected by a transversal are parallel if and only if the alternate interior angles are congruent.

Historic Parallel Postulate Given a line and a point not on the line, there is exactly one line passing through the point that is parallel to the given line.

Postulate 6.2 *SAS Congruence Postulate.* If two sides and an included angle of one triangle are congruent to the corresponding two sides and included angle of another triangle, then the two triangles are congruent.

Postulate 6.3 *ASA Congruence Postulate.* If two angles and an included side of one triangle are congruent to the corresponding two angles and included side of another triangle, then the two triangles are congruent.

Theorem 6.1 *Congruent Segment Bisector Theorem.* If two congruent segments are bisected, then the four resulting segments are congruent.

Theorem 6.2 Segment congruence is an equivalence relation.

Theorem 6.3 Supplements of congruent angles are congruent.

Theorem 6.4 Complements of congruent angles are congruent.

Theorem 6.5 Angle congruence is an equivalence relation.

Theorem 6.6 *Adjacent Angle Sum Theorem.* If two adjacent angles are congruent to another pair of adjacent angles, then the larger angles formed are congruent.

Theorem 6.7 *Adjacent Angle Portion Theorem.* If two angles, one in each of two pairs of adjacent angles, are congruent, and the larger angles formed are also congruent, then the other two angles are congruent.

Theorem 6.8 *Congruent Angle Bisector Theorem.* If two congruent angles are bisected, the four resulting angles are congruent.

Theorem 6.9 Triangle congruence is an equivalence relation.

Theorem 6.10 Circle congruence is an equivalence relation.

Theorem 6.11 Polygon congruence is an equivalence relation.

Theorem 6.12 *Alternate Exterior Angle Theorem.* Two lines intersected by a transversal are parallel if and only if the alternate exterior angles are congruent.

Theorem 6.13 *Corresponding Angle Theorem.* Two lines intersected by a transversal are parallel if and only if the corresponding angles are congruent.

Theorem 6.14 If a transversal is perpendicular to one of two parallel lines, then it is perpendicular to the other also.

Theorem 6.15 If two coplanar lines are perpendicular to the same line, then they are parallel to each other.

Theorem 6.16 The sum of the measures of the angles of any triangle is 180°.

Theorem 6.17 If two angles of one triangle are congruent to two angles of another triangle, then the third angles are also congruent.

Theorem 6.18 The acute angles of a right triangle are complementary.

Theorem 6.19 *SAA Congruence Theorem.* If two angles of a triangle and a side opposite one of the two angles are congruent to the corresponding angles and side of another triangle, then the two triangles are congruent.

Theorem 6.20 *Isosceles Triangle Theorem.* In an isosceles triangle the two base angles are congruent.

Theorem 6.21 If two angles of a triangle are congruent, then the sides opposite those angles are congruent, and the triangle is an isosceles triangle.

Theorem 6.22 A triangle is equilateral if and only if it is equiangular.

Theorem 6.23 *SSS Congruence Theorem.* If each side of one triangle is congruent to the corresponding side of a second triangle, then the two triangles are congruent.

Theorem 7.1 *HL Congruence Theorem.* If the hypotenuse and a leg of one right triangle are congruent to the hypotenuse and corresponding leg of another right triangle, then the two triangles are congruent.

Theorem 7.2 *LL Congruence Theorem.* If the two legs of one right triangle are congruent to the two legs of another right triangle, then the two triangles are congruent.

Theorem 7.3 *HA Congruence Theorem.* If the hypotenuse and an acute angle of one right triangle are congruent to the hypotenuse and corresponding acute angle of another right triangle, then the two triangles are congruent.

Theorem 7.4 *LA Congruence Theorem.* If a leg and one of the acute angles of a right triangle are congruent to the corresponding leg and acute angle of another right triangle, then the two triangles are congruent.

Theorem 7.5 Any point lies on the perpendicular bisector of a segment if and only if it is equidistant from the two endpoints.

Theorem 7.6 *Circumcenter Theorem.* The perpendicular bisectors of the sides of any triangle are concurrent at the circumcenter, which is equidistant from each vertex of the triangle.

Theorem 7.7 *Incenter Theorem.* The angle bisectors of the angles of a triangle are concurrent at the incenter, which is equidistant from the sides of the triangle.

Theorem 7.8 *Orthocenter Theorem.* The lines that contain the three altitudes are concurrent at the orthocenter.

Theorem 7.9 *Centroid Theorem.* The three medians of a triangle are concurrent at the centroid.

Theorem 7.10 *Exterior Angle Theorem.* The measure of an exterior angle of a triangle is equal to the sum of the measures of its two remote interior angles.

Theorem 7.11 *Exterior Angle Inequality.* The measure of an exterior angle of a triangle is greater than the measure of either remote interior angle.

Theorem 7.12 *Longer Side Inequality.* One side of a triangle is longer than another side if and only if the measure of the angle opposite the longer side is greater than the measure of the angle opposite the shorter side.

Theorem 7.13 *Hinge Theorem.* Two triangles have two pairs of congruent sides. If the measure of the included angle of the first triangle is larger than the measure of the other included angle, then the opposite (third) side of the first triangle is longer than the opposite side of the second triangle.

Theorem 7.14 *Triangle Inequality.* The sum of the lengths of any two sides of a triangle is greater than the length of the third side.

Theorem 7.15 The opposite sides of a parallelogram are congruent.

Theorem 7.16 SAS Congruence for Parallelograms

Theorem 7.17 A quadrilateral is a parallelogram if and only if the diagonals bisect one another.

Theorem 7.18 Diagonals of a rectangle are congruent.

Theorem 7.19 The sum of the measures of the four angles of every convex quadrilateral is 360°.

Theorem 7.20 Opposite angles of a parallelogram are congruent.

Theorem 7.21 Consecutive angles of a parallelogram are supplementary.

Theorem 7.22 If the opposite sides of a quadrilateral are congruent, then the quadrilateral is a parallelogram.

Theorem 7.23 A quadrilateral with one pair of parallel sides that are congruent is a parallelogram.

Midpoint Formula If M is the midpoint of $\overline{AB}$ where $A(x_1, y_1)$ and $B(x_2, y_2)$, then $M(\frac{x_1 + x_2}{2}, \frac{y_1 + y_2}{2})$.

Postulate 8.1 *Area Postulate.* Every region has an area given by a unique positive real number.

Postulate 8.2 *Congruent Regions Postulate.* Congruent regions have the same area.

Postulate 8.3 *Area of Square Postulate.* The area of a square is the square of the length of one side: $A = s^2$.

Postulate 8.4 *Area Addition Postulate.* If the interiors of two regions do not intersect, then the area of the union is the sum of their areas.

Theorem 8.1 The *area of a rectangle* is the product of its base and height: $A = bh$.

Theorem 8.2 The *area of a right triangle* is one-half the product of the lengths of the legs.

Theorem 8.3 The *area of a parallelogram* is the product of the base and the altitude: $A = bh$.

Theorem 8.4 The *area of a triangle* is one-half the base times the height: $A = \frac{1}{2}bh$.

Theorem 8.5 The *area of a trapezoid* is one-half the product of the altitude and the sum of the lengths of the bases: $A = \frac{1}{2}h(b_1 + b_2)$.

Theorem 8.6 The *area of a rhombus* is one-half the product of the lengths of the diagonals: $A = \frac{1}{2}d_1 d_2$.

Theorem 8.7 *Pythagorean Theorem.* In a right triangle, the sum of the squares of the lengths of the legs is equal to the square of the length of the hypotenuse: $a^2 + b^2 = c^2$.

Theorem 8.8 The *area of an equilateral triangle* is $\frac{\sqrt{3}}{4}$ times the square of the length of one side: $A = s^2 \frac{\sqrt{3}}{4}$.

Heron's Formula If $\triangle ABC$ has sides of lengths a, b, c, and semiperimeter s, then the area of the triangle is $A = \sqrt{s(s - a)(s - b)(s - c)}$.

Theorem 8.9 The central angles of a regular n-gon are congruent and measure $\frac{360°}{n}$.

Theorem 8.10 The *area of a regular polygon* is one-half the product of its apothem and its perimeter: $A = \frac{1}{2}ap$.

Theorem 8.11 The *apothem of an equilateral triangle* is one-third the length of the altitude: $a = \frac{1}{3}h$.

Theorem 8.12 The apothem of an equilateral triangle is $\frac{\sqrt{3}}{6}$ times the length of the side: $a = \frac{\sqrt{3}}{6}s$.

Theorem 8.13 The *area of a circle* is pi times the square of the radius: $A = \pi r^2$.

Theorem 8.14 The *surface area of a prism* is the sum of the lateral surface area and the area of the bases: $S = L + 2B$. The lateral surface area of a right prism is the product of its height and the perimeter of its base: $L = pH$.

Theorem 8.15 The *surface area of a cylinder* is the sum of the lateral surface area and the area of the bases: $S = L + 2B$. The lateral surface area of a right cylinder is the product of its circumference and height: $L = cH$.

Theorem 8.16 The *surface area of a pyramid* is the sum of the lateral surface area and the area of the base: $S = L + B$. For a regular pyramid, the lateral surface area is the sum of n equal triangular areas or $L = \frac{1}{2}pl$, and the total surface area is given by: $S = \frac{1}{2}p(L + a)$, where p is the perimeter of the base, L is the slant height, and a is the length of the apothem.

Theorem 8.17 The *surface area of a cone* is the sum of the lateral surface area and the area of the base: $S = L + B$; the lateral surface area of a circular cone is half the product of the circumference and slant height: $L = \frac{1}{2}cl$.

Theorem 8.18 The *surface area of a sphere* is 4π times the square of the radius: $S = 4\pi r^2$.

Theorem 8.19 The *surface area of a regular polyhedron* is the product of the number of faces and the area of one face: $S = nA$.

Postulate 9.1 *Chord Postulate.* If a line intersects the interior of a circle, then it contains a chord of the circle.

Postulate 9.2 *Arc Addition Postulate.* If B is a point on $\overparen{AB}$, then $m\overparen{AB} + m\overparen{BC} = m\overparen{AC}$.

Theorem 9.1 In a circle, if a radius is perpendicular to a chord of a circle, then it bisects the chord.

Theorem 9.2 In a circle or in congruent circles, if two chords are the same distance from the center(s), the chords are congruent.

Theorem 9.3 In a circle or in congruent circles, if two chords are congruent, then they are the same distance from the center(s).

Theorem 9.4 If a line is tangent to a circle, then it is perpendicular to the radius drawn to the point of tangency.

Theorem 9.5 *Law of Contradiction.* If an assumption leads to a contradiction, then the assumption is false and its negation is true.

Theorem 9.6 If a line is perpendicular to a radius at a point on the circle, then the line is tangent to the circle.

Theorem 9.7 Tangent segments extending from a given exterior point to a circle are congruent.

Theorem 9.8 *Major Arc Theorem.* $m\overparen{ACB} = 360 - m\overparen{AB}$

Theorem 9.9 Chords of congruent circles are congruent if and only if they subtend congruent arcs.

Theorem 9.10 In congruent circles, chords are congruent if and only if the corresponding central angles are congruent.

Theorem 9.11 In congruent circles, minor arcs are congruent if and only if their corresponding central angles are congruent.

Theorem 9.12 In congruent circles, two minor arcs are congruent if and only if the corresponding major arcs are congruent.

Theorem 9.13 The measure of an inscribed angle is equal to one-half the measure of its intercepted arc.

Theorem 9.14 If two inscribed angles intercept congruent arcs, then the angles are congruent.

Theorem 9.15 An angle inscribed in a semicircle is a right angle.

Theorem 9.16 The opposite angles of an inscribed quadrilateral are supplementary.

Theorem 9.17 The measure of an angle formed by two lines that intersect in the exterior of a circle is one-half the difference of the measures of the intercepted arcs.

Theorem 9.18 The measure of an angle formed by two lines that intersect in the interior of a circle is one-half the sum of the measures of the intercepted arcs.

Theorem 9.19 The measure of an angle formed by two lines that intersect at a point on a circle is one-half the measure of the intercepted arc.

Theorem 9.20 If the degree measure of an arc is θ and the circumference of the circle is c, then the length of the arc is given by $\frac{1}{c} = \frac{\theta}{360}$, or $1 = \frac{c\theta}{360}$.

Theorem 9.21 The area of a sector is given by the proportion $\frac{A}{\pi r^2} = \frac{\theta}{360}$ or $A = \frac{\pi r^2 \theta}{360}$ where A is the area of the sector of a circle with radius r and θ is the arc measure of the sector in degrees.

Postulate 10.1 *Space Separation Postulate.* Every plane separates space into three disjoint sets: the plane and two half-spaces.

Theorem 10.1 If the endpoints of a segment are equidistant from two other points, then every point between the endpoints is also equidistant from the two other points.

Theorem 10.2 A line perpendicular to two intersecting lines in a plane is perpendicular to the plane containing them.

Theorem 10.3 If a plane contains a line perpendicular to another plane, then the planes are perpendicular.

Theorem 10.4 If intersecting planes are each perpendicular to a third plane, then the line of intersection of the first two is perpendicular to the third plane.

Theorem 10.5 If $\overleftrightarrow{AB}$ is perpendicular to plane p at B, and $\overline{BC} \cong \overline{BD}$ in plane p, then $\overline{AC} \cong \overline{AD}$.

Theorem 10.6 Every point in the perpendicular bisecting plane of segment $\overline{AB}$ is equidistant from A and B.

Theorem 10.7 The perpendicular is the shortest segment from a point to a plane.

Theorem 10.8 Two lines perpendicular to the same plane are parallel.

Theorem 10.9 If two lines are parallel, then any plane containing exactly one of the two lines is parallel to the other line.

Theorem 10.10 A plane perpendicular to one of two parallel lines is perpendicular to the other line also.

Theorem 10.11 Two lines parallel to the same line are parallel.

Theorem 10.12 A plane intersects two parallel planes in parallel lines.

Theorem 10.13 Two planes perpendicular to the same line are parallel.

Theorem 10.14 A line perpendicular to one of two parallel planes is perpendicular to the other also.

Theorem 10.15 Two parallel planes are everywhere equidistant.

Theorem 10.16 Opposite edges of a parallelepiped are parallel and congruent.

Theorem 10.17 Diagonals of a parallelepiped bisect each other.

Theorem 10.18 Diagonals of a right rectangular prism are congruent.

Euler's Formula $V - E + F = 2$ where V, E, and F represent the number of vertices, edges, and faces of a convex polyhedron respectively.

Theorem 10.19 The intersection of a sphere and a secant plane is a circle.

Theorem 10.20 Two points on a sphere that are not on the same diameter lie on exactly one great circle of the sphere.

Theorem 10.21 Two great circles of a sphere intersect at two points that are endpoints of a diameter of the sphere.

Theorem 10.22 All great circles of a sphere are congruent.

Theorem 10.23 A secant plane of a sphere is perpendicular to the line containing the center of the circle of intersection and the center of the sphere.

Theorem 10.24 A plane is tangent to a sphere if and only if it is perpendicular to the radius at the point of tangency.

Postulate 11.1 *Volume Postulate.* Every solid has a volume given by a positive real number.

Postulate 11.2 *Congruent Solids Postulate.* Congruent solids have the same volume.

Postulate 11.3 *Volume of Cube Postulate.* The volume of a cube is the cube of the length of one edge: $V = e^3$.

Postulate 11.4 *Volume Addition Postulate.* If the interiors of two solids do not intersect, then the volume of their union is the sum of the volumes.

Postulate 11.5 *Cavalieri's Principle.* For any two solids, if all planes parallel to a fixed plane form sections having equal areas, then the solids have the same volume.

Theorem 11.1 The *volume of a rectangular prism* is the product of its length, width, and height: $V = lwH$.

Theorem 11.2 A cross section of a prism is congruent to the base of the prism.

Theorem 11.3 The *volume of a prism* is the product of the height and the area of the base: $V = BH$.

Theorem 11.4 The *volume of a cylinder* is the product of the area of the base and the height: $V = BH$. In particular, for a circular cylinder $V = \pi r^2 H$.

Theorem 11.5 The *volume of a pyramid* is one-third the product of the height and the area of the base: $V = \frac{1}{3}BH$.

Theorem 11.6 The *volume of a cone* is one-third the product of its height and base area: $V = \frac{1}{3}\pi r^2 H$.

Theorem 11.7 The *volume of a sphere* is four-thirds π times the cube of the radius: $V = \frac{4}{3}\pi r^3$.

Theorem 12.1 *Isometry Theorem.* Every isometry can be expressed as a composition of at most three reflections.

Postulate 13.1 *AA Similarity Postulate.* If two angles of one triangle are congruent to two angles of another triangle, then the two triangles are similar.

Theorem 13.1 *SSS Similarity Theorem*. If the three sides of one triangle are proportional to the corresponding three sides of another triangle, then the triangles are similar.

Theorem 13.2 *SAS Similarity Theorem*. If two sides of a triangle are proportional to the corresponding two sides of another triangle and the included angles between the sides are congruent, then the triangles are similar.

Theorem 13.3 Similarity of triangles is an equivalence ratio.

Theorem 13.4 An altitude drawn from the right angle to the hypotenuse of a right triangle separates the original triangle into two similar triangles, each of which is similar to the original triangle.

Theorem 13.5 In a right triangle, the altitude to the hypotenuse cuts the hypotenuse into two segments. The length of the altitude is the geometric mean between the lengths of the two segments of the hypotenuse.

Theorem 13.6 In a right triangle, the altitude to the hypotenuse divides the hypotenuse into two segments such that the length of a leg is the geometric mean between the hypotenuse and the segment of the hypotenuse adjacent to the leg.

Slopes of Perpendicular Lines If two distinct nonvertical lines are perpendicular, then their slopes are negative reciprocals.

Theorem 13.7 In similar triangles the lengths of the altitudes extending to corresponding sides are in the same ratio as the lengths of the corresponding sides.

Theorem 13.8 In similar triangles the lengths of the medians extending to corresponding sides are in the same ratio as the lengths of the corresponding sides.

Theorem 13.9 In similar triangles the lengths of the corresponding angle bisectors from the vertices to the points where they intersect the opposite sides are in the same ratio as the lengths of the corresponding sides of the triangles.

Theorem 13.10 In similar triangles the perimeters of the triangles are in the same ratio as the lengths of the corresponding sides.

Theorem 13.11 In similar triangles the ratio of the areas of the triangles is equal to the square of the ratio of the lengths of corresponding sides.

Theorem 13.12 If two chords intersect in the interior of a circle, then the product of the lengths of the segments of one chord is equal to the product of the lengths of the segments of the other chord.

Theorem 13.13 If two secants intersect in the exterior of a circle, then the product of the lengths of one secant segment and its external secant segment is equal to the product of the lengths of the other secant segment and its external secant segment.

Theorem 14.1 If the length of a leg of an isosceles right triangle (45-45 right triangle) is x, then the length of the hypotenuse is $x\sqrt{2}$.

Theorem 14.2 If the length of the leg opposite the 30° angle of a 30-60 right triangle is x, then the length of the leg opposite the 60° angle is $x\sqrt{3}$, and the length of the hypotenuse is $2x$.

Theorem 14.3 *Sum of Squares Identity*. For any angle x, $\sin^2 x + \cos^2 x = 1$.

Glossary

Acute angle An angle with a measure less than 90° or a triangle with three acute angles.

Adjacent angles Two coplanar angles that have a common side and a common vertex but no common interior points.

Alternate exterior angles Angles on opposite sides of the transversal and outside the other two lines.

Alternate interior angles Angles on opposite sides of the transversal and between the other two lines.

Altitude A perpendicular segment that extends from a vertex to the opposite side of a triangle, or the vertex to the plane of the base of a cone (or pyramid), or the parallel sides of a trapezoid, or the parallel bases of a cylinder (or prism).

Angle The union of two distinct rays with a common endpoint.

Angle bisector A ray that (except for its origin) is in the interior of an angle and forms congruent adjacent angles.

Apothem The perpendicular segment that joins the center with a side of a regular polygon.

Arc A curve that is a subset of a circle.

Arc measure The same measure as the degree measure of the central angle that intercepts the arc.

Area The number of square units needed to cover a region completely.

Between A point M is between A and B *(A-M-B)* if $AM + MB = AB$.

Biconditional A statement of the form "p if and only if q" $(p \leftrightarrow q)$, which means $p \rightarrow q$ and $q \rightarrow p$.

Bisector Any curve that intersects a segment only at the midpoint.

Central angle An angle that is in the same plane as a circle and whose vertex is the center of the circle.

Centroid The point of intersection of the medians of a triangle.

Chord A segment having both endpoints on a circle (or sphere).

Circle The set of all points that are a given distance from a given point in a given plane.

Circumcenter The point of intersection of the perpendicular bisectors of the sides of a triangle.

Circumference The distance around a circle (i.e., perimeter).

Circumscribed The outside figure when a polygon is positioned with all vertices on a circle or when a circle is tangent to each side of a polygon.

Closed A curve that begins and ends at the same point or a surface that has a finite size and divides the other points in space into an interior and an exterior.

Collinear points Points that lie on the same line.

Complementary angles The sum of two angle measures is 90°.

Concentric circles Circles with the same center but radii of different lengths.

Concurrent Lines that intersect at a single point.

Conditional A statement of the form "If p then q" ($p \rightarrow q$).

Cone The union of a region and all segments that connect the boundary of the region with a specific noncoplanar point.

Congruent angles Angles that have the same measure.

Congruent arcs Arcs on congruent circles that have the same measure.

Congruent circles Circles whose radii are congruent.

Congruent polygons Polygons that have three properties: 1) same number of sides, 2) congruent corresponding sides, and 3) congruent corresponding angles.

Congruent segments Segments that have the same length.

Congruent triangles Triangles in which corresponding angles and corresponding sides are congruent.

Construction A drawing made with the aid of only two instruments: an unmarked straightedge and a compass.

Convex The property that any two points of a set determine a segment contained in the set.

Corresponding angles Angles on the same side of the transversal and on the same side of their respective lines.

Cosine (cos A) The right triangle ratio adjacent over hypotenuse.

Cylinder The union of two regions of the same size and shape in different parallel planes, and the set of all segments that join corresponding points on the boundaries of the regions.

Diagonal A segment that connects two vertices of a polygon but is not a side of the polygon or that connects two vertices of a polyhedron that are not on the same face.

Diameter A chord that passes through the center of a circle or sphere.

Distance The distance between two points A and B is the absolute value of the difference of their coordinates. Distance between points A and B is denoted by AB, given by $AB = |a - b|$.

Edge One of the segments that defines a face of a polyhedron (or a line that divides half-planes).

Equiangular polygon A convex polygon in which all angles have the same degree measure.

Equilateral polygon A polygon in which all sides have the same length.

Equivalence relation A relation that is reflexive, symmetric, and transitive.

Exterior angle An angle that forms a linear pair with one of the angles of a triangle.

Face One of the polygonal regions that form the surface of a polyhedron (or either half-plane that defines a dihedral angle, or a plane that divides half-spaces).

Great circle The intersection of a sphere and a secant plane that contains the center of the sphere.

Greater than A real number a is greater than a real number b ($a > b$) if there is a positive real number c so that $a = b + c$.

Half-line The set of all points on a line on a given side of a given point of the line.

Incenter The point of intersection of the angle bisectors of a triangle.

Inscribed A polygon inscribed in a circle has all its vertices on the circle; an angle inscribed in a circle is one whose vertex is on a circle and whose sides each contain another point of the circle.

Isometry A transformation that preserves distance.

Isosceles triangle A triangle with at least two congruent sides.

Law of Deduction A method of proving a conditional statement by temporarily assuming the hypothesis and deducing the conclusion.

Linear pair A pair of adjacent angles whose noncommon sides form a straight angle (are opposite rays).

Measure of an angle The real number that corresponds to a particular angle.

Median of a triangle A segment extending from a vertex to the midpoint of the opposite side.

Midpoint The midpoint of $\overline{AB}$ is M if $A\text{-}M\text{-}B$ and $AM = MB$.

Obtuse An angle with a measure greater than 90° or a triangle with an obtuse angle.

Opposite Collinear rays with only the origin in common, or disjoint half-planes with a common edge, or (in a quadrilateral) sides with no common vertex or angles with no common side, or (in a hexahedron) faces with no common edge or edges on opposite faces not joined by an edge.

Orthocenter The point of intersection of the altitudes of a triangle.

Parallel Coplanar lines that do not intersect, planes that do not intersect, or a line and a plane that do not intersect.

Parallelepiped A hexahedron in which all faces are parallelograms.

Parallelogram A quadrilateral with two pairs of parallel opposite sides.

Perimeter The distance around a closed curve.

Perpendicular Two lines that intersect to form right angles, or two planes that intersect to form right dihedral angles, or a line that intersects a plane and is perpendicular to every line in the plane that passes through the point of intersection.

Platonic solids The five regular polyhedra.

Polygon A simple closed curve that consists only of segments.

Polyhedron A closed surface made up of polygonal regions

Prism A cylinder with polygonal regions as bases.

Pyramid A cone with a polygonal region as its base.

Radius A segment that connects a point on the circle or sphere with the center or that connects a vertex of a regular polygon to its center (plural: radii).

Ray The union of a half-line and its origin. It extends infinitely in one direction from a point.

Rectangle A parallelogram with four right angles.

Region The union of a simple closed curve and its interior.

Regular A polygon that is both equilateral and equiangular, a right pyramid or right prism with a regular polygon as base, or a polyhedron with congruent regular polygons as faces so that the same number of faces intersect at each vertex.

Remote interior angles The two angles of a triangle that do not form a linear pair with a given exterior angle.

Rhombus A parallelogram with four congruent sides.

Right An angle that measures 90°, a triangle with a right angle, a prism with lateral edges perpendicular to the base, or a pyramid with an altitude from the vertex that passes through the center of the base.

Scalene triangle A triangle with no congruent sides.

Secant A line that is in the same plane as the circle and intersects the circle in exactly two points.

Sector of a circle A region bounded by two radii and the intercepted arc.

Segment The set consisting of two points A and B and all the points in between.

Segment of a circle A region bounded by a chord and its intercepted arc.

Side One of the segments that forms a polygon or either ray that forms an angle.

Similar polygons Polygons having corresponding angles that are congruent and corresponding sides that are proportional.

Sine (sin A**)** The right triangle ratio opposite over hypotenuse.

Skew lines Lines that are not coplanar.

Slant height The distance on the surface of a cone or pyramid from the vertex to the base.

Solid The union of a closed surface and its interior.

Sphere A surface in space consisting of the set of all points at a given distance in space from a given point.

Square A rectangle with four congruent sides (or a rhombus with four congruent angles).

Supplementary angles The sum of two angle measures is 180°.

Symmetry An isometry (other than the identity) that maps a figure onto itself. A reflection results in line symmetry; a rotation, in rotational symmetry; a rotation of 180°, in point symmetry.

Tangent (tan A) The right triangle ratio opposite over adjacent.

Tangent line (or *tangent***)** A line in the plane of a circle that intersects the circle in exactly one point.

Transversal A line that intersects two or more distinct coplanar lines in two or more distinct points.

Trapezoid A quadrilateral with a pair of parallel opposite sides.

Triangle The union of segments that connect three non-collinear points.

Trichotomy property For any two real numbers a and b, exactly one of the following is true: $a = b$, $a < b$, or $a > b$.

Vertex A point at the intersection of segments or rays in angles, polygons, cones, prisms, pyramids, and polyhedra.

Vertical angles Angles adjacent to the same angle and forming linear pairs with it.

Volume The number of cubic units needed to fill up the interior of a solid completely.

Selected Answers

Chapter 1—Incidence Geometry

1.1
1. Answers will vary. **3.** Answers will vary.
5. Answers will vary. **7.** neither **9.** equivalent
11. $A \subseteq L$ **13.** $\emptyset \subseteq U$ **15.** $K = F$ **17.** $N \not\subseteq L$
19. $\{k, l, m\} \not\subseteq \{k, l, n\}$ **21.**
23.

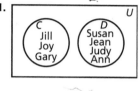

1.2
1. {dog, bird, cat, snail, rabbit, snake}

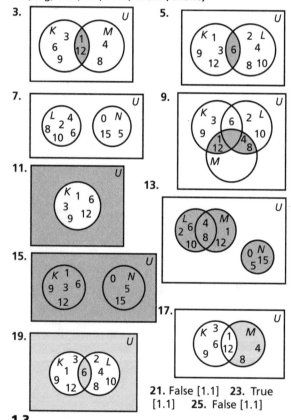

21. False [1.1] **23.** True [1.1] **25.** False [1.1]

1.3
1. not precise (not reversible) **3.** not precise (not reversible) **5.** not useful (defining numbers must precede defining sums) **7.** not concise (bad grammar) and not objective **9.** not useful (*Reredos* has not been defined.) **11.** $K \bullet$
13. $\longleftrightarrow$ l **15.** Point *F* lies on plane *n*.
17. Point *Q* lies outside line *l*. **19.** Answers will vary. A king is a male monarch. **21.** Answers will vary. A mosaic is a picture or pattern made from colored tiles.
35. disjoint: $A \cap B = \emptyset$ [1.2] **37.** intersecting: $A \cap B = \{1,11\}$ [1.2]

1.4
1. $\overleftrightarrow{KM}$, $\overleftrightarrow{KL}$ **3.** $\overleftrightarrow{KL}$, $\overleftrightarrow{MN}$, $\overleftrightarrow{KM}$, $\overleftrightarrow{LN}$ **5.** any three: $\{K, M\}$ $\{K, L\}$ $\{L, N\}$ $\{M, N\}$ **7.** $\overleftrightarrow{EH}$, $\overleftrightarrow{AH}$, $\overleftrightarrow{HC}$
9. point *C* **11.** points *F, A, B, H*. Answers will vary.
13. $\overleftrightarrow{EH}$, $\overleftrightarrow{HC}$, $\overleftrightarrow{DC}$. Answers will vary. **15.** no **17.** no
19. Logic is valid reasoning; it is step-by-step, principle-upon-principle thinking. **21.** Space is the set of all points. [1.3] **23.** $A = B$ [1.2]
25. [1.3]

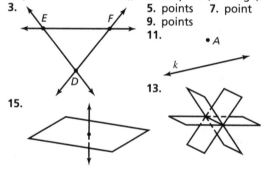

1.5
1. consistent (not contradictory), independent (streamlined), and complete (thorough)
3. **5.** points **7.** point **9.** points **11.** **13.** **15.**

17. Flat Plane Postulate **19.** Expansion Postulate **21.** False **23.** True **27.** True [1.1] **29.** False [1.2]

1.6
1. yes, Flat Plane Postulate **3.** one, Plane Postulate **5.** one, Theorem 1.2 **7.** point *D*; no, Theorem 1.1 **9.** no, definition of parallel lines
11. Answers will vary. $\overleftrightarrow{AB}$ and point *H* are contained in the plane determined by *A, B, E,* and *H*. **13.** $\overleftrightarrow{BH}$ and $\overleftrightarrow{GC}$ are parallel, and one plane determined by points *B, H, C,* and *G* passes through them.
15. $\overleftrightarrow{BG}$ and $\overleftrightarrow{CG}$ intersect at *G* and are contained in the plane determined by points *B, C,* and *G*.
17. no, parallel **19.** A plane is determined by three noncollinear points. Since each pair determines

a line, these three noncollinear points determine three lines. Thus there are at least three lines in a plane.   **27.** point, line, plane [1.3]   **29.** Answers will vary. Flat Plane Postulate [1.5]   **31.** No. Only four points are guaranteed in all of space so far. See Expansion Postulate. [1.5]

1.7
1.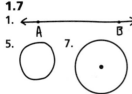
3.

5.   **7.**   **9.** sketch, drawing, and construction   **11.** drawing   **13.** sketch, drawing, construction

15. ―――――――――――
17. no, because there are no inch markings on a straightedge   **19.**
29. disjoint [1.2]
31. [1.4]   (a)

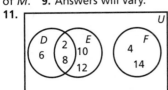
(b)

33. equivalent [1.1]

Chapter 1 Review
1. {California, Colorado, Connecticut} = $\{x \mid x =$ state names that begin with C}   **3.** K is a proper subset of L.   **5.** x is an element of B.   **7.** k is not an element of M.   **9.** Answers will vary.
11.

13. {2, 4, 8, 10, 12, 14}   **15.** ∅
17. $\{x \mid x$ is an even number and $x \neq 2, 6, 8\}$ or {4, 10, 12, 14, . . .}
19. U

21. Set A is equivalent to set C; $A = C$.   **23.** List three or four points to designate each plane (order doesn't matter): $ABC, CGH, BFG, FGH, ABF, ACH$ (*Note:* In the first three, D can replace any letter. The same is true for E in the last three).
25. $ABDC \parallel EFGH, ABFE \parallel CDGH, ACHE \parallel BDGF$
27. $\overleftrightarrow{ED}$ and $\overleftrightarrow{FC}$, $\overleftrightarrow{AB}$ and $\overleftrightarrow{BD}$. Answers will vary.
29. A construction is a drawing that uses only a straightedge and a compass.

Chapter 2—Subsets of Lines, Planes, and Space

2.1
1. $\overleftrightarrow{XY}$ **3.** $\overrightarrow{TS}$ **5.** 3; {C}, $\overrightarrow{CD}$, $\overrightarrow{CA}$ **7.** They have the same endpoint and go in the same direction.
9. $\overrightarrow{AB}$, $\overrightarrow{AD}$ **11.** $\overleftrightarrow{AB}$ **13.** ∅   **15.** $\overleftrightarrow{BC}$ **17.** $\overleftrightarrow{AD}$
19. light rays, laser beam. Answers will vary.
21. plane [1.3]   **23.** line [1.3]   **25.** point [1.3]

2.2
1. $\overleftrightarrow{KM}$ **3.** X   **5.** $\overrightarrow{LO}$ **7.** points L and M **9.** $\overrightarrow{FL}$ and $\overrightarrow{FM}$, $\overrightarrow{FK}$ and $\overrightarrow{FC}$ **11.** $\overline{LF}$, $\overline{BF}$, $\overline{FC}$, $\overline{FM}$. Answers will vary.   **13.** $\overleftrightarrow{AB}$ **15.** $\overset{\circ}{BC}$ **17.** ∅   **19.** $\overline{AD}$
21. $\overrightarrow{BA}$, $\overrightarrow{BC}$; opposite rays   **23.** $\overleftrightarrow{AX}$ **27.** Line Postulate (using B, C) [1.5]   **29.** the same line [1.1]

2.3
1. {D, B, E}, {A, B, C}   **3.** ∠CDG. Answers will vary.   **5.** points A, B, J **7.** $\overrightarrow{DC}$ (or $\overrightarrow{DF}$), $\overrightarrow{DG}$ (or $\overrightarrow{DI}$). Answers will vary.   **9.** s_1, s_2 **11.** c   **13.** ∅
15. s_1 **17.** D   **19.** B, C, D   **21.** {F}   **23.** ∠AFC, ∠AFD, ∠AFE   **27.** True [1.3]   **29.** True [1.3]

2.4
1. radius   **3.** circle   **5.** ∅   **7.** $\overline{LM}$, $\overline{XY}$, $\overline{AT}$ **9.** $\overset{\frown}{LA}$, $\overset{\frown}{AM}$. Answers will vary.   **11.** simple closed curve
13. closed curve   **15.** simple curve   **17.** curve
19. ∠B, ∠E, ∠BAC, ∠CAD, ∠DAE, ∠BAD, ∠CAE, ∠BAE, ∠BCA, ∠DCA, ∠CDA, ∠EDA   **21.** $\overline{AD}$
23. △XYZ   **25.** False   **27.** False   **29.** True
33. False [2.3]   **35.** False [2.1]

2.5
1. Concave octagon   **3.** Convex; not a polygon because it does not consist of segments   **5.** Convex quadrilateral   **7.** Concave; not a polygon because it is not simple   **9.** $\overline{AB}$, $\overline{BC}$, $\overline{CD}$, $\overline{DE}$, $\overline{EF}$, $\overline{FG}$, $\overline{GH}$, $\overline{HA}$; octagon
11.   **13.**   **15.** G, J   **17.** Any two: H, I, or J   **19.** yes; no
21. 3, 4, 5, 6, 7, 8, 9, 10
23. 0, 2, 5, 9, 14, 20, 27, 35   **29.** False [1.4]   **31.** False [1.4]

2.6
1. right circular cylinder   **3.** surface   **5.** regular triangular prism   **7.** circular cone   **9.** pentagonal pyramid, yes   **11.** △MAB, △MBC, △MCD, △MDE, △MEA; triangles   **13.** △ABC, △PQR; triangular
15. sewer pipes, drinking glass, soup can. Answers will vary.   **17.** ice-cream cone, party hat. Answers will vary.   **19.**
21. False. Pyramids are cones with polygonal bases. **23.** True
25. False. The base may be rectangular instead of a regular polygon.   **27.** Every closed surface separates space into three disjoint sets: the surface, the interior, and the exterior.   **31.** False [2.5]   **33.** True [2.5]
35. True [2.4]

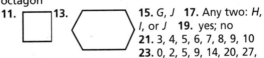

2.7
1. False   **3.** True   **5.** False   **7.** True   **9.** False
11. tetrahedron   **13.** dodecahedron   **15.** hexahedron   **17.** octahedron   **19.** points A, B, C, D, E, F
21. dodecahedron   **23.** $n + 1$, $n + 2$
27. convex [2.5]   **29.** convex [2.5]

Chapter 2 Review

1. $\vec{BA}$, $\vec{BC}$, $\vec{BD}$, $\vec{BE}$, $\vec{EB}$, $\vec{DB}$. Answers will vary.
3. $\angle ABC$, $\angle CBD$, $\angle DBE$, $\angle EBA$ **5.** $\overline{AD}$ **7.** $\vec{CE}$
9. yes **11.**  **13.**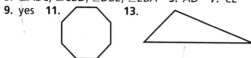

15. 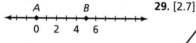 **17.** **19.** $\triangle LPO$, $\triangle LPM$, $\triangle MPN$, $\triangle NPO$, $\triangle LMO$, $\triangle MNO$, $\triangle LON$, $\triangle LMN$
21. M, N, P
23. any three: $\overline{TX}$, $\overline{TY}$, $\overline{TL}$, $\overline{TP}$, $\overline{TV}$ **25.** $\triangle PTX$ **27.** hexahedron **29.** octahedron

Chapter 3—Segments and Measurement

3.1

1.

> -6 -5 -4 -3 -2 -1 0 1 2 3 4 5 6

3. natural: 5, 8, $\frac{7}{1}$; whole: 0; integer: -3, -6, -5; rational: $\frac{1}{2}$, $\frac{2}{5}$, $\frac{1}{3}$; irrational: $\sqrt{3}$, $\sqrt{7}$, π
5. commutative property of multiplication
7. reflexive property of equality **9.** trichotomy
11. 8 **13.** 17 **15.** 18 **17.** True **19.** True
21. True **27.** $\varnothing$, line, plane [1.5] **29.** $\varnothing$, point, chord [2.4]

3.2

1., 3., 5.

> -7 -6 -5 -4 -3 -2 -1 0 1 2 3 4 5 6 7   C E   A

7. 6 **9.** -5 **11.** 9 **13.** 25 **15.** 17 **17.** A, C, and T are collinear, and $AC + CT = AT$; $8 + 12 = 20$
19. A, X, and F are collinear, and $XA + AF = XF$ since $5 + 3 = 8$. **21.** 59 **23.** 2659 **25.** $\frac{809}{15}$ or $53\frac{14}{15}$ or 53.93 **29.** yes [2.1] **31.** reflexive and transitive [3.1]

3.3

1. 10 **3.** 7 **5.** 10 **7.** 7 **9.** point E **11.** True
13. True **15.** False **17.** -9 and -3 **19.** 314
21. $\overline{XD} \cong \overline{FB}$ **27.** at point A (positive to right) [3.2]
29. [2.7]

> A   B
> 0 2 4 6

3.4

1. 37 **3.** 1 **5.** $2a + 2b$ **7.** $8c$
9. 31 cm **11.** $s = 18$m
13. $b = 4$, $a = 8$, $c = 10$ in.

	d	r	C in terms of π	C in decimal form
15.	12	6	12π	37.68
17.	9	4.5	9π	28.26
19.	6	3	6π	18.84

---

27. Polyhedron [2.7] **29.** -15 or 1 [2.2]

3.5

1. circumscribed **3.** neither **5.** either **7.** no
9. L, M, N, O, P **11.** points M, N, O
13. No; $\overline{MO}$ intersects $\odot L$ in more than one point.
15. 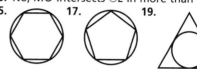 **17.** **19.**

23. False, should say "then $AB = BC$." [3.3]
25. True, they are the same segment [3.3]

3.6

1.

C'   D'

3.

A'   B'

5.

C'   D'   E'   F'

7.

ans.

C'   E'   D'   F'

9.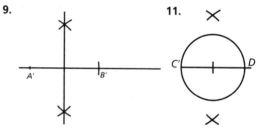

A'   B'

11.

C'   D'

13.

A'   B'   F'   P

15.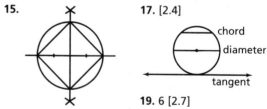

17. [2.4]

chord
diameter
tangent

19. 6 [2.7]

Chapter 3 Review

1. commutative property of addition **3.** associative property of multiplication **5.** trichotomy property **7.** 7 **9.** 6 **11.** -15 **13.** 0 **15.** 31
17. 15 units **19.** $C = 2\pi r = 34\pi \approx 107$ units

21.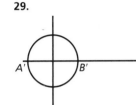

23. 64 units

25.

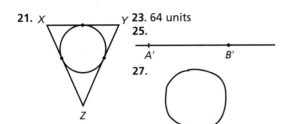

27.

29.

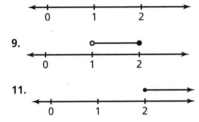

7.

9.

11.

13. $-1 < x < 1$

15. $x = -3$ or $x = 3$

17. $x < -3$ or $x > 3$

19. $\varnothing$ **21.** $\leq, \geq$; The others are not since $3 < 3$, $5 > 5$, and $2 \neq 2$ are all false. **23.** $<, >, \leq, \geq$; but $\neq$ is not transitive since $\sqrt{4} \neq 3$ and $3 \neq 2$, but $\sqrt{4} = 2$ **31.** A point separates a line into three disjoint sets: the point and two half-lines. [2.1]
33. Every point on a line can be placed in one-to-one correspondence with a real number. [3.2]
35. Every simple closed curve divides a plane into three disjoint sets: the curve, its interior, and its exterior. [2.4]

4.2
1. 23° **3.** 46° **5.** 180° **7.** 128° **9.** 82°
11. **13.**

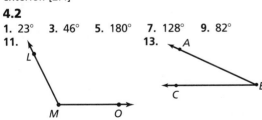

15. **17.**

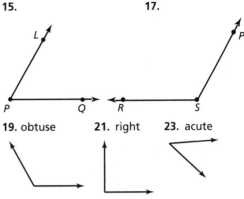

19. obtuse **21.** right **23.** acute

27. pyramids [2.6] **29.** circular cone [2.6]
31. sphere [2.6]

4.3
1. 45° **3.** 15° **5.** 90° **7.** 65° **9.** 135° **11.** 110°
13. 70° **15.** $m\angle 1 = m\angle 5$, or $\angle 1 \cong \angle 5$ **17.** 46°
19. $\angle GXC$ **21.** $m\angle NRQ$ **23.** $m\angle QRP$ **27.** point and two half-lines; Line Separation Postulate [2.2]
29. line and two half-planes; Plane Separation Postulate [2.3]

4.4
1. 125° **3.** $\angle GKJ \cong \angle HKI$; $\angle GKH \cong \angle JKI$ **5.** 35°, 125° **7.** $\angle EGF$ and $\angle FGB$; $\angle DGC$ and $\angle CGA$. Answers will vary. **9.** 50° **11.** $\angle EGF$ is acute, $\angle EGD$ is right, $\angle FGB$ is obtuse, $\angle FGC$ is straight. Answers will vary. **13.** Substitution (or transitive property of equality) **15.** $m\angle AXB + m\angle BXC = 180°$ **17.** $m\angle AXC = 180°$ **19.** Angle Addition Postulate **21.** Substitution (or transitive property of equality) **23.** Linear pairs are supplementary
25. Given **27.** Linear pairs form supplements
29. Transitive property of equality **31.** Definition of congruent angles **33.** Definition of congruent angles **35.** Substitution **37.** Multiplication property of equality **39.** Definition of right angles
41. If $a \leq b$, then $a + c \leq b + c$. [4.1] **43.** $a \leq a$ [4.1]
45. not symmetric [4.1]

4.5
1. right scalene **3.** obtuse scalene **5.** obtuse scalene **7.** $\angle H$ and $\angle J$, $\angle I$ and $\angle K$ **9.** any two: $\overline{PQ}$ and $\overline{QR}$, $\overline{QR}$ and $\overline{RS}$, $\overline{RS}$ and $\overline{SP}$, $\overline{SP}$ and $\overline{PQ}$
11. legs $\overline{MS}$, $\overline{ST}$, base $\overline{MT}$ **13.** congruent, 180 degrees **15.** $3^2 + 4^2 = 5^2$; Pythagorean theorem
17. rectangle **19.** square **21.** yes, no, yes
23. $QR = \sqrt{x^2 + 16}$ **29.** $67 - x$ [4.4]
31. $x + (2x + 15) = 180$ or $x = 2(180 - x) + 15$; 55°, 125° [4.4]

Chapter 4—Angles and Measurement
4.1
1. transitive property of inequality **3.** definition of greater than **5.** $\varnothing$

4.6

1. ∠PQR
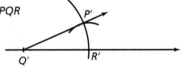

3. ∠XYZ
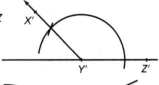

5.
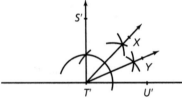

7.
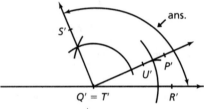

9.
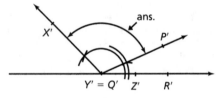

11.
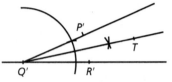

13.
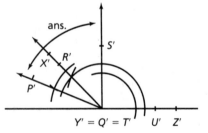

15.
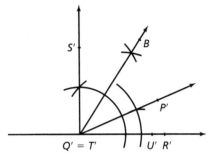

17.
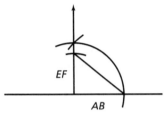

19.
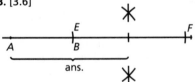

21. [3.3]
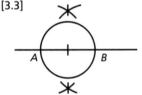

23. [3.6]
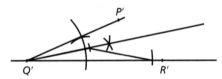

25. [3.6]
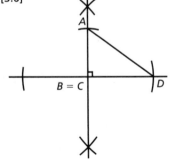

Chapter 4 Review

1. 112°, obtuse **3.** 90°; right **7.**
5. 80°; straight **9.** Answers will
vary. **11.** 68° **13.** $m\angle LMP$
15. given **17.** addition
property of inequality
19. transitive property of inequality
21.   **23.**

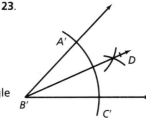

25. Definition of angle
bisector **27.** Angle
Addition Postulate
29. Distributive property
31. This passage describes the heavenly city. Since
the length, height, and width are equal, the shape
is a cube. The term "foursquare" stresses that the
angles at the corners are right angles.

Chapter 5—Preparing for Proofs

5.1

1. Jen
3.

	hang-gliding	rappelling	sky-diving (hockey)
Dean			
Kevin	C		
Frank (soccer)			

5.

	basketball	hockey	soccer (Frank)
hang-gliding (Kevin)			
rappelling			
skydiving			C

7. Answers will vary. **9.**
11. Mr. Bruckner is
Lindsey Kay's father.
13. There is only one
Toggle among the three
residents. **15.** {1, 2, 3, 9}
[1.2] **17.** ∅ [1.2]
19. True [1.1] **21.** True [1.2]

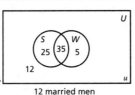

12 married men

5.2

1. true statement **3.** not a statement **5.** not a
statement **7.** false statement **9.** not a statement
11. not a statement **13.** All men are bald.
15. Prime numbers are not divisible by 2. **17.** All

students should study math. **19.** No men are bald.
21. 16, 17, 18, 20 **27.** True [4.5] **29.** False [2.7]

5.3

1. disjunction; true **3.** conjunction; false
5. disjunction; false **7.** disjunction; true
9. conjunction; true **11.** conjunction; false
13. conjunction; true
15.

p	$\sim p$	$p \wedge \sim p$
T	F	F
F	T	F

17.

a	b	$(a \wedge b)$	$\sim a$	$(a \wedge b) \vee \sim a$
T	T	T	F	T
T	F	F	F	F
F	T	F	T	T
F	F	F	T	T

19.

p	q	r	$(q \vee r)$	$p \wedge (q \vee r)$
T	T	T	T	T
T	T	F	T	T
T	F	T	T	T
T	F	F	F	F
F	T	T	T	F
F	T	F	T	F
F	F	T	T	F
F	F	F	F	F

23. True [4.5] **25.** True [2.4] **27.** False [2.7]

5.4

1. If I study my geometry, then I get good grades.
3. If two lines intersect, then they are not parallel.
5. If the sun shines, then the flowers bloom.
7. False **9.** True **11.** False (could be slick from
rain) **13.** If nothing was stolen, then no thief
came. **15.** If a quadrilateral is not a rectangle,
then it is not a trapezoid. **17.** If you study hard,
then you will get an A in geometry. If you get an
A in geometry, then you studied hard. **19.** $x \neq 10$,
or $x + 6 = 16$ **21.** The stoplight is not green, or
you can go. **23.** contrapositive **31.** True [1.6]
33. 20 faces [2.7]

5.5

1. valid; unsound **3.** valid; sound **5.** valid; sound
7. valid; sound **9.** valid; unsound **11.** B **13.** D
15. A **17.** invalid; unsound **19.** deductive, fallacy
of accident **21.** deductive, no fallacy (valid)
23. a nonrectangular parallelogram

27. [4.5]

29. Answers will vary. [2.5]

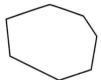

31. Answers will vary. [4.5]

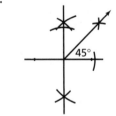

5.6
1. Two angles are vertical angles.   **3.** $\angle 1 \cong \angle 2$
5. $\angle A$ and $\angle B$ are not vertical angles   **7.** $A \rightarrow C$ by transitivity   **9.** assuming the converse
11. Modus ponens (steps 3 and 4)   **13.** modus ponens   **15.** transitivity   **17.** transitivity
19. $[(p \rightarrow q) \wedge p] \rightarrow q$ **25.** C [5.4]
27. Multiplication property of inequality [4.1]
29. Multiplicative inverse property [3.1]

5.7
1.

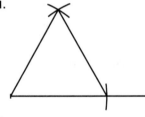

3.

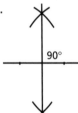

90°

5.

45°

7.

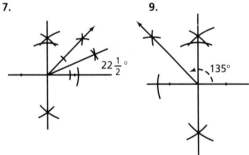

$22\frac{1}{2}°$

9.

135°

11.

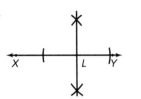

X L Y

13.

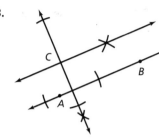

C B A

15. Radii of $\odot A$ are congruent
17. Definition of congruent segments

25.

27. $123°, 57°, 33°$ [4.4]

Chapter 5 Review
1. The parakeet Denise is owned by Christine.
3. false statement   **5.** not a statement   **7.** Girls do not have blond hair.   **9.** disjunction; true
11. disjunction; false   **13.** If a pen has ink, then it writes. A pen lacks ink, or it writes.   **15.** If two lines are parallel, then they do not intersect. Two lines are not parallel, or they do not intersect.
17. invalid, false   **19.** invalid, true   **21.** modus tollens   **23.** appeal to analogy   **25.** modus ponens
27.

A	B	$A \vee B$	$(A \vee B) \wedge A$	$[(A \vee B) \wedge A] \rightarrow B$
T	T	T	T	T
T	F	T	T	F
F	T	T	F	T
F	F	F	F	T

29. Counterexample: whale, octopus, coral, etc.

Chapter 6—Congruence
6.1
1. Given   **3.** Substitution property (step 2 into 1)
5. Addition property of equality   **7.** Expansion Postulate   **9.** Flat Plane Postulate   **11.** Given
13. Definition of betweenness   **15.** Definition of congruent segments   **17.** Law of Deduction
19.
   1. $LM = PQ$ 1. Given
   2. $MP = MP$ 2. Reflexive property of equality

3. $LM + MP = MP + PQ$ 3. Addition property of equality
4. $LM + MP = LP$;   4. Definition of $MP + PQ = MQ$ betweenness
5. $LP = MQ$ 5. Substitution

21.
1. $\overline{AB} \cong \overline{CD}$ 1. Given
2. $AB = CD$ 2. Definition of congruent segments
3. $CD = AB$ 3. Symmetric property of equality
4. $\overline{CD} \cong \overline{AB}$ 4. Definition of congruent segments

23. Segment congruence is an equivalence relation.
27. For any pair of points there is exactly one line containing them. [1.5]   **29.** If two points lie in a plane, then so does the line containing them. [1.5]

6.2
1. Given   **3.** Vertical Angle Theorem
5. Substitution (step 4 into 2)

7.
1. $\angle AXB$ and $\angle BXL$ are   1. Given supplementary; $\angle AXB \cong \angle BXL$
2. $m\angle AXB + m\angle BXL = 180$ 2. Definition of supplementary angles
3. $m\angle AXB = m\angle BXL$ 3. Definition of congruent angles
4. $m\angle AXB + m\angle AXB = 180$ 4. Substitution (step 3 into 2)
5. $2m\angle AXB = 180$ 5. Distributive property
6. $m\angle AXB = 90$ 6. Multiplication property of equality
7. $\angle AXB$ is a right angle   7. Definition of right angle
8. $\overline{AL} \perp \overline{BT}$ 8. Definition of perpendicular

9.
1. E is the midpoint of $\overline{AC}$ and $\overline{BD}$; $\overline{BE} \cong \overline{EC}$ 1. Given
2. $BE = EC$ 2. Definition of congruent segments
3. $BE = \frac{1}{2}BD$; $EC = \frac{1}{2}AC$ 3. Midpoint Theorem
4. $\frac{1}{2}BD = \frac{1}{2}AC$ 4. Substitution
5. $BD = AC$ 5. Multiplication property of equality
6. $\overline{AC} \cong \overline{BD}$ 6. Definition of congruent segments

11.
1. $\overrightarrow{OZ}$ bisects $\angle POY$ 1. Given
2. $\angle POZ \cong \angle ZOY$ 2. Definition of angle bisector
3. $\angle QOX \cong \angle ZOY$ 3. Vertical Angle Theorem
4. $\angle POZ \cong \angle QOX$ 4. Transitive property of congruent angles

13.
1. $m\angle ABC = m\angle ABC$ 1. Reflexive property of equality

2. $\angle ABC \cong \angle ABC$ 2. Definition of congruent angles

15.
1. $\angle ABC \cong \angle PQR$ $\angle DBC \cong \angle SQR$ 1. Given
2. $m\angle ABC = m\angle PQR$, $m\angle DBC = m\angle SQR$ 2. Definition of congruent angles
3. $m\angle ABC + m\angle DBC = m\angle PQR + m\angle SQR$ 3. Addition property of equality
4. $m\angle ABC + m\angle DBC = m\angle ABD$; $m\angle PQR + m\angle SQR = m\angle PQS$ 4. Angle Addition Postulate
5. $m\angle ABD = m\angle PQS$ 5. Substitution (step 4 into 3)
6. $\angle ABD \cong \angle PQS$ 6. Definition of congruent angles

17.
1. $\angle ABC \cong \angle DEF$ and $\overrightarrow{BR}$ bisects $\angle ABC$, $\overrightarrow{ES}$ bisects $\angle DEF$ 1. Given
2. $m\angle ABC = m\angle DEF$ 2. Definition of congruent angles
3. $\angle ABR \cong \angle RBC$, $\angle DES \cong \angle SEF$ 3. Definition of angle bisector
4. $m\angle ABC = 2m\angle ABR$, $m\angle DEF = 2m\angle DES$ 4. Angle measure is twice its bisected angles
5. $2m\angle ABR = 2m\angle DES$ 5. Substitution (step 4 into 2)
6. $m\angle ABR = m\angle DES$ 6. Multiplication property of equality
7. $\angle ABR \cong \angle DES$ 7. Definition of congruent angles
8. $\angle ABR$, $\angle RBC$, $\angle DES$, $\angle SEF$ are congruent   8. Transitive property of congruent angles (exercise 10)
9. If two congruent angles are bisected, then the four resulting angles are congruent   9. Law of Deduction

19.
1. $\angle 2$ and $\angle 3$ are supplementary angles   1. Given
2. $\angle 3$ and $\angle 4$ are a linear pair   2. Definition of linear pair
3. $\angle 3$ and $\angle 4$ are supplementary angles   3. Linear pairs are supplementary
4. $\angle 2 \cong \angle 4$ 4. Supplements of the same angle are congruent
5. $\angle 1 \cong \angle 2$ 5. Vertical Angle Theorem
6. $\angle 1 \cong \angle 4$ 6. Transitive property of congruent angles
7. $m\angle 1 = m\angle 4$ 7. Definition of congruent angles

21. [4.4]

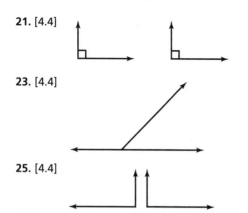

23. [4.4]

25. [4.4]

6.3
1. $\triangle ABC \cong \triangle PQL$ **3.** $\triangle LAM \cong \triangle VXP$
5. $\triangle PAU \cong \triangle HKT$ **7.** $\triangle DNA$ **9.** $\triangle DAN$
11. $\angle Q \cong \angle L$, $\angle M \cong \angle P$, $\angle N \cong \angle S$, $\overline{QM} \cong \overline{LP}$,
$\overline{MN} \cong \overline{PS}$, $\overline{QN} \cong \overline{LS}$ **13.** $\overline{BF} \cong \overline{KE}$, $\overline{FV} \cong \overline{EY}$,
$\overline{BV} \cong \overline{KY}$; $\angle B \cong \angle K$, $\angle F \cong \angle E$, $\angle V \cong \angle Y$
15. $BZ = BC$
17.
  1. $\triangle ABC \cong \triangle PQR$, $\triangle PQR \cong \triangle XYZ$ 1. Given
  2. $\overline{AB} \cong \overline{PQ}$, $\overline{BC} \cong \overline{QR}$, $\overline{AC} \cong \overline{PR}$, $\angle A \cong \angle P$,
     $\angle B \cong \angle Q$, $\angle C \cong \angle R$; $\overline{PQ} \cong \overline{XY}$, $\overline{QR} \cong \overline{YZ}$,
     $\overline{PR} \cong \overline{XZ}$, $\angle P \cong \angle X$, $\angle Q \cong \angle Y$, $\angle R \cong \angle Z$
     2. Definition of congruent angles
  3. $\overline{AB} \cong \overline{XY}$, $\overline{BC} \cong \overline{YZ}$, $\overline{AC} \cong \overline{XZ}$
     3. Transitive property of congruent segments
  4. $\angle A \cong \angle X$, $\angle B \cong \angle Y$, $\angle C \cong \angle Z$
     4. Transitive property of congruent angles
  5. $\triangle ABC \cong \triangle XYZ$ 5. Definition of congruent
     triangles
19.
  1. $\triangle ABC \cong \triangle PQR$ 1. Given
  2. $\angle A \cong \angle P$, $\angle B \cong \angle Q$, $\angle C \cong \angle R$ $\overline{AB} \cong \overline{PQ}$,
     $\overline{AC} \cong \overline{PR}$, $\overline{BC} \cong \overline{QR}$
     2. Definition of congruent triangles
  3. $\angle P \cong \angle A$, $\angle Q \cong \angle B$, $\angle R \cong \angle C$
     3. Symmetric property of congruent angles
  4. $\overline{PQ} \cong \overline{AB}$, $\overline{PR} \cong \overline{AC}$, $\overline{QR} \cong \overline{BC}$
     4. Symmetric property of congruent segments
  5. $\triangle PQR \cong \triangle ABC$ 5. Definition of congruent
     triangles
21. F [4.5] **23.** A [4.5] **25.** D and G [4.5]

6.4
1. 42 **3.** 120 **5.** 110 **7.** 120 **9.** $x^2 + 2x - 6$
11. Given **13.** Vertical Angle Theorem
15. Law of Deduction
17.
  1. $\angle 3 \cong \angle 6$; $\angle 3$ and $\angle 6$ are alternate interior
     angles 1. Given
  2. $a \parallel b$ 2. Parallel Postulate

3. $\angle 4 \cong \angle 5$ 3. Parallel Postulate
19.
  1. $\angle 2 \cong \angle 7$ 1. Given
  2. $p \parallel q$ 2. Alternate Exterior Angle Theorem
  3. $\angle 4 \cong \angle 8$ 3. Corresponding Angle Theorem
21.
  1. $\angle 2 \cong \angle 6$; $d \parallel c$ 1. Given
  2. $\angle 6 \cong \angle 14$ 2. Corresponding Angle Theorem
  3. $\angle 2 \cong \angle 14$ 3. Transitive property of
     congruent angles
23.
  1. $a \parallel b$; $\angle 16 \cong \angle 4$ 1. Given
  2. $\angle 16 \cong \angle 9$ 2. Alternate Exterior Angle
     Theorem
  3. $\angle 4 \cong \angle 9$ 3. Transitive property of
     congruent angles
  4. $c \parallel d$ 4. Parallel Postulate

27. A line in a plane
divides the plane into 3
disjoint sets: a line and
2 half-planes [2.3]

29. If M is the midpoint
of $\overline{AB}$, then $AM = \frac{1}{2}AB$.
[3.3]

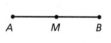

6.5
1. $\angle ADC$ **3.** $\overline{AE}$ **5.** 720 **7.**
$180(n - 2)$ **9.** 128.6 **11.**
$(n - 2)\frac{180}{n}$ **13.** 40 **15.** 29
17. 180 **19.** 97

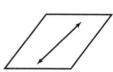

21.
  1. $\triangle ABC$ is a right triangle; $\angle C$ is the right angle
     1. Given
  2. $m\angle A + m\angle B + m\angle C = 180$ 2. The sum of the
     measures of the angles of a triangle is 180
  3. $m\angle C = 90$ 3. Definition of a right angle
  4. $m\angle A + m\angle B + 90 = 180$ 4. Substitution (step
     3 into 2)
  5. $m\angle A + m\angle B = 90$ 5. Addition property of
     equality
  6. $\angle A$ and $\angle B$ are complementary angles
     6. Definition of complementary angles
23.
  1. $\triangle ABC$ is an equiangular triangle 1. Given
  2. $m\angle A + m\angle B + m\angle C = 180$ 2. The sum of
     the measures of the angles of a triangle is 180
  3. $m\angle A = m\angle B = m\angle C$ 3. Definition of
     equiangular
  4. $m\angle A + m\angle A + m\angle A = 180$ 4. Substitution
     (step 3 into 2)
  5. $3m\angle A = 180$ 5. Distributive property

6. $m\angle A = 60$ 6. Multiplication property of
     equality
7. $m\angle B = 60$; $m\angle C = 60$ 7. Substitution
     (step 6 into 3)
27. Every angle corresponds to a real number x,
where $0 < x \leq 180$. [4.2] **29.** In any half-plane
bounded by $\overleftrightarrow{AC}$, for any real number x, where $0 <$
$x \leq 180$, there exists $\overrightarrow{AB}$ in the half-plane such that
$x = m\angle BAC$. [4.2]

6.6
1. SAS **3.** neither **5.** SAS **7.** SAS
9.
  1. $\overline{AC} \cong \overline{DF}$; $\overline{BC} \cong \overline{DE}$; $\angle EDF \cong \angle BCA$ 1. Given
  2. $\triangle ABC \cong \triangle FED$ 2. SAS
  3. $\angle A \cong \angle F$ 3. Definition of congruent
       triangles
11.
  1. $\triangle XYZ$ and $\triangle LMN$ are equiangular, $\overline{LN} \cong \overline{YZ}$
     1. Given
  2. $m\angle Y = 60°$, $m\angle Z = 60°$, $m\angle L = 60°$,
     $m\angle N = 60°$ 2. All angles of equiangular
     triangles measure $60°$
  3. $m\angle Z = m\angle L$; $m\angle Y = m\angle N$ 3. Transitive
     property of equality
  4. $\angle Z \cong \angle L$; $\angle Y \cong \angle N$ 4. Definition of
       congruent angles
  5. $\triangle XYZ \cong \triangle MNL$ 5. ASA
13.
  1. $\overleftrightarrow{AB} \parallel \overleftrightarrow{ED}$; $\overline{AC} \cong \overline{CD}$ 1. Given
  2. $\angle ACB \cong \angle DCE$ 2. Vertical Angle Theorem
  3. $\angle BAC \cong \angle EDC$ 3. Parallel Postulate
  4. $\triangle CAB \cong \triangle CDE$ 4. ASA
15.
  1. $\overleftrightarrow{AB} \parallel \overleftrightarrow{EF}$; $\overline{AB} \cong \overline{EF}$; $\overline{BC} \cong \overline{DE}$ 1. Given
  2. $\angle B \cong \angle E$ 2. Parallel Postulate
  3. $\triangle ABC \cong \triangle FED$ 3. SAS
  4. $\angle ACB \cong \angle FDE$ 4. Definition of congruent
       triangles
  5. $\overleftrightarrow{AC} \parallel \overleftrightarrow{DF}$ 5. Alternate Exterior Angle Theorem
17.
  1. $\triangle ABD \cong \triangle DEA$; $\overline{EF} \cong \overline{BC}$ 1. Given
  2. $\overline{BD} \cong \overline{AE}$ 2. Definition of congruent triangles
  3. $\angle CBD \cong \angle FEA$ 3. Proved in exercise 16
  4. $\triangle BCD \cong \triangle EFA$ 4. SAS
25. $8 - x$ [3.2] **27.** 6π [3.4] **29.** $2x + 4$ [2.4]

6.7
1. $m\angle B = m\angle C = 80°$ **3.** $m\angle P = m\angle Q = 65°$,
$m\angle R = 50°$
5.
  1. $\angle CAB \cong \angle CBA$, $\angle ADB \cong \angle BEA$ 1. Given

  2. $\overline{AB} \cong \overline{AB}$ 2. Reflexive property of
       congruent segments
  3. $\triangle EAB \cong \triangle DBA$ 3. SAA
  4. $\angle EAB \cong \angle DBA$ 4. Definition of congruent
       triangles
7.
  1. $\angle A \cong \angle B$ 1. Given
  2. Draw $\overleftrightarrow{CD} \perp \overline{AB}$ 2. Auxiliary perpendicular line
  3. $\angle CDA$ and $\angle CDB$ are right angles
     3. Definition of perpendicular lines
  4. $\angle CDA \cong \angle CDB$ 4. All right angles are
       congruent
  5. $\overline{CD} \cong \overline{CD}$ 5. Reflexive property of congruent
       segments
  6. $\triangle ADC \cong \triangle BDC$ 6. SAA
  7. $\overline{AC} \cong \overline{CB}$ 7. Definition of congruent triangles
  8. $\triangle ABC$ is an isosceles triangle
     8. Definition of isosceles triangle
9.
  1. equiangular $\triangle ABC$ 1. Given
  2. $m\angle A = m\angle B = m\angle C$ 2. Definition of
       equiangular triangle
  3. $\angle A \cong \angle B \cong \angle C$ 3. Definition of congruent
       angles
  4. $\overline{AB} \cong \overline{AC}$; $\overline{AB} \cong \overline{BC}$; $\overline{AC} \cong \overline{BC}$ 4. If two
     angles of a triangle are congruent, then the
     opposite sides are congruent (exercise 7)
  5. $\triangle ABC$ is equilateral   5. Definition of
       equilateral triangle
11.
  1. $\angle 1 \cong \angle 4$; $\angle 2 \cong \angle 3$ 1. Given
  2. $m\angle 1 = m\angle 4$; $m\angle 2 = m\angle 3$
     2. Definition of congruent angles
  3. $m\angle 1 + m\angle 2 = m\angle 3 + m\angle 4$
     3. Addition property of equality
  4. $m\angle 1 + m\angle 2 = m\angle PMN$;   4. Angle Addition
     $m\angle 3 + \angle 4 = m\angle ONM$ Postulate
  5. $m\angle PMN = m\angle ONM$ 5. Substitution
       (step 4 into 3)
  6. $\angle PMN \cong \angle ONM$ 6. Definition of congruent
       angles
  7. $\overline{MN} \cong \overline{MN}$ 7. Reflexive property of
       congruent segments
  8. $\triangle MPN \cong \triangle NOM$ 8. ASA
13.
  1. $\triangle MPN \cong \triangle NOM$ 1. Given
  2. $\angle PMN \cong \angle ONM$, $\angle 2 \cong \angle 3$
     2. Definition of congruent triangles
  3. $\angle 1 \cong \angle 4$ 3. Adjacent Angle Portion
       Theorem (Theorem 6.7)

15.
1. $\overline{LM} \cong \overline{LN}$, P midpoint of $\overline{LM}$, O midpoint of $\overline{LN}$
   1. Given
2. $\overline{LP} \cong \overline{LO}$, $\overline{MP} \cong \overline{NO}$ 2. Congruent Segment
   Bisector Theorem
3. $\angle L \cong \angle L$ 3. Reflexive property of congruent
   angles
4. $\triangle MLO \cong \triangle NLP$ 4. SAS
5. $\angle 1 \cong \angle 4$ 5. Definition of congruent
   triangles
6. $\angle PQM \cong \angle OQN$ 6. Vertical Angle Theorem
7. $\triangle PQM \cong \triangle OQN$ 7. SAA

17.
1. $\angle PQT \cong \angle SRT$ 1. Given
2. $\angle PQT$ and $\angle TQR$ are supplementary angles;
   $\angle SRT$ and $\angle TRQ$ are supplementary angles
   2. Linear pairs are supplementary
3. $\angle TQR \cong \angle TRQ$ 3. Supplements of congruent
   angles are congruent
4. $\triangle TQR$ is an isosceles triangle   4. If two angles
   of a triangle are congruent, sides opposite are
   congruent and the triangle is isosceles

19.
1. $\overline{BD} \cong \overline{CD}$; $\angle 1 \cong \angle 4$ 1. Given
2. $\triangle BDC$ is an isosceles triangle
   2. Definition of isosceles triangle
3. $\angle 2 \cong \angle 3$ 3. Isosceles Triangle Theorem
4. $m\angle 1 = m\angle 4$; $m\angle 2 = m\angle 3$
   4. Definition of congruent angles
5. $m\angle 1 + m\angle 2 = m\angle 3 + m\angle 4$
   5. Addition property of equality
6. $m\angle 1 + m\angle 2 = m\angle ABC$;
   $m\angle 3 + m\angle 4 = m\angle ACB$
   6. Angle Addition Postulate
7. $m\angle ABC = m\angle ACB$ 7. Substitution
   (step 6 into 5)
8. $\angle ABC \cong \angle ACB$ 8. Definition of congruent
   angles
9. $\triangle ABC$ is an isosceles triangle   9. If two angles
   of a triangle are congruent, then the opposite
   sides are congruent and the triangle is
   isosceles

23. any three: equality of real numbers; biconditionals of statements; congruence of segments; congruence of angles; congruence of circles; congruence of triangles; congruence of polygons [3.1]
25. For any point P and for any line l, there is a line m so that m is parallel to l and P lies on m [5.2]; Historic Parallel Postulate. [6.4]

6.8
1. 75   **3.** 40   **5.** definition of congruent triangles

7. AAA

9. SAA
11.
1. $\angle M \cong \angle P$; $\angle MNO \cong \angle PON$ 1. Given
2. $\overline{NO} \cong \overline{NO}$ 2. Reflexive property of
   congruent segments
3. $\triangle MNO \cong \triangle PON$ 3. SAA

13.
1. $\overline{AB} \cong \overline{CD}$; $\overline{AD} \cong \overline{CB}$ 1. Given
2. $\overline{BD} \cong \overline{BD}$ 2. Reflexive property of congruent
   segments
3. $\triangle ABD \cong \triangle CDB$ 3. SSS

15.
1. $\overline{DC} \cong \overline{BA}$; $\angle BAC \cong \angle DCA$ 1. Given
2. $\overline{AC} \cong \overline{AC}$ 2. Reflexive property of
   congruent segments
3. $\triangle BAC \cong \triangle DCA$ 3. SAS
4. $\angle BCA \cong \angle DAC$ 4. Definition of
   congruent triangles
5. $\overleftrightarrow{AD} \parallel \overleftrightarrow{BC}$ 5. Parallel Postulate

17.
1. $\angle 3 \cong \angle 4$; $\overline{HK} \cong \overline{JK}$ 1. Given
2. $\overline{IK} \cong \overline{IK}$ 2. Reflexive property of congruent
   segments
3. $\triangle HKI \cong \triangle JKI$ 3. SAS
4. $\angle 1 \cong \angle 2$ 4. Definition of congruent triangles
5. $\overrightarrow{IK}$ bisects $\angle HIJ$ 5. Definition of angle bisector

19.
1. $\overline{VW} \cong \overline{XY}$; $\overline{UV} \cong \overline{ZY}$; $\overline{UX} \cong \overline{ZW}$ 1. Given
2. VW = XY   2. Definition of congruent
   segments
3. VW + WX = WX + XY   3. Addition property
   of equality
4. VW + WX = VX; WX + XY = WY
   4. Definition of betweenness
5. VX = WY   5. Substitution (step 5 into 4)
6. $\overline{VX} \cong \overline{WY}$ 6. Definition of congruent
   segments
7. $\triangle UVX \cong \triangle ZYW$ 7. SSS
8. $\angle VXU \cong \angle YWZ$ 8. Definition of congruent
   triangles
9. $\angle UXY$ and $\angle UXV$ are supplementary; $\angle ZWV$
   and $\angle ZWY$ are supplementary 9. Linear pairs
   are supplementary
10. $\angle ZWV \cong \angle UXY$ 10. Supplements of
    congruent angles are congruent
21. E [5.2]   **23.** C [5.3]   **25.** D [5.3]

Chapter 6 Review

1.

 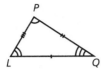

3. 115 **5.** 115 **7.** 60

9.
1. $l \parallel n; p \parallel q$ 1. Given
2. $\angle 12 \cong \angle 13$ 2. Parallel Postulate
3. $\angle 13 \cong \angle 5$ 3. Corresponding Angle Theorem
4. $\angle 12 \cong \angle 5$ 4. Transitive property of congruent angles

11.
1. $\angle ABC \cong \angle XYZ$; $\angle ACB \cong \angle XZY$; $\overline{BC} \cong \overline{YZ}$ 1. Given
2. $\triangle ABC \cong \triangle XYZ$ 2. ASA

13.
1. $\overline{HN} \cong \overline{KL}$; $\angle H \cong \angle K$; $\overline{HI} \cong \overline{JK}$ 1. Given
2. $\overline{IJ} \cong \overline{IJ}$ 2. Reflexive property of congruent segments
3. $HI = JK$; $IJ = IJ$ 3. Definition of congruent segments
4. $HI + IJ = IJ + JK$ 4. Addition property of equality
5. $HI + IJ = HJ$; $IJ + JK = IK$ 5. Definition of betweenness
6. $HJ = IK$ 6. Substitution (step 5 into 4)
7. $\overline{HJ} \cong \overline{IK}$ 7. Definition of congruent segments
8. $\triangle HNJ \cong \triangle KLI$ 8. SAS

15.
1. $\overline{WY} \cong \overline{WU}$; $\overline{VW} \cong \overline{XW}$ 1. Given
2. $\angle W \cong \angle W$ 2. Reflexive property of congruent angles
3. $\triangle UWX \cong \triangle YWV$ 3. SAS
4. $\angle U \cong \angle Y$ 4. Definition of congruent triangles

17. 55 **19.** 120 **21.** $\frac{8(180)}{10} = 144°$

23.
1. P is center of $\odot P$ 1. Given
2. $\overline{PC} \cong \overline{PA}$ 2. Radii of a circle are congruent
3. $\triangle PCA$ is isosceles 3. Definition of isosceles triangle
4. $\angle PCA \cong \angle PAC$ 2. Isosceles Triangle Theorem

25.
1. $\overline{LI} \cong \overline{KI}$; $\angle HLI$ is complementary to $\angle LIH$; $\angle JKI$ is complementary to $\angle KIJ$; $\angle LIH \cong \angle KIJ$ 1. Given
2. $\angle HLI \cong \angle JKI$ 2. Complements of congruent angles are congruent
3. $\triangle LIH \cong \triangle KIJ$ 3. ASA

27.
1. $\angle VUP \cong \angle WUQ$; $\angle PUX \cong \angle QUX$; $\overline{VU} \cong \overline{WU}$

1. Given
2. $m\angle VUP + m\angle PUX = m\angle VUX$; $m\angle WUQ + m\angle QUX = m\angle WUX$ 2. Angle Addition Postulate
3. $m\angle VUP = m\angle WUQ$; $m\angle PUX = m\angle QUX$ 3. Definition of congruent angles
4. $m\angle VUP + m\angle PUX = m\angle WUQ + m\angle QUX$ 4. Addition property of equality
5. $m\angle VUX = m\angle WUX$ 5. Substitution (step 2 into 4)
6. $m\angle VUX \cong m\angle WUX$ 6. Definition of congruent angles
7. $\overline{UX} \cong \overline{UX}$ 7. Reflexive property of congruent segments
8. $m\triangle UVX \cong m\triangle UWX$ 8. SAS

29.
1. hexagon $ABCDEF$ with P equidistant from all vertices 1. Given
2. $PA = PB = PC = PD = PE = PF$ 2. Definition of equidistant
3. $\overline{PA} \cong \overline{PB} \cong \overline{PC} \cong \overline{PD} \cong \overline{PE} \cong \overline{PF}$ 3. Definition of congruent segments
4. $AB = BC = CD = DE = EF = AF$ 4. Definition of regular polygon
5. $\overline{AB} \cong \overline{AF} \cong \overline{BC} \cong \overline{CD} \cong \overline{DE} \cong \overline{EF}$ 5. Definition of congruent segments
6. $\triangle PFA \cong \triangle PAB \cong \triangle PBC \cong \triangle PCD \cong \triangle PDE \cong \triangle PEF$ 6. SSS

Chapter 7—Triangles and Quadrilaterals

7.1
1. See pp. 262-65 for HL, LL, and HA

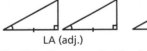 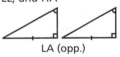

LA (adj.)   LA (opp.)

3. ASA **5.** none **7.** HL **9.** $\triangle MNH \cong \triangle QPH$; LA
11. LA (opposite case)
1. $\triangle RST$ and $\triangle UVW$ are right triangles; $\overline{RS} \cong \overline{UV}$; $\angle T \cong \angle W$ 1. Given
2. $\angle S$ and $\angle V$ are right angles 2. Definition of right triangles
3. $\angle S \cong \angle V$ 3. All right angles are congruent
4. $\triangle RST \cong \triangle UVW$ 4. SAA

13.
1. $\angle P$ and $\angle Q$ are right angles; $\overline{PR} \cong \overline{QR}$ 1. Given
2. $\overline{RT} \cong \overline{RT}$ 2. Reflexive property of congruent segments
3. $\triangle PRT \cong \triangle QRT$ 3. HL
4. $\overline{PT} \cong \overline{QT}$ 4. Definition of congruent triangles

15.
1. $\overline{WY} \perp \overline{XZ}$; $\angle X \cong \angle Z$ 1. Given

2. ∠XYW and ∠ZYW are right angles
   2. Definition of perpendicular
3. △XYW and △ZYW are right triangles
   3. Definition of right triangle
4. $\overline{WY} \cong \overline{WY}$ 4. Reflexive property of
   congruent segments
5. △XYW ≅ △ZYW   5. LA

17.
1. ∠XYW ≅ ∠ZYW; $\overline{XY} \cong \overline{ZY}$ 1. Given
2. ∠XYW and ∠ZYW are right angles
   2. Congruent angles in a linear pair are right
   angles
3. △XYW and △ZYW are right triangles
   3. Definition of right triangle
4. $\overline{YW} \cong \overline{YW}$ 4. Reflexive property of
   congruent segments
5. △XYW ≅ △ZYW   5. LL

19.
1. △XYW and △ZYW are right triangles;
   $\overline{XY} \cong \overline{ZY}$ 1. Given
2. $\overline{WY} \cong \overline{WY}$ 2. Reflexive property of congruent
   segments
3. △XYW ≅ △ZYW   3. LL
4. $\overline{XW} \cong \overline{ZW}$ 4. Definition of congruent triangles

23. 90° [4.3]   **25.** 45° [4.3]

7.2

1.   **3.**

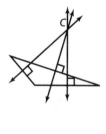

5.

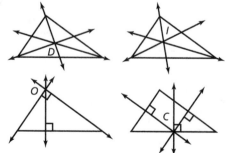

7.

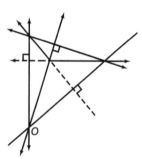

9. 10
11. incenter, inscribed circle
13. 8
15.

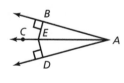

1. $\overline{XL}$ is both a median
   and an altitude of
   △MXN
   1. Given
2. L is the midpoint of $\overline{MN}$ 2. Definition of
   median
3. LM = LN   3. Definition of midpoint
4. $\overline{LM} \cong \overline{LN}$ 4. Definition of congruent segments
5. $\overleftrightarrow{MN} \perp \overleftrightarrow{XL}$ 5. Definition of altitude
6. ∠MLX and ∠NLX are right angles
   6. Definition of perpendicular
7. △MLX and △NLX are right triangles
   7. Definition of right triangle
8. $\overline{XL} \cong \overline{XL}$ 8. Reflexive property of congruent
   segments
9. △MLX ≅ △NLX   9. LL

17.
1. $\overrightarrow{AC}$ bisects ∠BAD; E is
   any point on the bisec-
   tor   1. Given
2. $\overline{EB}$ and $\overline{ED}$ are the
   perpendiculars to the
   sides of the angle   2.
   Auxiliary lines
3. ∠EBA and ∠EDA are right angles
   3. Definition of perpendicular
4. △AEB and △AED are right triangles
   4. Definition of right triangle
5. $\overline{AE} \cong \overline{AE}$ 5. Reflexive property of congruent
   segments

636   ANSWER KEY

6. ∠BAE ≅ ∠DAE   6. Definition of angle
                   bisector
7. △AEB ≅ △AED   7. HA
8. $\overline{EB}$ ≅ $\overline{ED}$ 8. Definition of congruent triangles
9. EB = ED   9. Definition of congruent segments
23. $\overrightarrow{AB}$ [2.2]   25. $\overrightarrow{AB}$ [2.2]

7.3
1. 50, 65, 115   3. 40, 40, 100   5. 80, 100, 20, 60
7. 120°   9. 100°   11. 120°   13. 39°
15. m∠EDA = 2m∠DAC
17.
1. △ABC with right angle at C   1. Given
2. 90 = m∠A + m∠B   2. Exercise 16
3. 90 > m∠A, 90 > m∠B   3. Definition of
                            greater than
4. ∠A and ∠B are acute angles
   4. Definition of acute angle
19.
1. m∠ABE > m∠ACE   1. Exterior Angle
                        Inequality
2. m∠ACE > m∠CED   2. Exterior Angle
                        Inequality
3. m∠ABE > m∠CED   3. Transitive property of
                        inequality
25. b ∥ c since they are perpendicular to the same
line [6.4]   27. l ∥ n by the Corresponding Angle
Theorem

7.4
1. AC, AB, BC   3. NM, LM, LN   5. ∠F, ∠H, ∠G
7. ∠Z, ∠X, ∠Y   9. m∠Z > m∠F   11. AC > PR
13.
1. △ABC is a right
   triangle   1. Given
2. m∠B = 90   2.
   Definition of right
   angle
3. ∠A, ∠C acute angles
   3. Exercise 17, section
   7.3
4. m∠A < 90, m∠C < 90
   4. Definition of acute angle
5. m∠B > m∠A; m∠B > m∠C
   5. Substitution (step 2 into 4)
6. $\overline{AC}$ is the longest side   6. Longer Side
                                  Inequality
15.
1. AB > AC, AC > BC   1. Given
2. AB > BC   2. Transitive property of inequality
3. m∠C > m∠A   3. Longer Side Inequality
17.
1. $\overline{MB}$ is base of isosceles △BCM; M is midpoint
   of $\overline{AB}$ 1. Given
2. m∠BMC = m∠B   2. Isosceles Triangle Theorem

3. m∠BMC > m∠A   3. Exterior Angle Theorem
4. m∠B > m∠A   4. Substitution (step 2 into 3)
19.
1. △ABC is obtuse with its obtuse angle at A
   1. Given
2. Extend $\overrightarrow{AB}$ to D so that D-A-B   2. Auxiliary line
3. ∠BAC and ∠DAC are supplementary
   3. Linear pairs are supplements
4. m∠DAC + m∠BAC = 180
   4. Definition of supplementary
5. m∠BAC = 180 − m∠DAC
   5. Addition property of equality
6. m∠BAC > 90   6. Definition of obtuse angle
7. 180 − m∠DAC > 90   7. Substitution (step 5
                            into 6)
8. 90 > m∠DAC   8. Addition property of
                   inequality
9. m∠DAC > m∠C   9. Exterior Angle Inequality
10. 90 > m∠C   10. Transitive property of
                   inequality
11. m∠C < m∠BAC   11. Transitive property of
                      inequality
12. BC > AB   12. Longer Side Inequality
13. BC > AC   13. Repeat steps 9-12 for ∠B
21. 143 [6.4]   23. 180 − x [4.4]   25. 185 − 3x [6.5]

7.5
1. triangle exists   3. no triangle: a + b = c
(collinear)   5. no triangle: a = b + c (collinear)
7. B-A-C   9. impossible

11. isosceles triangle
13. feasible and efficient
15. not desired: total distance is
greater than 20 feet
17. feasible and efficient
19.
1. △DEF with interior point P   1. Given
2. DF + FG > DG; GE + PG > PE   2. Triangle
                                    Inequality
3. DF + FG + GE + PG > DG + PE   3. Addition
   of inequalities
4. DP + PG = DG; FG + GE = FE   4. Definition
   of betweenness
5. DF + FE + PG > DP + PG + PE
   5. Substitution (step 4 into 3)
6. DF + FE > DP + PE   6. Addition property of
                          inequality
23. Definition of perpendicular [4.3]   25. Vertical
Angle Theorem [4.4]

7.6
1. 48°   3. 77°   5. 132°

7. lines not parallel   **9.** all acute, but none
complementary

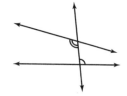

11.
1. Parallelogram *ABCD*   1. Given
2. Draw diagonals $\overline{AC}$ and $\overline{BD}$ 2. Line Postulate
3. $\overleftrightarrow{AB} \parallel \overleftrightarrow{CD}$, $\overleftrightarrow{AD} \parallel \overleftrightarrow{BC}$ 3. Definition of parallelogram
4. $\angle ABE \cong \angle CDE$, $\angle EAB \cong \angle ECD$
   4. Parallel Postulate
5. $\overline{AB} \cong \overline{CD}$ 5. Opposite sides of parallelogram
   congruent
6. $\triangle ABE \cong \triangle CDE$ 6. ASA
7. $\overline{BE} \cong \overline{ED}$, $\overline{AE} \cong \overline{EC}$ 7. Definition of congruent
   triangles
8. $BE = ED$; $AE = EC$ 8. Definition of congruent
   segments
9. $\overline{AC}$ bisects $\overline{BD}$, $\overline{BD}$ bisects $\overline{AC}$
   9. Definition of segment bisector

13.
1. Rectangle *ABCD*   1. Given
2. $\angle ABC$ and $\angle BCD$ are right angles
   2. Definition of rectangle
3. $\overline{AB} \cong \overline{CD}$ 3. Opposite sides of a parallelogram
   congruent
4. $\overline{BC} \cong \overline{BC}$ 4. Reflexive property of congruent
   segments
5. $\triangle ABC \cong \triangle DCB$ 5. LL (or SAS)
6. $\overline{AC} \cong \overline{BD}$ 6. Definition of congruent triangles

15.
1. *ABCD* is a parallelogram with diagonals $\overline{AC}$
   and $\overline{BD}$ 1. Given
2. $\overline{AB} \cong \overline{CD}$, $\overline{AD} \cong \overline{BC}$
   2. Opposite sides of a parallelogram congruent
3. $\overline{BD} \cong \overline{BD}$, $\overline{AC} \cong \overline{AC}$ 3. Reflexive property of
   congruent segments
4. $\triangle DAB \cong \triangle BCD$; $\triangle ABC \cong \triangle CDA$ 4. SSS
5. $\angle DAB \cong \angle BCD$; $\angle ABC \cong \angle CDA$
   5. Definition of congruent triangles

17.
1. $\overline{AB} \cong \overline{CD}$, $\overline{AD} \cong \overline{BC}$ 1. Given
2. $\overline{BD} \cong \overline{BD}$ 2. Reflexive property of congruent
   segments
3. $\triangle ABD \cong \triangle CDB$ 3. SSS
4. $\angle ABD \cong \angle CDB$; $\angle ADB \cong \angle CBD$
   4. Definition of congruent triangles
5. $\overleftrightarrow{AD} \parallel \overleftrightarrow{BC}$; $\overleftrightarrow{AB} \parallel \overleftrightarrow{CD}$ 5. Parallel Postulate

6. *ABCD* is a parallelogram   6. Definition of
   parallelogram
7. If a quadrilateral has congruent opposite sides,
   then it is a parallelogram. 7. Law of Deduction

19.
1. *ABCD* and *DEFG* are parallelograms; $\overline{AG}$ and
   $\overline{EC}$ bisect each other at *D*   1. Given
2. *D* is the midpoint of $\overline{AG}$ and $\overline{EC}$
   2. Definition of segment bisector
3. $AD = DG$; $ED = DC$ 3. Definition of midpoint
4. $\overline{AD} \cong \overline{DG}$; $\overline{ED} \cong \overline{DC}$ 4. Definition of
   congruent segments
5. $\angle ADC \cong \angle GDE$ 5. Vertical Angle Theorem
6. $ABCD \cong GFED$ 6. SAS for parallelograms

25. definition of congruent segments [6.1]
27. converse of Isosceles Triangle Theorem [6.7]
29. radii of a circle congruent [6.3]

7.7
1.

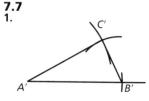

3.

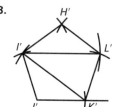

5.

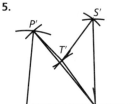

7.

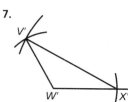

9.   **11.**

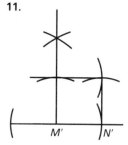

13.

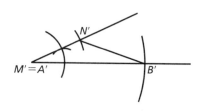

15.

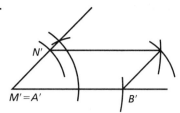

17.

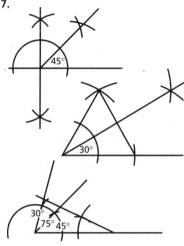

19. SSS and definition of congruent triangles
25. definition of congruent triangles [6.3]
27. Answers will vary. [6.2]   **29.** opposite angles congruent [7.6]

Chapter 7 Review

1.
  1. Rectangle *ABCD* with diagonal $\overline{BD}$
    1. Given
  2. ∠*A* and ∠*C* are right angles
    2. Definition of rectangle
  3. △*ABD* and △*CDB* are right triangles
    3. Definition of right triangle
  4. $\overline{AB} \cong \overline{DC}$; $\overline{AD} \cong \overline{BC}$
    4. Opposite sides of parallelogram congruent
  5. △*ABD* ≅ △*CDB*   5. LL

3.
  1. $\overline{TU} \perp \overline{VU}$; $\overline{TS} \perp \overline{SV}$ 1. Given
  2. ∠*TSV* and ∠*TUV* are right angles
    2. Definition of perpendicular
  3. $\overline{ST} \cong \overline{TU}$ 3. Radii of a circle are congruent

  4. $\overline{VT} \cong \overline{VT}$ 4. Reflexive property of congruent segments
  5. △*VTS* and △*VTU* are right triangles
    5. Definition of right triangle
  6. △*VTS* ≅ △*VTU*   6. HL

5.

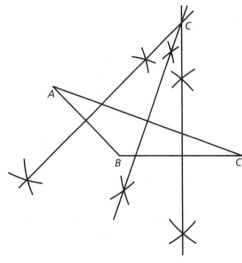

7.

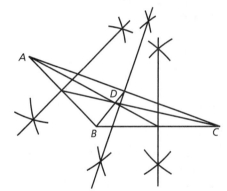

9. They are equal.   **11.** Concurrent lines are lines that intersect at a common point.

13.
  1. ABCD and PQRS are rhombi; $\overline{AB} \cong \overline{PQ}$; ∠*A* ≅ ∠*P*   1. Given
  2. $\overline{AB} \cong \overline{AD}$; $\overline{PQ} \cong \overline{PS}$ 2. Definition of rhombus
  3. $\overline{AD} \cong \overline{PS}$ 3. Transitive property of congruent segments
  4. ABCD ≅ PQRS   4. SAS for parallelograms

15. longest—$\overline{AC}$; shortest—$\overline{BC}$; Longer Side Inequality   **17.** Exterior Angle Inequality
19. Hinge Theorem   **21.** impossible, segment lengths must be positive   **23.** equilateral triangle

(all sides same length) **25.** right triangle (Pythagorean theorem) **27.** collinear, $\angle BAC$ is a straight angle and B-A-C **29.** $\angle B \cong \angle C$ (larger than $\angle A$ in #24-26 when $x < 7$)

Chapter 8—Area

8.1
1. $A = 225$ sq. yd. **3.** $A = 144$ sq. cm
5. $s = 5.7$ m **7.** $A = 72$ sq. ft. **9.** $A = 26.16$ sq. cm **11.** $b = 5.47$ yd. **13.** 52 sq. units **15.** 108 sq. units **17.** Answers will vary. **19.** 405 tiles
21. $\sqrt{35}$ sq. units **23.** $x^2 - 49$ sq. units
27. 124 [3.4] **29.** $0° < x < 60°$ [6.5]

8.2
1. $A = bh$ **3.** $A = \frac{1}{2}bh$ **5.** $A = \frac{1}{2}h(b_1 + b_2)$
7. 35 sq. units **9.** 28 sq. units **11.** 153 sq. units **13.** 72 sq. units **15.** 175.5 sq. units
17. In a parallelogram $b_1 = b_2 = b$, $A = \frac{1}{2}$ $(b_1 + b_2)h = \frac{1}{2}(b + b)h = \frac{1}{2}(2b)h = bh$
19.

1. Trapezoid ABCD with diagonal $\overline{AC}$ 1. Given

2. Area $ABCD$ = Area $\triangle ABC$ + Area $\triangle ADC$
2. Area Addition Postulate

3. Area $\triangle ABC = \frac{1}{2}h(BC)$; Area $\triangle ADC = \frac{1}{2}h(AD)$
3. Area of Triangle

4. Area $ABCD = \frac{1}{2}h(BC) + \frac{1}{2}h(AD)$
4. Substitution

5. Area $ABCD = \frac{1}{2}h(BC + AD)$ 5. Distributive property

21. $(x + 5)(x - 3) = 0$; $x = -5$ or $x = 3$
23. $x(x + 3) = \frac{22}{9}$; $\frac{2}{3}$ inch
27. [2.7] **29.** [2.7]

8.3
1. 4 **3.** $\sqrt{85}$ **5.** $\sqrt{5}$ **7.** $2\sqrt{13}$ **9.** $\sqrt{x^2 + 25}$
11. no **13.** yes **15.** $\frac{5\sqrt{39}}{2}$ sq. inches
17. $h = \sqrt{119}$ units; $A = 15\sqrt{119}$ sq. units
19. $\frac{25\sqrt{3}}{4}$ **21.** $h = 4$, $A = 36$ sq. in.
23. $\triangle CDA \cong \triangle CDB$ by HL since the equilateral triangle has congruent sides and the altitude is congruent to itself. Thus, $\overline{AD} \cong \overline{BD}$ and D is midpoint of AB; $(AD)^2 + h^2 = s^2$ by Pythagorean

theorem (on $\triangle ACD$). So $(\frac{1}{2}s)^2 + h^2 = s^2$ and $h^2 = s^2 - \frac{1}{4}s^2 = \frac{3}{4}s^2$ or $h = \frac{\sqrt{3}}{2}s$.
29. 21 [8.1] **31.** $7 < c < \sqrt{74}$ [7.4–7.5] **33.** 198 cm [8.1]

8.4
1. $\angle AHG$, $\angle FHE$, $\angle EHD$, $\angle DHC$, $\angle CHB$, $\angle BHA$, $\angle GHF$ **3.** 84 sq. units **5.** 2784 sq. units **7.** 64 sq. units **9.** 6, $6\sqrt{2}$, 144 **11.** 11.7, 233
13. $\sqrt{3}$, $2\sqrt{3}$, $9\sqrt{3}$
15.

$A = \frac{1}{2}bh = \frac{1}{2}sh$ and
$A = \frac{1}{2}ap = \frac{1}{2}a$ $(3s) = \frac{3as}{2}$.
Therefore: $\frac{1}{2}sh = \frac{3}{2}as$ or
$sh = 3as$ or $h = 3a$ or
$a = \frac{h}{3}$.

17. Since the radii are congruent, the triangles are isosceles. The base angles are congruent by the Isosceles Triangle Theorem.
19. $A_{\text{hexagon}} = 6A_{\text{triangle}}$ (where the triangles are equilateral)

$$= 6\,(s^2\frac{\sqrt{3}}{4}) \qquad = 3\,\frac{\sqrt{3}}{2}\,s^2$$

21. prism [2.7] **23.** skew [1.4]
25. linear pairs [4.4]

8.5
1. 18π, 18, 81π **3.** 4π, 2, 4π **5.** 1.6π, 0.8, 0.64π **7.** 15, 30, 225π **9.** 10π, 5, 10 **11.** $144 - 36\pi$ sq. units (≈ 30.9) **13.** 128π sq. units (≈ 402.1) **15.** 21π sq. units (≈ 66.0) **17.** $9 \cdot 4 - \frac{1}{2}\pi \cdot 3^2 = 36 - \frac{9}{2}\pi$ (≈ 21.9) **19.** $A_{\text{outer}} - A_{\text{inner}} = \pi x^2 - \pi y^2 = \pi(x^2 - y^2)$ **21.** B is between A and C if $\overrightarrow{BC} \cap \overrightarrow{BA} = \{B\}$ and A, B, C are collinear (or) if $AB + BC = AC$ [2.2] **23.** $\triangle ABC \cong \triangle DEF$ if $\overline{AB} \cong \overline{DE}$, $\overline{AC} \cong \overline{DF}$, $\overline{BC} \cong \overline{EF}$, $\angle A \cong \angle D$, $\angle B \cong \angle E$, $\angle C \cong \angle F$ [6.3]
25. a segment joining a vertex of the triangle to the midpoint of the opposite side [7.2]

8.6
1. $L = 1200$ sq. units **3.** $L = 200$ sq. units, $S = 287.5$ sq. units **5.** $L = 1104$ sq. units, $S = 1104 + 192\sqrt{3}$ sq. units ≈ 1437 sq. units **7.** $L = 3604$ sq. units, $S = 3604 + 504 = 4108$ sq. units **9.** $L = 294\pi$ sq. meters, Top = 49π sq. meters (no paint needed for bottom), Total = 343π sq. meters ≈ 1077.6 sq. meters **11.** 420π sq. in. ≈ 1319 sq. in. **13.** 15" x 15" x 15" **15.** $270\sqrt{3}\pi$ sq. ft ≈ 1469 sq. ft.
17.

1. $S = pH + 2B$ 1. Surface area of prism
2. $B = s^2$ 2. Area of Square Postulate
3. $p = 4s$ 3. Perimeter of square

4. $H = s$ 4. Definition of cube
5. $S = 4s \cdot s + 2s^2 = 6s^2$ 5. Substitution

21. set of all points a given distance from a given point [2.4]   **23.** two angles for which the sum of angle measures is 180 [4.4]   **25.** the center of the circumscribed circle [7.2]

8.7

1. 724.5, 1017.45   **3.** 12, 432, 1008   **5.** 8, 72, 238.3
7. 12, 108   **9.** 3.2, 3, 184.4   **11.** 108π, 189π
13. 3, 27π **15.** 28, 128π **17.** $16\frac{1}{2}$" x $16\frac{1}{2}$"
19.
  1. $S = L + B$, $L = \frac{1}{2}pl$ 1. Area of pyramid
  2. $B = s^2$ 2. Area of Square Postulate
  3. $p = 4s$ 3. Perimeter of square
  4. $L = \frac{1}{2}(4s)l = 2sl$ 4. Substitution (step 3 into 1)
  5. $S = 2sl + s^2$ 5. Substitution (steps 2 and 4 into 1)
  6. $S = s(2l + s)$ 6. Distributive property
21. Using the right triangle, shown dotted, the height and radius are equal to half the diagonal of the square $h = r = 3\sqrt{2}L$ (2 cones) $=$ $2\pi rl = 2\pi(3\sqrt{2})(6) = 36\pi\sqrt{2}\,\text{m}^2$ **25.** B [2.2]
27. A [2.1]

8.8

1. 64π sq. units   **3.** 20π sq. units
5. $4\sqrt{5}$ sq. yd. ≈ 8.94 **7.** dodecahedron, 12, 30, 20
9. octahedron, 8, 12, 6   **11.** 80 sq. inches
13. $180\sqrt{3}$ sq. cm   **15.** 4620 sq. ft.   **17.** 18 cm
19.
  1. $S = 4\pi r^2$ 1. Surface area of sphere
  2. $d = 2r$ 2. Diameter is twice radius
  3. $r = \frac{d}{2}$ 3. Multiplication property of equality
  4. $S = 4\pi\left(\frac{d}{2}\right)^2 = 4\pi\frac{d^2}{4} = \pi d^2$ 4. Substitution (step 3 into 1)

29. $\varnothing$ [2.8]   **31.** edge (segment) [2.7]   **33.** simple closed curve [2.6]

Chapter 8 Review

1. 153 sq. units   **3.** 7 sq. units   **5.** 72 sq. units
7. 15 sq. units   **9.** $81\sqrt{3}$ sq. units   **11.** 102 sq. units   **13.** 81π sq. units   **15.** $39 + \frac{1}{2}\pi$ **17.** 2904 $\sqrt{3}$ sq. units   **19.** 140 sq. units   **21.** $L = 704$ sq. units; $S = 832$ sq. units   **23.** $L = 648\pi$ sq. units; $S = 810\pi$ sq. units   **25.** $L = 78\pi$ sq. units; $S = 114\pi$ sq. units   **27.** The triangles are isosceles (radii congruent), and the Isosceles Triangle Theorem guarantees congruent base angles.

29. For each triangular face, $h = \frac{\sqrt{3}}{2}s$ and

$a = \frac{\sqrt{3}}{6}s$. By Pythagorean theorem: $H^2 + a^2 = h^2$
$$H^2 = h^2 - a^2 = \left(\frac{\sqrt{3}}{2}s\right)^2 - \left(\frac{\sqrt{3}}{6}s\right)^2$$

$= \frac{3}{4}s^2 - \frac{3}{36}s^2 = \frac{2}{3}s^2$, $H = \frac{\sqrt{6}}{3}s$

Chapter 9—Circles

9.1

1. $\odot P$, P, $\overline{AF}$ **3.** $\overline{PA}$, $\overline{PN}$, $\overline{PF}$ **5.** 12 units   **7.** 6 units
9. 5 units   **11.** $\sqrt{33}$ units   **13.** $4\sqrt{7}$ units
15. 22 units
17.
  1. $\odot D$ with $\overleftrightarrow{DB} \perp \overleftrightarrow{AC}$; $\overleftrightarrow{DF} \perp \overleftrightarrow{GE}$; $\overline{BD} \cong \overline{DF}$
    1. Given
  2. $BD = DF$ 2. Definition of congruent segments
  3. $\overline{AC} \cong \overline{EG}$ 3. Chords equidistant from center are congruent
  4. $\overline{AD} \cong \overline{DE} \cong \overline{CD} \cong \overline{DG}$ 4. Radii of a circle are congruent
  5. $\triangle ADC \cong \triangle GDE$ 5. SSS
  6. $\angle CAD \cong \angle DGE$ 6. Definition of congruent triangles
19.
  1. $\odot D$ with $\overleftrightarrow{DB} \perp \overleftrightarrow{AC}$; $\overleftrightarrow{DF} \perp \overleftrightarrow{GE}$; $\overline{AC} \cong \overline{GE}$
    1. Given
  2. $BD = FD$ 2. Congruent chords are equidistant from the center
  3. $\overline{BD} \cong \overline{FD}$ 3. Definition of congruent segments
  4. $\overline{CD} \cong \overline{DG}$ 4. Radii of a circle are congruent
  5. $\angle DBC$ and $\angle DFG$ are right angles
    5. Definition of perpendicular
  6. $\triangle DBC$ and $\triangle DFG$ are right triangles
    6. Definition of right triangle
  7. $\triangle DBC \cong \triangle DFG$ 7. HL
21.
  1. $\odot M$ with $\overleftrightarrow{MQ} \perp \overleftrightarrow{PR}$; $\overleftrightarrow{MT} \perp \overleftrightarrow{PS}$; $\angle RPM \cong \angle SPM$ 1. Given
  2. $\angle MQP$ and $\angle MTP$ are right angles
    2. Definition of perpendicular
  3. $\triangle MQP$ and $\triangle MTP$ are right triangles
    3. Definition of right triangle
  4. $\overline{MP} \cong \overline{MP}$ 4. Reflexive property of congruent segments
  5. $\triangle MQP \cong \triangle MTP$ 5. HA
  6. $\overline{PQ} \cong \overline{PT}$ 6. Definition of congruent triangles
  7. $PQ = PT$ 7. Definition of congruent segments
  8. $\overline{MQ}$ bisects $\overline{PR}$ (at midpoint Q); $\overline{MT}$ bisects $\overline{PS}$ (at midpoint T)   8. Radius perpendicular to chord bisects chord (definition of segment bisector)
  9. $PQ = \frac{1}{2}PR$; $PT = \frac{1}{2}PS$ 9. Midpoint Theorem
  10. $\frac{1}{2}PR = \frac{1}{2}PS$ 10. Substitution (step 9 into 7)
  11. $PR = PS$ 11. Multiplication property of equality

12. $\overline{PR} \cong \overline{PS}$ 12. Definition of congruent segments

23.
1. $\overleftrightarrow{AB} \parallel \overleftrightarrow{CD}; \overline{CX} \cong \overline{XD}; \overleftrightarrow{PQ} \perp \overline{CD}$ 1. Given
2. $\overline{SP} \perp \overleftrightarrow{AB}$ 2. Line perpendicular to one of two parallel lines is perpendicular to the other
3. $\overline{SP}$ bisects $\overline{AB}$ 3. A radius perpendicular to a chord bisects the chord.

27. $m\angle ABC = m\angle DEF$ [4.2]   29. $\overline{AB} \cong \overline{DE}$, $\overline{AC} \cong \overline{DF}$, $\overline{BC} \cong \overline{EF}$, $\angle A \cong \angle D$, $\angle B \cong \angle E$, $\angle C \cong \angle F$ [6.3]

9.2
1. 90   3. $2\sqrt{65}$ units   5. 90   7. $3\sqrt{13}$ 9. $5\sqrt{11}$ units   11. (e), q 13. b, c, l 15. $\odot R, \odot S$
17.
19.

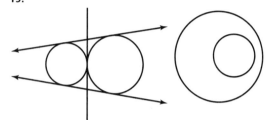

21.

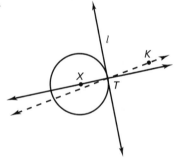

1. $\overleftrightarrow{KT} \perp l$; l intersects $\odot X$ at T 1. Given
2. $\overleftrightarrow{KT}$ does not contain X 2. Assumption
3. Draw $\overline{XT}$ 3. Auxiliary line
4. $\overline{XT} \perp l$ 4. Tangent is perpendicular to radius at point of tangency
5. $\overleftrightarrow{XT} = \overleftrightarrow{KT}$ 5. There is exactly one line perpendicular to a given line at a given point T
6. $\overleftrightarrow{KT}$ contains the center X 6. Law of Contradiction (compare steps 2 and 5)

23.
1. $\odot M$ and $\odot N$ with common tangents $\overleftrightarrow{XW}$ and $\overleftrightarrow{XZ}$ 1. Given
2. $\overline{XY} \cong \overline{XV}; \overline{XZ} \cong \overline{XW}$ 2. Tangent segments congruent

3. $XY = XV; XZ = XW$ 3. Definition of congruent segments
4. $XY + YZ = XZ; XV + VW = XW$ 4. Definition of betweenness
5. $XY + YZ = XV + VW$ 5. Substitution (step 4 into 3)
6. $YZ = VW$ 6. Addition property of equality ($XY = XV$ by step 3)
7. $\overline{YZ} \cong \overline{VW}$ 7. Definition of congruent segments

27. Law of Deduction [5.6]
29. [5.4]

p	q	$\sim q$	$q \wedge \sim q$	$p \rightarrow (q \wedge \sim q)$
T	T	F	F	F
T	F	T	F	F
F	T	F	F	T
F	F	T	F	T

$\sim p$	$[p \rightarrow (q \wedge \sim q)] \rightarrow \sim p$	[5.4]
F	T	
F	T	
T	T	
T	T	

9.3
1. Answers will vary.   3. $\widehat{AFC}, \widehat{ADC}$ 5. 130   7. 90
9. $m\widehat{ABC} = 180$ 11. 80   13. They are congruent since congruent central angles subtend congruent chords.

15.
1. $\odot U; \overline{XY} \cong \overline{YZ} \cong \overline{ZX}$ 1. Given
2. $\overline{XY} \cong \overline{YZ} \cong \overline{ZX}$ 2. Chords are congruent when arcs are congruent
3. $\triangle XYZ$ is an equilateral triangle
3. Definition of equilateral triangle

17.
1. $\odot O$; E is the midpoint of $\overline{BD}$ and $\overline{AC}$; $\overline{BE} \cong \overline{AE}$
1. Given
2. $BE = AE$ 2. Definition of congruent segments
3. $BE = ED; AE = EC$ 3. Definition of midpoint
4. $ED = EC$ 4. Substitution (step 3 into 2)
5. $BE + ED = AE + EC$ 5. Addition property of equality
6. $BE + ED = BD; AE + EC = AC$ 6. Definition of betweenness
7. $BD = AC$ 7. Substitution (step 6 into 5)
8. $\overline{BD} \cong \overline{AC}$ 8. Definition of congruent segments

19.

1. $\overline{AB} \cong \overline{CD}$; $\odot K \cong \odot P$ 1. Given
2. Draw radii $\overline{AK}, \overline{BK}, \overline{CP}, \overline{DP}$ 2. Line Postulate
3. $\overline{AK} \cong \overline{BK} \cong \overline{CP} \cong \overline{DP}$ 3. Radii of congruent circles are congruent
4. $\triangle AKB \cong \triangle CPD$ 4. SSS
5. $\angle AKB \cong \angle CPD$ 5. Definition of congruent triangles
6. If two chords in congruent circles are congruent, then the central angles that intercept them are congruent   6. Law of Deduction

25. The measure of an exterior angle of a triangle is always greater than the measure of either remote interior angle. [7.3]   **27.** If $a = b + c$, where $c > 0$, then $a > b$. [7.3]

9.4
1. 30°   **3.** 50°   **5.** 28°   **7.** 305°   **9.** 35°   **11.** 79°
13. 60°   **15.** $m\angle MNO = 106°$, $m\angle MLO = 74°$
17.
1. $\odot K$ with inscribed $\angle ABC$
   1. Given
2. K lies in the interior of $\angle ABC$ 2. Given for Case 2
3. Draw diameter BD
   3. Auxiliary line
4. $m\angle ABD = \frac{1}{2} m\widehat{AD}$; $m\angle DBC = \frac{1}{2} m\widehat{DC}$
   4. Proved in Case 1
5. $m\angle ABC = m\angle ABD + m\angle DBC$ 5. Angle Addition Postulate
6. $m\angle ABC = \frac{1}{2} m\widehat{AD} + \frac{1}{2} m\widehat{DC}$ 6. Substitution (step 4 into 5)
7. $m\angle ABC = \frac{1}{2} (m\widehat{AD} + m\widehat{DC})$ 7. Distributive property
8. $m\widehat{AD} + m\widehat{DC} = m\widehat{AC}$ 8. Arc Addition Postulate
9. $m\angle ABC = \frac{1}{2} m\widehat{AC}$ 9. Substitution (step 8 into 7)

19.
1. $\angle ABC$ and $\angle PQR$ are inscribed angles and $\widehat{AC} \cong \widehat{PR}$ 1. Given
2. $m\angle ABC = \frac{1}{2}m\widehat{AC}$; $m\angle PQR = \frac{1}{2}m\widehat{PR}$
   2. Inscribed angle measures half of intercepted arc
3. $m\widehat{AC} = m\widehat{PR}$ 3. Definition of congruent arcs
4. $m\angle ABC = \frac{1}{2}m\widehat{PR}$ 4. Substitution (step 3 into 2)
5. $m\angle ABC = m\angle PQR$ 5. Transitive property of equality
6. $\angle ABC \cong \angle PQR$ 6. Definition of congruent angles
7. If two inscribed angles intercept congruent

arcs, then the angles are congruent
7. Law of Deduction

21.
1. $\odot R$; $m\angle TSU = \frac{1}{2}m\widehat{UV}$ 1. Given
2. $m\angle TSU = \frac{1}{2}m\widehat{TU}$ 2. Inscribed angle measures half of intercepted arc
3. $\frac{1}{2} m\widehat{UV} = \frac{1}{2}m\widehat{TU}$ 3. Transitive property of equality
4. $m\widehat{UV} = m\widehat{TU}$ 4. Multiplication property of equality
5. $\widehat{UV} \cong \widehat{TU}$ 5. Definition of congruent arcs

23.
1. $\widehat{ABC}$ is a semicircle
   1. Given
2. $m\widehat{ABC} = 180$ 2. Definition of semicircle
3. $m\widehat{ABC} + m\widehat{ADC} = 360°$
   3. Arc Addition Postulate
4. $m\widehat{ADC} = 180$ 4. Addition property of equality
5. $m\angle ABC = \frac{1}{2} m\widehat{ADC}$ 5. Inscribed angle measures half of intercepted arc
6. $m\angle ABC = \frac{1}{2} (180) = 90$ 6. Substitution (step 4 into 5)
7. $\angle ABC$ is a right angle   7. Definition of right angle

27. definition of perpendicular [4.3]   **29.** Exterior Angle Theorem [7.3]   **31.** 180° in a triangle [6.5]

9.5
1. the angle formed measures half the difference of the measures of the intercepted arcs.   **3.** half the sum of the measures of the intercepted arcs
5. Theorem 9.19   Theorem 9.13

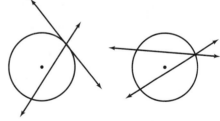

7. $x = 78$ **9.** $x = 14$ **11.** $x = 68$ **13.** $x = 64$
15. $x = 45$ **17.** Two tangents cannot have the same point of tangency. (Otherwise both tangents would be perpendicular to the same radius at the same point.)
19.
1. $\overleftrightarrow{DC}$ is a tangent and $\overleftrightarrow{AC}$ is a secant to $\odot B$
   1. Given
2. Draw $\overleftrightarrow{AD}$ 2. Auxiliary line

3. $m\angle ADE = m\angle CAD + m\angle 1$ 3. Exterior Angle Theorem

4. $m\angle ADE - m\angle CAD = m\angle 1$ 4. Addition property of equality

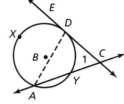

5. $m\angle ADE = \frac{1}{2}m\widehat{AXD}$
   5. Measure of angle formed by secant and tangent at point of tangency

6. $m\angle CAD = \frac{1}{2}m\widehat{DY}$ 6. Inscribed angle measures half of intercepted arc

7. $\frac{1}{2}m\widehat{AXD} - \frac{1}{2}m\widehat{DY} = m\angle 1$ 7. Substitution (steps 5 and 6 into 4)

8. $\frac{1}{2}(m\widehat{AXD} - m\widehat{DY}) = m\angle 1$ 8. Distributive property

25. perpendicular bisectors of each other [7.6]

27. 30 [8.2]

9.6

1. $\frac{10\pi}{3}$ 3. 6   5. 2   7. 10   9. $300°$

11.	$\frac{11\pi}{18}$	$\frac{11\pi}{18} + 4$
13.	120	$\frac{8\pi}{3} + 8$
15.	240	
17.	$\frac{65\pi}{4}$	$\frac{13\pi}{6} + 30$

19. 196,350 sq. ft. (4.5 acres)

21. $\frac{8\pi}{3} - 4\sqrt{3}$ sq. units

23. $50\pi - 50$ sq. units

33. A [9.5]   35. E [2.2]

9.7

1.
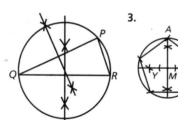

3.

5.
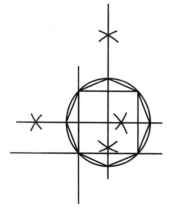

7. Incenter Theorem

9., 11.
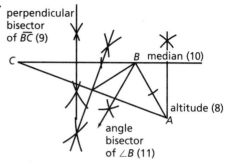

13.
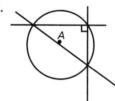

15.
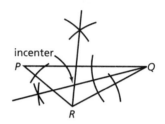

17.
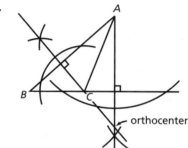

units   9. $\angle BRP$ or $\angle QPR$

11.
1. $\odot P$; $\angle PDC$ is a right angle   1. Given
2. $\overline{DP} \perp \overline{AC}$ 2. Definition of perpendicular
3. $\overline{DP}$ bisects $\overline{AC}$ 3. Radius perpendicular to chord bisects the chord
4. D is the midpoint of $\overline{AC}$ 4. Definition of segment bisector
5. $AD = DC$ 5. Definition of midpoint
6. $\overline{AD} \cong \overline{DC}$ 6. Definition of congruent segments

19. Answer with number of sides. 3, 4, 5, 6, 8, 10, 12, 16, 20, 24

21. [4.6]   23. [3.6]

25. [5.7]

Chapter 9 Review

1. O   3. $\overline{RB}$, $\overline{RP}$, $\overline{AB}$, or $\overleftrightarrow{PQ}$ 5. $\overline{CD}$ 7. 6

7. $\overline{BD} \cong \overline{BD}$ 7. Reflexive property of congruent segments
8. $\angle ADB$ and $\angle CDB$ are right angles
   8. Definition of perpendicular
9. $\triangle ADB$ and $\triangle CDB$ are right triangles
   9. Definition of right triangle
10. $\triangle ADB \cong \triangle CDB$ 10. LL
11. $\angle ABD \cong \angle CBD$ 11. Definition of congruent triangles
12. $\overrightarrow{BP}$ bisects $\angle ABC$ 12. Definition of angle bisector

13.
1. $\overleftrightarrow{AB}$ and $\overleftrightarrow{AC}$ are tangent to $\odot X$; $\overrightarrow{AX}$ bisects $\angle BAC$ 1. Given
2. $\angle BAD \cong \angle DAC$ 2. Definition of angle bisector
3. $\overline{AB} \cong \overline{AC}$ 3. Tangent segments from same point are congruent
4. $\overline{AD} \cong \overline{AD}$ 4. Reflexive property of congruent segments
5. $\triangle BAD \cong \triangle CAD$ 5. SAS
6. $\angle BDA \cong \angle CDA$ 6. Definition of congruent triangles
7. $m\angle BDA = m\angle CDA$ 7. Definition of congruent angles
8. $\angle BDA$ and $\angle CDA$ are supplementary angles
   8. Linear pairs are supplementary
9. $m\angle BDA + m\angle CDA = 180$ 9. Definition of supplementary angles
10. $m\angle BDA + m\angle BDA = 180$ 10. Substitution (step 7 into 9)
11. $2m\angle BDA = 180$ 11. Distributive property
12. $m\angle BDA = 90$ 12. Multiplication property of equality
15. 40   **17.** 18 units   **19.** 180; semicircle   **21.** 96
23. 130   **25.** $\frac{2\pi}{3}$ units   **27.** $12 + \frac{2\pi}{3}$ units
29.

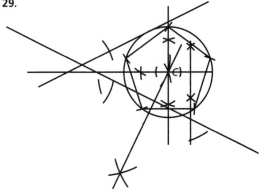

Chapter 10—Space
10.1
1.

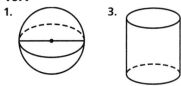

3.

5.

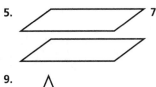

7.

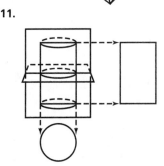

9.

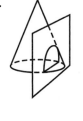

11.

13.

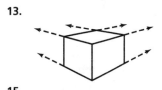

15.

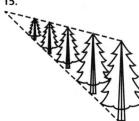

17. Answers will vary.
21. False [1.5]
23. False [3.3]
25. True [1.6]

10.2
1. $\angle A\text{-}CD\text{-}E$, $\angle F\text{-}CD\text{-}E$, $\angle F\text{-}CD\text{-}A$
3. $\angle P\text{-}LM\text{-}R$, $\angle R\text{-}LM\text{-}U$, $\angle U\text{-}NO\text{-}T$, $\angle T\text{-}NO\text{-}V$, $\angle V\text{-}NO\text{-}S$, $\angle S\text{-}NO\text{-}V$, $\angle U\text{-}LM\text{-}Q$, $\angle Q\text{-}LM\text{-}P$, $\angle U\text{-}NO\text{-}V$, $\angle P\text{-}LM\text{-}U$, $\angle Q\text{-}LM\text{-}R$, $\angle T\text{-}NO\text{-}S$ **5.** $\overleftrightarrow{PQ}$
7. right dihedral angle   **9.** $\angle X\text{-}LM\text{-}Y$, excluding $\overleftrightarrow{LM}$
11. empty set   **13.** $s_1 \cup s_2$ **15.** empty set   **17.** No, it divides space into 3 disjoint sets but it does not have finite size (its surface area is infinite).

19. *a.*

b.

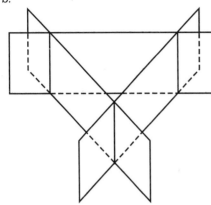

21. True [4.2] **23.** True [4.4] **25.** False [6.4]

10.3

1.

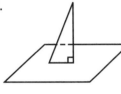

3.

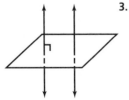

5.

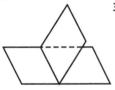

7. $m\angle A\text{-}BC\text{-}X +$
$m\angle X\text{-}BC\text{-}D =$
$m\angle A\text{-}BC\text{-}D$
9. two dihedral angles the sum of whose measures is 90°
11. congruent (dihedral angles)

13. form a dihedral linear pair **15.** so is the other
19. SAS, ASA [6.6] **21.** HL (SSA) [7.1] **23.** LA [7.1]

10.4

1.

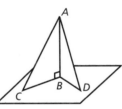

3.

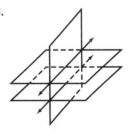

5.

7.

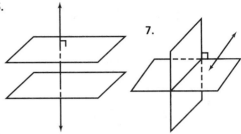

9.

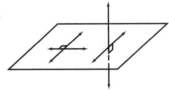

11.

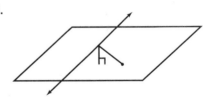

13.

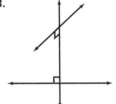

15.

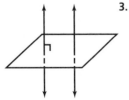

19. False [2.7] **21.** False [2.7] **23.** True [2.7]

10.5

1. True **3.** True **5.** True **7.** 10 faces, 20 edges, 12 vertices **9.** 10 faces, 18 edges, 10 vertices (nonagonal pyramid) **11.** 10 faces, 16 edges, 8 vertices **13.** 5 sides, 15 edges **15.** 9 faces, 14 vertices, 21 edges **17.** 12 vertices, 6 sides, 18 edges **19.** The diagonals bisect each other and are congruent.
21.
  1. A parallelepiped 1. Given

2. *ABCD, EFGH, ABFE, CDHG, BCGF,* and *ADHE* are parallelograms   2. Definition of parallelepiped

3. $\overline{AB} \cong \overline{CD}$, $\overline{CD} \cong \overline{GH}$, $\overline{GH} \cong \overline{EF}$, $\overline{AD} \cong \overline{BC}$, $\overline{BC} \cong \overline{GF}$, $\overline{GF} \cong \overline{HE}$, $\overline{AE} \cong \overline{BF}$, $\overline{BF} \cong \overline{CG}$, $\overline{CG} \cong \overline{DH}$ 3. Opposite sides of a parallelogram are congruent

4. $\overline{AB} \cong \overline{GH}$, $\overline{CD} \cong \overline{EF}$, $\overline{AD} \cong \overline{GF}$, $\overline{BC} \cong \overline{HE}$, $\overline{AE} \cong \overline{CG}$, $\overline{BF} \cong \overline{DH}$ 4. Transitive property of congruent segments

25. Using the marked congruence and vertical angles, the triangles are congruent by SAS. By definition (of congruent triangles), the desired angles are congruent. [6.6]   **27.** Use the disjunction truth table to evaluate the statement ($F \vee T = T$). [5.3]

10.6

1. ∅, a point, a circle

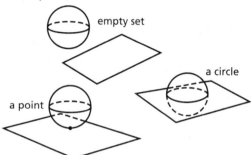

empty set

a circle

a point

3.

5. $\frac{7\pi}{9} \approx 2.4$ units

7. $\frac{14\pi}{45} \approx 1.0$ units

9. $\frac{8\pi}{3} \approx 8.4$ units

11. $\frac{43\pi}{2} \approx 67.5$ units   **13.** 790 ≈ 2480.6 mi.

15. $\frac{37,525\pi}{18} \approx 6546.0$ mi.

17.

1. Sphere *S* with center *C* and two great circles
   1. Given
2. Plane *m* and plane *n* contain *C* and the great circles respectively   2. Definition of great circle
3. *m* ∩ *n* is a line *l*   3. Plane Intersection Postulate
4. *l* contains a diameter $\overline{AB}$, with points *A* and *B* as the only points of the sphere
   4. Definition of diameter (since *C* is in both planes)
5. *l* intersects each great circle at *A* and *B*
   5. A line in a plane of a circle that intersects the interior of the circle intersects the circle at exactly two points
6. The intersection of the two great circles is {*A, B*}   6. Definition of intersection

21. 7 [8.2]   **23.** 20 [8.2]   **25.** $\frac{9\sqrt{3}}{2}$ [8.3]

10.7

1. Philadelphia, Pennsylvania   **3.** Cape Town, South Africa   **5.** Paris, France   **7.** Monterrey, Mexico   **9.** Alexandria, Egypt   **11.** 13° north, 122° east   **13.** 42° north, 14° east   **15.** 72° south, 70° west   **17.** 22° north, 117° east   **19.** 33° north, 8° west   **21.** North and South Poles   **23.** 21° S, 158° W, Avarua, Rarotonga in the Cook Islands
25. 21° N, 22° E, Libya   **27.** 196 million sq. mi. [8.8]
29. SAS [7.6]   **31.** congruent edges [10.5]

10.8

1. 90°   **3.** The circle at 45° N latitude is not a great circle, so it is not a line in the system.

5. yes

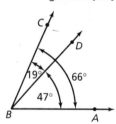

yes   no, two points

7. no, in exactly 2 points
9. more (180° even without ∠*P*)
15. valid, unsound deductive argument [5.6]

17. *m*∠*ABC* = *m*∠*ABD* + *m*∠*CBD* (since 66° = 47° + 19°) [4.2]

Chapter 10 Review

1. True   **3.** True   **5.** False   **7.** 12; ∠*A-BC-D* is one of the possible answers

9.

11.

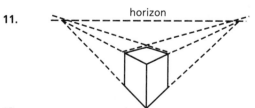

horizon

13.
8 vertices, 12 edges, 6 faces   **15.** $\frac{16\pi}{9} \approx 5.6$ units

17.

19. Lima, Peru
21. 38° S, 156° W
23. The plane is the surface of the sphere. The points are the points of the sphere. The lines are the great circles.

25.
1. Sphere S with center C and chord $\overline{AB}$
   1. Given
2. A great circle E contains A and B 2. Two points on a sphere lie on a great circle
3. Draw $\overleftrightarrow{CP}$ perpendicular to $\overline{AB}$ ($P \in S$)
   3. Auxiliary line
4. $\overleftrightarrow{CP}$ bisects $\overline{AB}$ 4. The radius of a circle perpendicular to a chord of the circle bisects the chord

27.
1. $n \perp \overleftrightarrow{AB}$ at A, $m \perp \overleftrightarrow{AB}$ at B 1. Given
2. Assume n and m intersect in at least one point C 2. Assumption
3. $\angle ABC$ and $\angle BAC$ are right angles   3. Definition of perpendicular
4. $\triangle ABC$ is a right triangle with a right angle at A 4. Definition of right triangle
5. $\angle BAC$ is acute   5. In a right triangle, two angles are acute
6. Planes n and m are parallel   6. Law of Contradiction (see steps 3 and 5)

29.
1. $\overleftrightarrow{AB} \parallel \overleftrightarrow{CD}$, $m \cap n = \overleftrightarrow{AB}$, $n \cap p = \overleftrightarrow{CD}$, $m \cap p = \overleftrightarrow{EF}$ 1. Given
2. $\overleftrightarrow{AB}$ intersects $\overleftrightarrow{EF}$ at Q 2. Assumption
3. $Q \in n$ since $\overleftrightarrow{AB} \subset n$, $Q \in p$ since $\overleftrightarrow{EF} \subset p$ 3. Flat Plane Postulate
4. $Q \in n \cap p$ 4. Definition of intersection (step 3)
5. $Q \in \overleftrightarrow{CD}$ 5. Substitution (step 1 into 4)
6. $\overleftrightarrow{AB} \parallel \overleftrightarrow{EF}$ 6. Law of Contradiction (steps 1, 2, and 5)
7. $\overleftrightarrow{CD} \parallel \overleftrightarrow{EF}$ 7. Two lines parallel to same line are parallel

Chapter 11—Volume

11.1
1. 64 cu. units   **3.** 70 cu. units   **5.** x^3 cu. units
7. 1200 boxes   **9.** $250\sqrt{2}$ **11.** $12\sqrt{39}$ **13.** 153
15. $x^3 \frac{\sqrt{3}}{9}$ **17.** 29,376 g   **21.** Area Postulate and Congruent Regions Postulate [8.1]
23. $A = 6A_{\text{triangle}} = 6s^2 \frac{\sqrt{3}}{4} = 6 \cdot 16^2 \frac{\sqrt{3}}{4}$

$= 384\sqrt{3} = 665.1$ [8.4]
25. $S = 2lw + 2lH + 2wH$ [8.6]

11.2
1. $81\frac{\sqrt{3}}{2}$ cu. units   **3.** 192 cu. units   **5.** 216 cu. units   **7.** 14,924 cu. units   **9.** 480 cu. units
11. 2288 cu. units   **13.** 576 cu. units   **15.** 576 cu. ft.
17. $H = 9$ cm; $V = 468$ cm^3 **19.** $V = s^2H$, $V = \frac{1}{2}bhH$ **23.** $A = s^2$ [8.1]   **25.** $V = e^3$ [11.1]

11.3
1. 441π cu. units   **3.** 240π cu. units   **5.** 2299π cu. units   **7.** 1500π cu. ft.   **9.** $8424 - 2106\pi \approx 1808$ cu. in.   **11.** $S = 9116\pi$ m^2; $V = 116{,}487\pi$ m^3
13. 16.75 in.   **15.** $\frac{1}{3}$(170) ≈ 56.7 cu. ft. or 2.1 cu. yds.   **17.** 15,707.96 cu. ft.   **25.** circle [11.2]
27. triangle [11.2]   **29.** circle [10.6]

11.4
1. 672 cu. units   **3.** $3371\frac{2}{3}\pi$ cu. units   **5.** 3264 cu. units   **7.** 4060 cu. in.   **9.** 2700 cm^3 **11.** $H = 9$ in.; $r = 4.5$ in.; $V = 60.75\pi \approx 190.76$ cu. in. using $\pi \approx 3.14$ **13.** $V = \frac{1}{3} \cdot 27 \cdot 18 - \frac{1}{3} \cdot 3 \cdot 6 = 156$ cm^3
15. $V = 91{,}152{,}065$ cu. ft.; $L = 921{,}388$ sq. ft.
17.
1. Circular cone of radius r and height H
   1. Given
2. Area of base $B = \pi r^2$ 2. Area of circle
3. $V = \frac{1}{3} BH$ 3. Volume of Prism (exercise 16)
4. $V = \frac{1}{3} \pi r^2 H$ 4. Substitution
19. False [10.5]   **21.** 384 cu. ft. [11.2]
23. Law of Deduction [5.6]

11.5
1. $7776\pi \approx 24{,}429$ cu. ft.   **3.** 27 cu. yd.   **5.** $\frac{4}{3} \approx 1.3$ cu. units   **7.** $135 + 63\sqrt{5} \approx 275.9$ mm^3 **9.** 5461.3 π cm$^3 \approx 17{,}157$ cm^3 **11.** 106π cu. in. ≈ 333 cu. in.
13. 6π in. ≈ 18.8 in.   **15.** 23 mm   **17.** 1,648,703 (or 1,647,867.2 gal. using $\pi \approx 3.14$)   **19.** 170.67 π cu. in. ≈ 536.17 cu. in.   **21.** 818,833.3 $\pi \approx$ 2,572,440.7 cu. ft.   **25.** scalene triangle [4.5]
27. segment [9.6]

11.6
1. base region and altitude   **3.**

5.

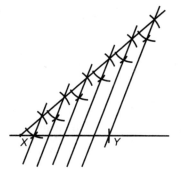

7.

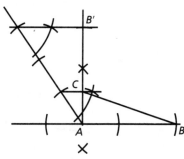

9.

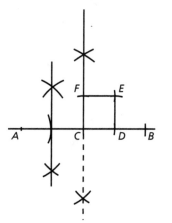

11.

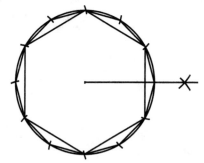

13.

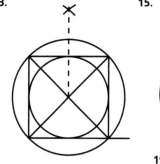

15.

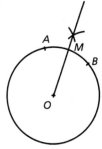

19. [5.8]

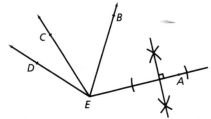

21. [6.9]

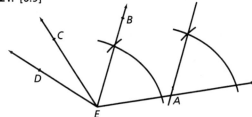

23. [3.7]

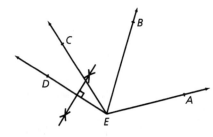

Chapter 11 Review

1. $261\frac{1}{3}\pi$ cu. units ≈ 821 cu. units **3.** 30.2 cu. units **5.** 972π cu. units ≈ 3053.6 cu. units **7.** 729 cu. units **9.** $9\frac{\sqrt{2}}{4} \approx 3.2$ cu. units **11.** $V = 1408$ cu. units **13.** $V = 120\pi$ cu. units ≈ 377 cu. units **15.** $V = 18\sqrt{2}$ cu. units ≈ 25.5 cu. units **17.** $125\frac{\sqrt{2}}{3}$ cu. units ≈ 58.9 cu. units **19.** $S = 696$ sq. units **21.** The number of cubic units needed to fill a solid **23.** $10\pi \approx 31.4$ cu. in. **25.** $3\pi^2 \approx 29.6$ sq. in. **27.** $S = (2\pi a)(2\pi r) = 4\pi^2 ar$

29.

1. Cube   1. Given
2. $r = \frac{\sqrt{3}}{2}e$ 2. Exercise 28
3. $e = \frac{2}{\sqrt{3}}r$ 3. Multiplication property of equality
4. $V = e^3$ 4. Volume of Cube Postulate
5. $V = \left(\frac{2}{\sqrt{3}}r\right)^3 = \frac{8}{3}\sqrt{3}\,r^3 = 8\,\frac{\sqrt{3}}{9}r^3$
   5. Substitution (step 3 into 4)

Chapter 12—Transformations and Symmetry

12.1

1.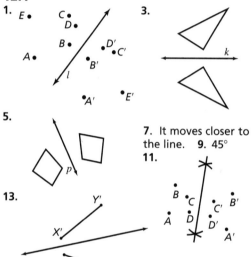

3.

5.

7. It moves closer to the line.   **9.** 45°

11.

13.

15. Connect the preimage and the image points; their perpendicular bisector will be the line of reflection.

	Figure	Surface Area	Volume
19.	Sphere	$4\pi r^2$ [8.8]	$\frac{4}{3}\pi r^3$ [11.5]
21.	Cone	$\pi r l + \pi r^2$ [8.7]	$\frac{1}{3}\pi r^2 H$ [11.4]

12.2

1.

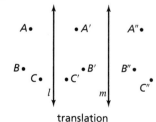

translation

3.

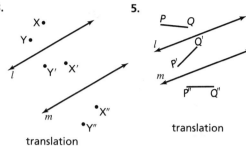

translation

5.

P   Q

translation

7.

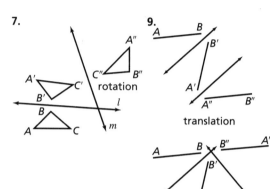

rotation

9.

A   B

translation

rotation

11.

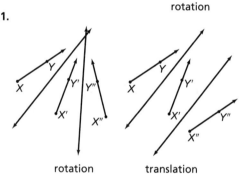

rotation   translation

13. 40°   **15.** identity transformation   **17.** rotation; P, 80°   **23.** $(1.1)^{12}$, 3.14, π, $\frac{22}{7}$, $\sqrt{10}$, $\sqrt[3]{32}$ [3.1]

	Perimeter	Area
25.	$2\pi r$ [3.4]	πr^2 [8.5]
27.	ns [3.4]	$\frac{1}{2}ap$ (or $\frac{1}{2}ans$) [8.4]

12.3

1. ≈ 2.6   **3.** 45   **5.** 8   **7.** 82   **9.** 3, enlargement (or expansion)   **11.** 1, identity transformation
13. $\frac{y}{x}$ **15.** 1.67

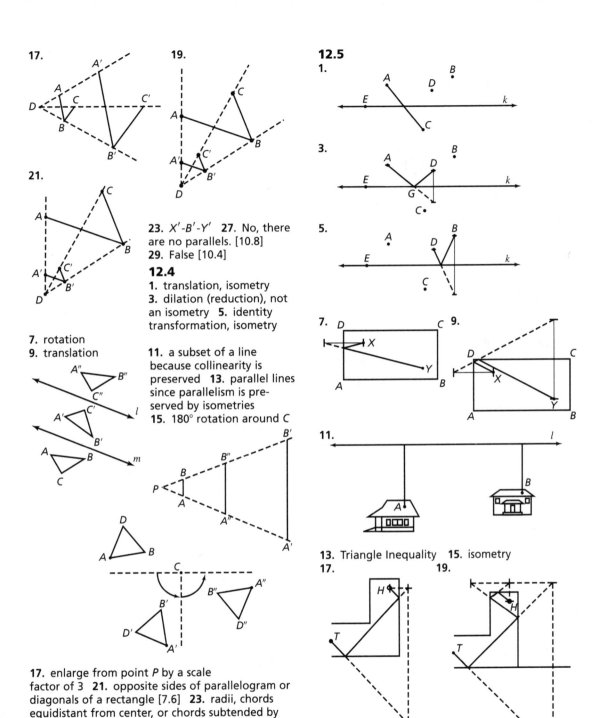

17.

19.

21.

23. X'-B'-Y' **27.** No, there are no parallels. [10.8]
29. False [10.4]

12.4
1. translation, isometry
3. dilation (reduction), not an isometry **5.** identity transformation, isometry

7. rotation
9. translation

11. a subset of a line because collinearity is preserved **13.** parallel lines since parallelism is preserved by isometries
15. 180° rotation around C

17. enlarge from point P by a scale factor of 3 **21.** opposite sides of parallelogram or diagonals of a rectangle [7.6] **23.** radii, chords equidistant from center, or chords subtended by congruent arcs [9.1] **25.** bisections of congruent segments [6.1]

12.5
1.

3.

5.

7. **9.**

11.

13. Triangle Inequality **15.** isometry
17.   **19.**

21. center [12.4] **23.** infinitely many [12.4]

25. It's the identity. [12.4]

12.6
1. one axis (horizontal)   **3.** no axes
5. one axis (vertical)
7. no axes

9. exercises 2, 4, 6, 7
11. exercises 2, 4, 7
13. 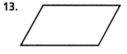   **15.**   **17.**

19. Yes, because it will also have 180° symmetry.
21. {1, −3}, C ⊆ A [1.1]   **23.** B = {x | x is a power of 2} or B = {x | x = 2^n, where n is a whole number} [1.1]   **25.** A ∪ C = {−3, −2, −1, 0, 1} = A, B ∪ C = {−3, 1, 2, 4, 8, 16, . . .} [1.2]

Chapter 12 Review
1.   **3.**

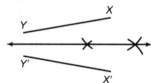

5. rotation   **7.** translation

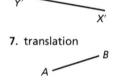

9. 38°   **11.**

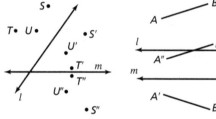

13. house   **15.** pinwheel   **17.** window   **19.** 17
21. similar   **23.** neither   **25.** Draw the perpendicular from the vertex to the hypotenuse.
27. distance

29.
1. $\overline{KI} \perp \overline{HJ}$; a dilation with scale factor k and center O maps H onto H', I onto I', J onto J', and K onto K' 1. Given
2. $\angle KIJ$ is a right angle   2. Definition of perpendicular
3. $m\angle KIJ = 90°$ 3. Definition of right angle
4. $m\angle KIJ = m\angle K'I'J'$ 4. Dilations preserve angle measure
5. $m\angle K'I'J' = 90°$ 5. Substitution (step 3 into 4)
6. $\angle K'I'J'$ is a right angle   6. Definition of right angle
7. $\overline{K'I'} \perp \overline{H'J'}$ 7. Definition of perpendicular

Chapter 13—Similarity
13.1
1. 50   **3.** 7   **5.** $\frac{1}{4}$ **7.** $\frac{1}{4}$ **9.** $m = 16$ units, $n = 12$ units   **11.** $x = 4\frac{1}{2}$; $y = 7\frac{1}{2}$ units   **13.** $\angle L \cong \angle R$; $\angle P \cong \angle S$; $\angle Q \cong \angle T$; $\frac{RS}{LP} = \frac{ST}{PQ} = \frac{RT}{LQ}$ **15.** yes
17.
1. $\triangle ABC \cong \triangle LMN$ 1. Given
2. $\angle A \cong \angle L$; $\angle B \cong \angle M$; $\angle C \cong \angle N$; $\overline{AB} \cong \overline{LM}$; $\overline{BC} \cong \overline{MN}$; $\overline{AC} \cong \overline{LN}$ 2. Definition of congruent triangles
3. $AB = LM$, $BC = MN$, $AC = LN$ 3. Definition of congruent segments
4. $\frac{LM}{AB} = \frac{AB}{AB} = 1$, $\frac{MN}{BC} = \frac{BC}{BC} = 1$, $\frac{LN}{AC} = \frac{AC}{AC} = 1$ 4. Multiplication property of equality
5. $\frac{LM}{AB} = \frac{MN}{BC} = \frac{LN}{AC} = 1$ 5. Transitive property of equality
6. $\triangle ABC \sim \triangle LMN$ 6. Definition of similarity
7. If two triangles are congruent, then they are also similar   7. Law of Deduction
19. Perimeter of $\triangle ABC = 14$ units; $\triangle XYZ = 42$ units; yes; they are in the same ratio as the sides.
21. [12.3]

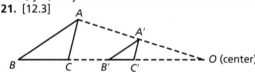

23. 1 [12.3]   **25.** [12.3]

13.2
1. yes   **3.** yes   **5.** no
7. yes, AA   **9.** yes, SSS
11. yes, AA
13.
1. $WXYZ$ is a parallelogram   1. Given
2. $\overleftrightarrow{ZY} \parallel \overleftrightarrow{WX}$; $\overleftrightarrow{ZW} \parallel \overleftrightarrow{YX}$ 2. Definition of parallelogram

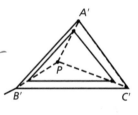

3. $\angle ZYW \cong \angle YWX$; $\angle ZWY \cong \angle XYW$
   3. Parallel Postulate
4. $\triangle WXY \sim \triangle YZW$ 4. AA

15.
1. $\triangle MNP \sim \triangle QOP$ 1. Given
2. $\angle PQO \cong \angle PMN$ 2. Definition of similar
   triangles
3. $\overleftrightarrow{MN} \parallel \overleftrightarrow{QO}$ 3. Corresponding Angle Theorem

17.
1. $\triangle ABC \sim \triangle DEF$ 1. Given
2. $\angle A \cong \angle D$, $\angle B \cong \angle E$ 2. Definition of similar
   triangles
3. $\angle D \cong \angle A$, $\angle E \cong \angle B$ 3. Symmetric property
   of congruent angles
4. $\triangle DEF \sim \triangle ABC$ 4. AA

19.
1. $\overleftrightarrow{DB} \perp \overleftrightarrow{DE}$; $\overleftrightarrow{DB} \perp \overleftrightarrow{AB}$ 1. Given
2. $\triangle ABC \sim \triangle EDC$ 2. Exercise 18
3. $\frac{DE}{AB} = \frac{CE}{AC}$ 3. Definition of similar triangles

21. equivalence relation [3.1]   **23.** yes [12.2]
25. yes, the Volume Postulate and Congruent
Solids Postulate [11.1]

13.3
1. $x = 34$ **3.** $x = 6$ **5.** $x = 4$ **7.** $10\sqrt{3}$ units
9. $2\sqrt{39}$ units   **11.** $2\sqrt{6}$ units   **13.** $4\sqrt{11}$ units
15. $6\sqrt{3}$ units

17.
1. $\triangle ABC$ is a right triangle with altitude $\overline{CD}$
   1. Given
2. $\frac{AD}{CD} = \frac{CD}{BD}$ 2. Altitude is geometric mean of
   two parts of hypotenuse
3. $(CD)^2 = (AD)(BD)$ 3. Multiplication property
   of equality

23. similar, $k = 2$ [13.1]   **25.** not similar [13.1]
27. similar, $k = \frac{t}{s}$ [13.1]

13.4
1. 3 units   **3.** 5 units   **5.** 4 units   **7.** 40 units   **9.** 45
sq. units   **11.** Answers will vary. $\frac{AB}{MN} = \frac{AE}{MQ}$

13. Answers will vary. $\frac{AC}{MO} = \frac{BF}{NR}$

15. Answers will vary; $\frac{AC}{MO} = \frac{\text{perimeter } \triangle ABC}{\text{perimeter } \triangle MNO}$

17.
1. $\triangle ABC \sim \triangle EFG$; $\overline{CD}$ and $\overline{GH}$ are medians
   1. Given
2. $\angle B \cong \angle F$; $\frac{EF}{AB} = \frac{FG}{BC}$ 2. Definition of similar
   triangles
3. D and H are the midpoints of $\overline{AB}$ and $\overline{EF}$
   3. Definition of median
4. $DB = \frac{1}{2}AB$; $HF = \frac{1}{2}EF$ 4. Midpoint Theorem
5. $AB = 2DB$; $EF = 2HF$ 5. Multiplication
   property of equality

6. $\frac{2HF}{2DB} = \frac{FG}{BC}$ 6. Substitution (step 6 into 2)
7. $\frac{HF}{DB} = \frac{FG}{BC}$ 7. Cancellation property
8. $\triangle CDB \sim \triangle GHF$ 8. SAS
9. $\frac{GH}{CD} = \frac{FG}{BC}$ 9. Definition of similar triangles

19.
1. $\triangle ABC \sim \triangle EFG$ with altitudes $\overline{CD}$ and $\overline{GH}$
   respectively   1. Given
2. $\frac{EF}{AB} = \frac{GH}{CD} = k$ 2. The altitudes of similar triangles are in the same ratio as the corre-
   sponding sides
3. $A_1 = $ Area $\triangle ABC = \frac{1}{2}(AB)(CD)$; $A_2 = $ Area
   $\triangle EFG = \frac{1}{2}(EF)(GH)$ 3. Area of a triangle
4. $\frac{A_2}{A_1} = \dfrac{\frac{1}{2}(EF)(GH)}{\frac{1}{2}(AB)(CD)}$ 4. Multiplication property
   of equality
5. $\frac{A_2}{A_1} = \frac{(EF)(GH)}{(AB)(CD)} = \frac{EF}{AB} \cdot \frac{GH}{CD}$
   5. Cancellation property
6. $\frac{A_2}{A_1} = k \cdot k = k^2$ 6. Substitution (step 2
   into 5)

23. hexagonal region, convex [2.5]   **25.** half-plane,
convex [2.5]   **27.** angle and its exterior, concave
[2.5]

13.5
1. 2.5 miles   **3.** 17.5 miles   **5.** 1.2 inches   **7.** 480 ft.
9. 9,750 ft.   **11.** 555 ft.   **13.** 5.3 in.   **15.** 70 yd. or
210 ft.   **17.** 450 ft., buy 150 ft. more   **21.** D [2.4]
23. A [2.4]   **25.** C [2.4]

13.6
1. $\frac{20}{3}$ units (or $6\frac{2}{3}$)   **3.** 4 units   **5.** 4 units
7. 63 units   **9.** $x = 32.5$ **11.** $x = 2$ **13.** $x = 2$
15. $r = \frac{17}{4}$ (or 4.25)   **17.** $\frac{AB}{MN} = \frac{AP}{NP} = \frac{BP}{MP}$ (or
$\frac{t}{s} = \frac{r}{c} = \frac{v}{r}$)   **19.** $\frac{1}{c} = \frac{v}{1}$, so $v = \frac{1}{c}$ **21.** simple
closed [2.4]   **23.** closed [2.4]   **25.** simple [2.4]

13.7
1. $\frac{1}{3}$ **3.** $\frac{1}{2}$ **5.** $\frac{26.5 \text{ mm}}{15.5 \text{ mm}} = 1.71$ **7.** $\frac{35.5 \text{ mm}}{10.5 \text{ mm}} = 3.0$
9. $\frac{27 \text{ mm}}{16 \text{ mm}} = 1.69$ **11.** 6 and 10   **15.** Answers will
vary.   **17.** Answers will vary.   **21.** SAS and SSS
[13.2]   **23.** AA covers all three [13.2]

25. SSA, proportional sides, congruent angles, but triangles not similar [13.2]

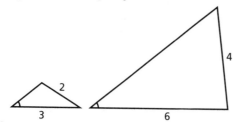

Chapter 13 Review

1. $\frac{1}{2}$

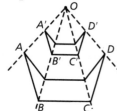

3. 4 units **5.** 15 units
7. $\frac{10}{3}$ (or $3\frac{1}{3}$) **9.** similar: SSS **11.** not similar; figures are not the same shape

13.
1. $\triangle ABF \sim \triangle EBC$; $\overline{GB} \perp \overline{AF}$; $\overline{BD} \perp \overline{CE}$
   1. Given
2. $\overline{BG}$ and $\overline{BD}$ are altitudes to $\overline{AF}$ and $\overline{CE}$ respectively 2. Definition of altitude
3. $\frac{FB}{CB} = \frac{GB}{DB}$ 3. Altitudes proportional to sides

15. yes **17.** no **19.** $4\sqrt{3}$ units **21.** $4\sqrt{5}$ units
23. 4 units **25.** 13 units **27.** 44 ft. **29.** Answers will vary.

Chapter 14—Trigonometry

14.1

1. $\frac{6\sqrt{61}}{61}$ **3.** $\frac{4\sqrt{6}}{11}$ **5.** $\frac{3\sqrt{13}}{13}$

7.   **9.**

11. $\sin A = \frac{4}{5}$; $\sin B = \frac{3}{5}$; $\cos A = \frac{3}{5}$; $\cos B = \frac{4}{5}$; $\tan A = \frac{4}{3}$; $\tan B = \frac{3}{4}$ **13.** $\sin A = \frac{2\sqrt{5}}{5}$; $\sin B = \frac{\sqrt{5}}{5}$; $\cos A = \frac{\sqrt{5}}{5}$; $\cos B = \frac{2\sqrt{5}}{5}$; $\tan A = 2$; $\tan B = \frac{1}{2}$ **15.** $\tan A = 2\sqrt{2}$, $\sin A = \frac{2\sqrt{2}}{3}$ **17.** $\cos A = \frac{\sqrt{5}}{3}$, $\tan A = \frac{2\sqrt{5}}{5}$ **19.** $\frac{\sqrt{7}}{4}$ **21.** $\frac{\sqrt{21}}{2}$

27. [13.2]
1. $m\angle A = m\angle B$, $\angle C$ and $\angle D$ are right angles
   1. Given
2. $\angle A \cong \angle B$ 2. Definition of congruent angles
3. $\angle C \cong \angle D$ 3. All right angles congruent
4. $\triangle ACE \sim \triangle BDF$ 4. AA

29. [14.1] $\frac{BF}{AE} = \frac{BD}{AC}$, $(BF)(AC) = (BD)(AE)$, $\frac{AC}{AE} = \frac{BD}{BF}$, $\cos A = \cos B$, $\frac{BD}{AC} = \frac{FD}{EC}$, $(BD)(EC) = (FD)(AC)$, $\frac{EC}{AC} = \frac{FD}{BD}$, $\tan A = \tan B$

14.2

1. $AB = 18\sqrt{2}$ units, $AC = 18$ units **3.** $AB = 12\sqrt{2}$ units, $AC = 12$ units **5.** $AC = BC = 2\sqrt{2}$ units **7.** $XZ = 6\sqrt{3}$ units, $YZ = 6$ units **9.** $XY = 4$ units, $XZ = 2\sqrt{3}$ units **11.** 45-45 **13.** 30-60 **15.** neither **17.** $\frac{\sqrt{2}}{2}$ **19.** $\sqrt{3}$ **21.** $\frac{1}{2}$ **25.** side is 20 units [8.3] **27.** radius is $\sqrt{5}$ [8.5]

14.3

1. 0.6561 **3.** 7.1154 **5.** 0.5150 **7.** 29° **9.** 12° **11.** cosine, DF $\approx$ 19.1 **13.** sine, EF $\approx$ 5.8 **15.** complement, $m\angle E = 90° - 17° = 73°$ **17.** $m\angle Z = 51°$; $m\angle X = 39°$; $y = 19.2$ units **19.** $n = 4.5$ units; $l = 2.2$ units; $m\angle N = 64°$ **21.** $m\angle A = 26°$; $m\angle B = 64°$; $b = 30.5$ units **23.** $m\angle B = 43°$; $a = 19.3$ units; $c = 26.4$ units **25.** $m\angle A = 12°$; $a = 5.8$ units; $b = 27.4$ units **27.** similar, AA [13.2] **29.** similar, theorem on altitude of right triangle [13.3] **31.** congruent, SAS [6.6]

14.4

1.
$$\frac{\sec x}{\csc x} = \frac{\sec x}{\csc x}$$
$$= \frac{\frac{1}{\cos x}}{\frac{1}{\sin x}}$$
$$= \frac{1}{\cos x} \cdot \frac{\sin x}{1}$$
$$= \frac{\sin x}{\cos x}$$
$$= \tan x$$

3.
$$\cot x = \cot x$$
$$= \frac{1}{\tan x}$$
$$= \frac{1}{\frac{\sin x}{\cos x}}$$
$$= 1 \cdot \frac{\cos x}{\sin x}$$
$$= \frac{\cos x}{\sin x}$$

5.
$$\cos x \csc x = \cos x \csc x$$
$$= \cos x \left(\frac{1}{\sin x}\right)$$
$$= \frac{\cos x}{\sin x}$$
$$= \cot x$$

7.
$$\cot^2 x + 1 = \cot^2 + 1$$
$$= \frac{\cos^2 x}{\sin^2 x} + 1$$
$$= \frac{\cos^2 x + \sin^2 x}{\sin^2 x}$$
$$= \frac{1}{\sin^2 x}$$
$$= \csc^2 x$$

9.
$$\cot x + \tan x = \cot x + \tan x$$
$$= \frac{\cos x}{\sin x} + \frac{\sin x}{\cos x}$$
$$= \frac{\cos^2 x + \sin^2 x}{\sin x \cos x}$$
$$= \frac{1}{\sin x \cos x}$$
$$= \frac{1}{\sin x} \cdot \frac{1}{\cos x}$$
$$= \csc x \sec x$$

11.
$$\sin x \sec x \cot x = \sin x \sec x \cot x$$
$$= \sin x \cdot \frac{1}{\cos x} \cdot \frac{1}{\tan x}$$
$$= \frac{\sin x}{\cos x} \cdot \frac{1}{\tan x}$$
$$= \tan x \cdot \frac{1}{\tan x}$$
$$= 1$$

13. $1 - \sin x \cos x \tan x = 1 - \sin x \cos x \tan x$
$$= 1 - \sin x \cos x \cdot \frac{\sin x}{\cos x}$$
$$= 1 - \sin^2 x$$
$$= \cos^2 x$$

15.
$$\csc x - \cos x \cot x = \csc x - \cos x \cot x$$
$$= \frac{1}{\sin x} - \cos x \cdot \frac{\cos x}{\sin x}$$
$$= \frac{1}{\sin x} - \frac{\cos^2 x}{\sin x}$$
$$= \frac{1 - \cos^2 x}{\sin x}$$
$$= \frac{\sin^2 x}{\sin x}$$
$$= \sin x$$

17.
$$\sec^2 x\,(1 - \sin^2 x) = \sec^2 x\,(1 - \sin^2 x)$$
$$= \sec^2 x\,(\cos^2 x)$$
$$= \frac{1}{\cos^2 x} \cdot \cos^2 x$$
$$= \frac{\cos^2 x}{\cos^2 x}$$
$$= 1$$

21. F [12.2]   **23.** H [12.1]   **25.** G [13.1]

14.5
1. 14.3 ft.   **3.** 1026.6 ft.   **5.** 60° angle   **7.** 59°
9. 376.8 ft.   **11.** 17 feet   **13.** 1142.5 ft.   **15.** about 30 mi. east (157,689 ft.)   **17.** 54.7 ft.   **19.** 34°
21. 40° [8.4]   **23.** $\frac{25\sqrt{3}}{4} \approx 10.8$ [8.3]   **25.** 98.4 [8.4]

14.6
1. 72°, 36°   **3.** 4.1, 5.1   **5.** 36°, 18°
7. $r = \frac{s}{2\sin 18°}$
9. $s = 24\sin 18° \approx 7.4$
11. $a = \frac{s}{2\tan 22\frac{1}{2}°} = \frac{s}{2}\cot 22.5°$
$$r = \frac{s}{2\sin 22\frac{1}{2}°} = \frac{s}{2}\csc 22.5°$$
13. $7 = \frac{s}{2} \cdot 22.5$
$$s = 14\tan 22.5 \approx 5.8$$
27. edge of 4 in. [11.1]   **29.** edge of 6 inches [11.5]
31. $\tan 9° = \frac{2}{9}$, $a \approx 12.6$, $B = 504$ sq. units
$V = 1848$ cu. units [11.4]

Chapter 14 Review
1. $\sin A = \frac{7\sqrt{149}}{149}$; $\sin B = \frac{10\sqrt{149}}{149}$; $\cos A = \frac{10\sqrt{149}}{149}$; $\cos B = \frac{7\sqrt{149}}{149}$; $\tan A = \frac{7}{10}$; $\tan B = \frac{10}{7}$
3. $\sin A = \frac{12\sqrt{193}}{193}$; $\tan A = \frac{12}{7}$
5. $AC = 6\sqrt{2}$ units; $BC = 6\sqrt{2}$ units
7. $MO = 13\sqrt{3}$ units; $NO = 13$ units   **9.** 1.000
11. 0.0698   **13.** 3.078   **15.** 28°   **17.** $c = 9\sqrt{13}$ units ≈ 32.4; $m\angle B = 34°$; $m\angle C = 56°$

19.
$$\sec x - \cos x = \sec x - \cos x$$
$$= \frac{1}{\cos x} - \cos x$$
$$= \frac{1}{\cos x} - \frac{\cos^2 x}{\cos x}$$
$$= \frac{1 - \cos^2 x}{\cos x}$$
$$= \frac{\sin^2 x}{\cos x}$$
$$= \sin x \cdot \frac{\sin x}{\cos x}$$
$$= \sin x \tan x$$

21. 3763.7 ft.   **23.** $\cot x = \frac{1}{\tan x}$ **25.** $c = x\sqrt{2}$, $m\angle A = m\angle B = 45°$, $m\angle C = 90°$ **27.** $\sin 45° = \frac{\sqrt{2}}{2}$, $\cos 45° = \frac{\sqrt{2}}{2}$, $\tan 45° = 1$ **29.** 0.3657

Index

Photo Credits

The following agencies and individuals have furnished materials to meet the photographic needs of this textbook. We wish to express our gratitude to them for their important contribution.

Andersen Windows, Inc.

Aramco World Magazine

Artemis Images

Art Resource

The Biltmore Company

BJU Press

Cedar Point

Stephen Christopher

George R. Collins

Corel Corporation

Creation Science Foundation, Ltd.

Susan Day

Dr. E. R. Degginger

Digital Stock

Jack Dill

Eastman Chemical Division

M. C. Escher Company

The Field Museum

Ford Motor Company

Gaffney Board of Public Works

Greater Quebec Area Tourism and Convention Bureau

Greek National Tourist Office

Hemera Technologies, Inc.

IBM Corporation

iStockphotos, Inc.

Brian D. Johnson

Kenya Tourist Office

KVLY-TV

Fred E. Mang Jr.

Colonel Kemp Moore

National Aeronautics and Space Administraion (NASA)

National Park Service

National Radio Astronomy Observatory

Naval Historical Foundation

Nikon, Inc.

PhotoDisc, Inc.

Planet Art

J. Norman Powell

Providence/Warwick Conventions & Visitors Bureau

Dr. Margene Ranieri

Ted Rich

Kay Shaw Photography

Six Flags Over Georgia

Smithsonian Institution

Ron Tagliapietra

Texas Water Development Board

United States Department of Agriculture (USDA)

Unusual Films

U.S. Forest Service

U.S. Geological Survey (USGS)

U.S. Naval Observatory

U.S. Navy

USS *Alabama* Battleship Commission

Harry Ward

Ward's Natural Science Establishment

Cover

Unusual Films: shell, turtle; George R. Collins: mimosa; PhotoDisc, Inc.: bridge, bubbles, wood grain, staircase; NASA: galaxy

Front Matter

Jack Dill vii; Greek National Tourist Office xii (top); Dr. Margene Ranieri xii (bottom); *Aramco World* Magazine xiii; PhotoDisc, Inc. xiv; Unusual Films xv

Chapter 1

Unusual Films xvi-1, 2, 27; Corel Corporation 3; PhotoDisc, Inc. 6; National Zoological Park, Smithsonian Institution 9; Digital Stock 11, 32; The Field Museum (#GEO85827c); © 2005 Hemera Technologies, Inc. All Rights Reserved. 22; Brian D. Johnson 25

Chapter 2

Harry Ward 40-41; Unusual Films, Courtesy of Six Flags Over Georgia 42; George R. Collins 43; Digital Stock 45, 68; PhotoDisc, Inc. 48, 54; Planet Art 52; Corel Corporation 53 (both); Kay Shaw Photography 55; Susan Day/Daybreak Imagery 56; Greater Quebec Area Tourism and Convention Bureau 62; M. C. Escher's *Moebius Strip II* © 2005 M. C. Escher Company - Baarn - Holland. All rights reserved 70; National Park Service 73

Chapter 3

PhotoDisc, Inc. 84 (top), 95, 106; Corel Corporation 84 (bottom); Jack Dill 89; Digital Stock 97, 99; Scala/Art Resource, NY, Pinacoteca di Brera, Milan, Italy 109

Chapter 4

PhotoDisc/Getty Images 116-17; Colonel Kemp Moore 118; The Biltmore Company 121 (top); Used with Permission from the Biltmore Company, *A Guide to Biltmore Estate* 121 (bottom); PhotoDisc, Inc. 135, 143; Digital Stock 146

Chapter 5

Courtesy of IBM Corporation 158-59; PhotoDisc, Inc. 160, 166, 172, 176, 189; M. C. Escher's *Ascending and Descending* ©2005 M. C. Escher Company - Baarn - Holland. All rights reserved 199

Chapter 6

Stephen Christopher 206-7; Brian Johnson 208, 220 (top); PhotoDisc, Inc. 214, 233, 240; Eastman Chemical Division 220 (bottom), 245; BJU Press files 226; Artemis Images 238; Digital Stock 239; Corel Corporation 252

Chapter 7

M. C. Escher's *Transitional System IA-IA* © 2005 M. C. Escher Company - Baarn - Holland. All rights reserved 260-61; Andersen Windows, Inc. 262, 273, 281, 298; PhotoDisc, Inc. 267; Jack Dill 279, 280, 291; Digital Stock 287

Chapter 8

Courtesy of the Texas Water Development Board 308-9; Unusual Films 311; Ward's Natural Science Establishment 316 (top), 355; © 2005 Dr. E. R. Degginger 316 (bottom), 333; Digital Stock 323; George R. Collins 332; Courtesy Six Flags Over Georgia 338; PhotoDisc, Inc. 348

Chapter 9

USDA 364-65; Corel Corporation 366, 389; Ford Motor Company 374; PhotoDisc, Inc. 381, 388, 394; Fred E. Mang Jr., Courtesy of National Park Service 390; Brian Johnson 399; Courtesy of Six Flags Over Georgia 405

Chapter 10

NASA 414-15, 428, 434, 440, 448; Corel Corporation 416; PhotoDisc, Inc. 420; J. Norman Powell 423; U. S. Naval Observatory 439; USDA 451

Chapter 11

PhotoDisc, Inc. 458-59, 471 (top), 488, 489; Unusual Films 461; © 2005 iStockphotos, Inc. 464; USDA 471 (bottom); Corel Corporation 475; National Radio Astronomy Observatory 481; Gaffney Board of Public Works 482; Ward's Natural Science Establishment 490; Unusual Films 495

Chapter 12

Creation Science Foundation, Ltd. 498-99; Kenya Tourist Office 500; PhotoDisc, Inc. 501, 515, 525; Cedar Point, photo by Dan Feicht 506; M. C. Escher's *Circle Limit 3* © 2005 M. C. Escher Company - Baarn - Holland. All rights reserved 512; J. Norman Powell 522; Brian Johnson 526

Chapter 13

Unusual Films 534-35; Ron Tagliapietra 536; Planet Art 540; Alinari/Art Resource, NY, S. Maria delle Grazi, Milan, Italy 541; Jack Dill 548, 566; Digital Stock 556, 563, 573 (top), 575; Réunion des Musées Nationaux/Art Resource, NY, Louvre, Paris, France 571, 574 (detail, bottom); Corel Corporation 572; USS *Alabama* Battleship Commission 573 (middle); Naval Historical Foundation 573 (bottom); U. S. Navy 574 (top)

Chapter 14

U. S. Forest Service 580-81; PhotoDisc, Inc. 588, 594; Unusual Films 592; USGS 593 (top); Nikon, Inc. 593 (bottom); Providence/Warwick Conventions & Visitors Bureau 604; KVLY-TV 605; BJU Press files 609

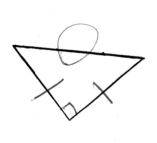